Springer Remote Sensing/Photogrammetry

The Springer Remote Sensing/Photogrammetry series seeks to publish a broad portfolio of scientific books, aiming at researchers, students, and everyone interested in the broad field of geospatial science and technologies. The series includes peer-reviewed monographs, edited volumes, textbooks, and conference proceedings. It covers the entire area of Remote Sensing, including, but not limited to, land, ocean, atmospheric science and meteorology, geophysics and tectonics, hydrology and water resources management, earth resources, geography and land information, image processing and analysis, satellite imagery, global positioning systems, archaeological investigations, and geomorphological surveying.

More information about this series at http://www.springer.com/series/10182

Lei Yan · Bin Yang · Feizhou Zhang ·
Yun Xiang · Wei Chen

Polarization Remote Sensing Physics

北京大学出版社
PEKING UNIVERSITY PRESS

Springer

Lei Yan
Earth and Space Science
Peking University
Beijing, China

Feizhou Zhang
Peking University
Beijing, China

Wei Chen
China University of Mining
and Technology, Beijing
Beijing, China

Bin Yang
College of Electrical
and Information Engineering
Hunan University
Changsha, China

Yun Xiang
Hebei Academy of Meteorological Sciences
Shijiazhuang, China

ISSN 2198-0721 ISSN 2198-073X (electronic)
Springer Remote Sensing/Photogrammetry
ISBN 978-981-15-2885-9 ISBN 978-981-15-2886-6 (eBook)
https://doi.org/10.1007/978-981-15-2886-6

Jointly published with Peking University Press
The print edition is not for sale in China (Mainland). Customers from China (Mainland) please order the print book from: Peking University Press.

This Springer imprint is published by the registered company Springer Nature Singapore Pte Ltd.
The registered company address is: 152 Beach Road, #21-01/04 Gateway East, Singapore 189721, Singapore

Foreword By Huilin Jiang

The polarized signal is an important component of the surface and atmospheric system reflection signals and is an important signal source for retrieving surface and atmospheric information using remote sensing information. However, the current research on optical remote sensing mostly focuses on non-polarized remote sensing, and there are few studies on polarization remote sensing, which neglects an important data source in remote sensing inversion.

In response to this, the author of this book conducted a long-term systematic study from four aspects, including the physical basis of polarization remote sensing, the polarization remote sensing of ground object parameter inversion, the polarization remote sensing of atmospheric parameter inversion, and the application of polarization remote sensing.

This book contains not only a theoretical breakthrough, but a series of practical experiments. Based on the data analysis of more than 300,000 sets of field experiments conducted by the author's research team since 1978, the theoretical system of optical polarization remote sensing is established, and the main material in this book stems from the theoretical achievements and experimental accumulation of the research team led by the first author over the past 40 years.

Polarized remote sensing technology, with its objectivity, stability, and repeatability, should take the leading responsibility in the China Geoscience Research. The book provides a new vision for the practitioners in related fields to solve their own problems by means of polarization remote sensing and can be used as a guide for polarized remote sensing research to understand polarization remote sensing theory.

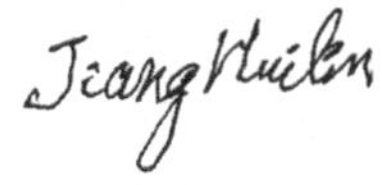

Huilin Jiang
Changchun University of Science and Technology
Academician of Chinese Academy of Engineering

Foreword By Junhao Chu

Polarization is an important part of the surface reflection signal, but optical polarization remote sensing still lacks in-depth research and application. The international astronomical community mainly uses polarization methods for the study of the sun, stars, and planets. Optical polarization is expected to play a role in the remote sensing science.

From the optics point of view, the solar electromagnetic wave is the only carrier that runs through the whole process of remote sensing, including the atmosphere, the surface and the observation instrument. By exploring the polarization parameters from the optics point of view are we able to make the whole remote sensing process have a unified physical dimension and lay the foundation for the integrated automation model construction of remote sensing.

This book mainly analyzes the two geoscience inversion bottlenecks: the intensity of the ground reflection spectrum and the attenuation error effect of the atmospheric window. The author comprehensively summarize the physical principle of polarization remote sensing, and use polarization remote sensing to solve the problem of target detection under the over-bright or over-dark background.

This book is a reference book for practitioners in the field of optical polarization to learn the application of polarized optical technology in the field of remote sensing. It also provides a new technical solution for practitioners in the field of remote sensing. It can be used as a guide for polarization remote sensing scientific research and a teaching material to understand polarization remote sensing theory, techniques and methods.

Junhao Chu
Junhao Chu
Shanghai Institute of Technical Physics
Chinese Academy of Sciences
Academician of Chinese
Academy of Sciences

Foreword By Jouni Peltoniemi

Polarization is expected to play a role in the burgeoning science of remote sensing, however, the realization of application goals requires the establishment, interpretation, and breakthrough of a sound theoretical basis. So far, there are no books systematically and comprehensively introducing the mechanism, theory, and application of optical polarization remote sensing, either at home or abroad. The publication of this book fills the theoretical and experimental gaps in this cognate area.

In this book, Prof. Yan has contributed to the polarimetric effects of three elements of remote sensing processing-chain, i.e. ground, atmosphere and instruments, as well as the three-element-related integrated modeling, observation methods and instrumentation. He has established a series of polarization inversion models of three-element high-resolution quantitative remote sensing, adding an additional dimension to the domain of resolutions; besides spatial, spectral, radiometric and temporal resolution, also the degree of polarization can now be considered. He has led to new theoretical and experimental results in the areas of representation, capture and analysis of polarized electromagnetic radiation. He has developed the theory and mechanism of a new earth polarization field. The field contains not only the "line", i.e. the sky polarization distribution, but also the "central point", i.e. atmospheric polarization neural point. The polarization field makes a new atmospheric window for ground observation, and reveals information of atmospheric pollution and many other meteorological characteristics.

This monograph is the first in the world to integrate the theory, methodology and applications of polarimetric remote sensing, providing good foundation for polarimetric remote sensing method and instrumentations.

Jouni Peltoniemi
Finnish Geospatial Research
Institute FGI, Research manager
of space geodesy

Preface

Polarization is an important part of the surface reflection signal, but polarization information has not received the attention it deserves in remote sensing. At present, the US NASA has recognized the significance of polarization remote sensing and has developed a new type of polarization sensor; France has already used polarization remote sensing detectors, and the international astronomy community mainly uses polarization methods for the study of the sun, stars, and planets. Polarization remote sensing will inevitably play a huge role in the ascending remote sensing discipline. However, there are no books at home and abroad that systematically introduce the theory and application of polarization remote sensing. Therefore, the author analyzes the current status and trends of polarization remote sensing research at home and abroad and summarizes the research results of his team for more than 20 years since 1988 to complete the full text of this book.

This book introduces five characteristics of polarization remote sensing: multi-angle reflection physical features (Chap. 2), multispectral chemical properties (Chap. 3), roughness and density structure properties (Chap. 4), characteristics of info-background high contrast ratio filter (Chap. 5), and the features of radiative energy transfer (Chap. 6). This book introduces important connotations of atmospheric polarized remote sensing: the nature and physical characteristics of the full-sky polarization pattern (Chap. 7), regional characteristics of atmospheric polarization neutral points and decoupling of ground-atmospheric parameters (Chap. 8), and multi-angle stereoscopic observation chromatography of atmospheric particles under full-sky polarization vector field (Chap. 9). This book introduces new areas in polarization: bionic polarization automatic navigation based on earth's polarization vector field (Chap. 10), polarization remote sensing methods for advanced space detection and a global change study (Chap. 11).

The first author is in charge of all the concepts, the establishment of the first- and second-level titles, the guidance of each part and all chapters, the summary of the five characteristics of the polarization remote sensing of the ground, the extraction of the three regular characteristics of the atmospheric polarization remote sensing, and the opening of the five new frontier international cooperation fields of polarization remote sensing. The second author Bin Yang and the third author Feizhou

Zhang are in charge of the fifth and sixth chapters. The fourth author Yun Xiang is in charge of the third chapter, cooperating with the first author for the manuscripts of the second to fourth chapters. The fifth author Wei Chen is in charge of the ninth chapter. The first author is in charge of the other chapters, checks the manuscripts of each chapter and reviews for more than three times.

This book is a crystallization of the creative achievements accumulated by the first author's team for a long period of time concentrated on basic research, experiments, and teaching. Special thanks should be given to Professor Hu Zhao for the fourth chapter of the book, Professor Taixia Wu for the eighth chapter of the book, and Professor Guixia Guan for the tenth chapter of the book. The first author expresses his gratitude to all the contributors with sincere respect. The first author expresses his gratitude to the relevant state departments for their financial support.

Finally, I would also like to thank Qingxi Tong, Academician of the Academy of Sciences, Xianlin Liu, Academician of the Academy of Engineering, and Xiaowen Li, Academician of the Academy of Sciences for their review, suggestions, and recommendations.

This book has been written since 2015. Although it has been revised, screened, and refined in the past 4 years, due to the limited knowledge of the author, it is still difficult to avoid mistakes. Please give advice and point out my error, my email is lyan@pku.edu.cn.

October 2019

Dr. Lei Yan
Written at Weiming Lake
Peking University

Afterword

The final draft of the book is a mark of the miles of my life. It has been 17 years since I started to pay attention to polarization remote sensing. It has been 14 springs and autumns to actually launch basic theoretical research on polarization remote sensing.

Science is a stream, a trickle; little by little, drop hollows the stone

Since 1998, when I finished my first monograph "structural control of resources, environment and ecological giant system" from the field of electronic and electrical automation to the field of geosciences, I began to comprehensively examine my "territory" of remote sensing. After more than 2 years of tracking and observation of polarization remote sensing research carried out by Professor Yunsheng Zhao of Northeast Normal University, based on my background of optoelectronics and electromagnetics, I realized that polarization, the unique role of this physical phenomenon, can become a virgin land to be cultivated urgently for remote sensing and optical observation and is expected to become an indispensable supplement branch of conventional remote sensing and optical imaging observation. In 2001, I sent doctoral students to Northeast Normal University for the first time for three months, which verified that the field experiment data can be reproduced, which means that there are rules to explore; at the same time, it was found that the measurement instrument data can only be recorded and read manually, and then input into the computer processing and analysis, the efficiency and reliability are affected. Under the traction of the polarization remote sensing demand of Zhao's team, our team compiled the instrument automatically computer interface. Thus, the bottleneck of automatic data acquisition and computer analysis has been completely solved, making more than 100000 sets of original data accumulated manually in the past 13 years, and more than 300000 sets of measurement data, including more than 200000 sets of field joint measurement data, plus nearly 100000 sets of flight data of our team. Thus, the strong data onto automatic acquisition and analysis and the common field experiment foundation are established for the study of polarization remote sensing theory.

Secondly, in 2002, Yunsheng Zhao and I went south to Hefei, researcher team of Anhui Institute of Optics and Mechanics, Chinese Academy of Sciences, the research unit of polarization remote sensing instrument, held the long-term research strategy seminar of polarization remote sensing for the first time, and determined that the Northeast Normal University focuses on the acquisition and experimental analysis of field data outside polarization remote sensing, and the Institute of Optics and Safety, Chinese Academy of Sciences focuses on the development and indicators of polarization remote sensing instrument It is verified that Peking University focuses on the establishment of polarization remote sensing theory, mechanism interpretation, link work under the guidance of the theory of instrument indicators and ground feature parameters based on the data and instrument requirements of the outfield features; and recommends that Peking University team be responsible for convening all parties to carry out the continuous annual consultation mechanism to focus on attracting the domestic scientific community to participate in polarization remote sensing research. This consultation system has been maintained in the weakest period and has gradually diversified forms and developed continuously: In 2004, our team supported the integration of Xingfa Gu teacher's team of Remote Sensing Application Institute of Chinese Academy of Sciences (CAS) and polarization remote sensing instrument team of Anhui Optics and Precision Mechanism Institute of CAS with all the achievements of the first doctoral dissertation on polarization remote sensing mechanism in China, and applied to the state for project approval successfully. In 2005, we and Northeast Normal University; in 2006, we successfully attracted the domestic excellent team of polarization optical devices, the team of Professor Jianyi Yang and Yubo Li from Zhejiang University to join in supporting polarization remote sensing observation and research, and in 2008, we won the first team of polarization remote sensing devices jointly applied by the National Natural Science Foundation of China Objective: In 2008, he established academic contacts and talent exchanges with Zhongquan Qu team of Yunnan Observatory of Chinese Academy of Sciences on polarization observation and remote sensing polarization observation and sent each other graduate students to study the complementary mapping of surface atmospheric polarization remote sensing and planetary polarization observation; in 2007, he conducted bionic polarization navigation research with Professor Jinkui Chu, academician of Liding Wang of Dalian University of Technology and obtained the results in 2009: the first bionic polarization navigation project jointly applied by our team. Thus, the combination and cross of multiple teams of polarization remote sensing in China has been realized, which has established the basis of truth-seeking team cooperation beyond the economic interest relationship for the full implementation of polarization remote sensing in China and won the first prize of Beijing Science and Technology Progress Award (natural science category) in 2012 with the achievement of multiple team cooperation "systematic theory, methods and several discoveries of polarization remote sensing."

In terms of project support, due to the lack of national projects in terms of polarization, we have to find solutions respectively. Yunsheng Zhao and I am mainly national funds. In the aspect of aviation polarization data acquisition, Xingfa Gu's team gave us free help. In the world's first flight experiment aimed at the atmospheric polarization neutral point area, Beijing Xingtiandi Technology Co,

Ltd, the leader of the national aviation remote sensing industry alliances, has borne all the expenses of the flight experiment in Zhuhai free of charge. Chunfeng Zhang, the general manager, and Chao Zhang, the chairman of the board, know that such a flight will provide the experimental basis of us to overcome the atmospheric attenuation before coming. They have arranged a half month flight of us without hesitation. These have become an inexhaustible driving force for our continuous research.

Science is the fountainhead, moisten the soil, and nurture later learning

I remember that when Hu Zhao, the first polarization doctoral student, defended against 2003, strict experts judged his polarization remote sensing theory and density conversion theory, which basically failed to pass. Hu Zhao was so aggrieved that he wanted to give up. I talked to him for a month and decided to continue. We encouraged each other and persevered, so that he spent another year perfecting his thesis, and finally conquered all the judges and became the first doctor in the basic theory research of polarization remote sensing in China. His polarization non-contact estimation of the surface density of the Qilian Mountains and the moon is within 15% of the actual measurement error of the sample, which makes us deeply feel the important role of polarization remote sensing, and thus has the theoretical construction of the fourth chapter of this book. On this basis, Hu Zhao successfully applied for NSFC after graduation and achieved the deepening results.

The second Ph. D student Yun Xiang used hyperspectral polarization remote sensing for ground feature research. More than 30 rock density measurements and geological samples were measured. The accuracy was within the allowable range of engineering error, which built up our confidence, and thus the theoretical construction of Chap. 3 of this book. She was pregnant and ran around Changchun and Wuhan to prove the figure of polarization theory with various geotechnical samples, which was always engraved on the way of seeking polarization remote sensing.

Under their leadership, a group of postdoctoral and master's students have carried out arduous research on methods of remote sensing of ground objects, thus accumulating the second chapter of this book.

On the premise of breakthrough in the study of ground object polarization, I have noticed that atmospheric attenuation is the biggest error source of remote sensing. In the field of atmosphere, we use polarization scattering transmission attenuation to estimate aerosol thickness and atmospheric effect. Why can't we use polarization method to try to explore the law forbidden area of remote sensing such as atmospheric window? So I turned to Taixia Wu, the third doctoral student in polarized remote sensing, to the study of atmospheric attenuation. Half a year after his visit to the USA, he emailed me that he found that there was an area in the atmosphere where the polarization attenuation effects were zero. So it was difficult to measure the atmospheric effect on polarization. I asked him how he studied the atmosphere. He said that because the atmospheric attenuation affects the inversion of surface features. I said that if the polarized light carries the ground feature information, isn't it the new, low attenuation atmospheric window that we expect to observe or obtain the polarized ground feature information in the area where the polarization effect is zero? This makes his doctoral project all deepen to the study of the neutral point region of atmospheric polarization. This is the basis of the theoretical breakthrough in Chap. 8 of this book. After graduation, he cooperated with us and applied to NSFC.

The fourth doctoral student Guixia Guan's polarization research started from bionic polarization navigation and found a large number of laws of atmospheric polarization pattern, which became the basis of theoretical breakthrough in Chap. 10 of this book. After graduation, she cooperated with us to preside over the application for in-depth research of NSFC, and all her graduate students joined the team of Peking University.

Based on the work of Taixia and Guixia, group of our master's and undergraduate students have been engaged in the observation of atmospheric polarization pattern, which has become the basis for the theoretical construction of Chap. 7 of this book.

Wei Chen, the fifth doctoral student, was sent to Zhanqing Li, an atmospheric remote sensing scientist and an atmospheric scientist at the University of Maryland, U.S., on the basis of the theoretical results of the atmospheric polarization model map and the atmospheric polarization neutral point region. Under the influence of Liang's team, we have achieved a breakthrough in the method of three-dimensional chromatography of atmospheric polarization effect and large-scale characterization of atmospheric particle polarization remote sensing, which has become the core theory in Chap. 9 of this book.

Bin Yang, the sixth doctoral student, and Mr. Yunsheng Zhao have carried out polarization remote sensing research on surface water, soil, and vegetation and cooperated with the international famous global biomass remote sensing inversion team to carry out relevant research, which has made very positive progress, thus forming the fifth and sixth chapters of this book together with Mr. Zhao. On this basis, he led the research team to prove the theory of polarization "strong light weakening" and "weak light strengthening" and solved the long-standing problem of polarization remote sensing.

Science is manna and nutrients, besten the fruit, comb the harvest

Since 2007, my students and I have to go out and come in every year to combine our polarization research results with international cutting-edge and hot issues, and attract the attention of peers at home and abroad. Our theory has begun to explain and guide the application of technology and methods in the field of related disciplines. The 7th International Workshop on solar, 2013 At the world's highest 3-year Global Conference on astronomical polarization observation of stars and planets, we were invited to make a report on the polarization remote sensing earth observation conference to lead the complementary and conversion of polarization observation means in the field of astronomy and remote sensing; we proposed the global comparison and objective enlightenment of polarization vector field, gravity field and geomagnetic field in the whole sky to determine the third earth The breakthrough of the whole field on the observation means of different astronomical problems may be reported in the relevant organizations and conferences of atmospheric observation, environment, geology and geophysics, and our observation instrument "polarized remote sensing imaging" System "won the 2013 Geneva International Invention Gold Medal. It also exchanged visits with Professor Ansundi, chairman of Singapore Optical Society and Optical Research Center of Nanyang University of Technology and established the development plan of remote sensing and multioptical polarization instruments. Changyong Cao, director of NOAA National Calibration Center, USA, invited our Chinese team to join in, hoping to prove the irreplaceable role of remote sensing radiance reference to the moon by means of polarization observation, so as to realize the world problem of polarization "strong light weakening" filtering saturation and "weak light strengthening" highlighting extremely small fluctuation of radiance. At present, our Chinese team of scientists has been continuously involved in the observation and research for 5 years; Boston University famous remote sensing in the USA, Professor Yuri, an expert in vegetation inversion, and a team of European and American scientists and another group of international ecologists argued on PNAs in the USA about the positive–negative correlation between global vegetation biomass and C, N content, pointing out that polarized remote sensing is expected to become a unique screening method, and our Chinese scientists are invited to become the backbone to jointly correct and improve the basic theory of remote sensing inversion; polarized remote sensing objects. The systematic theoretical outlook of theory makes the USA, Britain, Ireland, Hungary, Austria, Australia, Singapore, Australia, etc., become the international arena for long-term or multiple interactive discussions, so as to achieve significant progress in polarization remote sensing and wider acceptance of polarization remote sensing methods and theories. This leads to the challenging and rich fruits of Chap. 11.

I am awed by all those who have participated in this arduous and fundamental exploration of natural laws and scientific methods. On the basis of what I have mentioned above, I put forward two bottleneck problems existing in the remote sensing observation, namely the limitation of the intensity of the electromagnetic spectrum reflected by the ground object at both ends and the attenuation error effect of the atmospheric window, and extracted the breakthrough progress of the three parts of the polarized remote sensing: first, the basis of "weak light enhancement" and "strong light weakening" of the polarized remote sensing of the ground object, so as to extend the detection area at both ends of the remote sensing, and explore Based on the analysis of the laws in the past ten years, I have summed it up into five characteristics of polarization remote sensing: multi-angle reflection physical characteristics (Chap. 2), multispectral chemical characteristics (Chap. 3), roughness and density structure characteristics (Chap. 4), information background high contrast filtering characteristics (Chap. 5), radiation transmission energy characteristics (Chap. 4) The six chapter. Secondly, the polarization method of atmospheric attenuation is accurately described, the law is discovered, and the new theoretical basis of atmospheric window is explored. The important connotation of atmospheric polarization remote sensing is obtained: the law of all sky polarization pattern and physical characteristics (Chap. 7), the regional law of atmospheric polarization neutral point and the separation of ground air parameters (Chap. 8), and the law of all sky polarization vector field gas particle multi-angle observation of solid chromatography (Chap. 9). Thirdly, the latest development of polarization remote sensing application is given to prove the objectivity, uniqueness, stability, and repeatability of polarization remote sensing physics, so as to bear China's leading responsibility in the theory, method, and application of international polarization remote sensing physics. Therefore, it becomes the first chapter of the book: the physical basis of polarization remote sensing; and connects other chapters in series.

The above lament is for the feeling that the collection is in awe of all the contributors and grateful for the national basic research support projects.

The first Author Dr. Lei Yan
Originally written at a 30 hours flight to Santiago, Chile, November 11, 2014 (first time to be invited to present a polarization remote sensing report at the South American International Conference)
Revised at Weiming Lake, Peking University, October 4, 2019

Contents

Part I
Introduction

Chapter 1
Physical Rationale of Polarized Remote Sensing

Polarization, together with intensity, frequency, and phase, are the four main physical features of remote sensing using electromagnetic waves. This chapter discusses the physical rationale of remote sensing. For the two bottleneck problems we have in remote sensing, namely (1) restriction from the two ends of electromagnetic spectrum reflected by land objects and (2) atmospheric window attenuation difference, this chapter provides three breakthroughs: (1) the basis from which to support "bright light attenuation" and "weak light intensification" in the area of polarized remote sensing of land objects, this can greatly expand the detection range at both the dark and the bright ends of remote sensing data inversion; (2) accurate description of polarization methods and research on the rules of atmospheric attenuation, as well as exploration of the basis for new atmosphere window theory of atmosphere polarized remote sensing.

1.1 Polarization of Light

The development of remote sensing science and technology is now at a new stage. To be specific, theoretically it is shifting from simple description to quantitative research, from just describing the relationship between measured radiation values and phenomena on the land to quantifying their relationship around bidirectional reflectance and radiation, from forward radiative transfer models to quantitative inversion of radiative transfer models, and from focusing on separate wave ranges (for example, only research on visible, thermal infrared or microwave spectrum) to integrating wave spectrum ranges. From the view of technical development, it evolved from a single wave range to multiple wave ranges, angles, polarization, multi-temporal as well as multi-modal, and from single sensor to multiple sensor integration. Polarized remote sensing started to develop against this background and marks a relatively new remote sensing area which needs to be further studied.

L. Yan et al., *Polarization Remote Sensing Physics*, Springer Remote Sensing/Photogrammetry, https://doi.org/10.1007/978-981-15-2886-6_1

Polarization is defined as the asymmetry of vibration direction relative to spreading direction (Yao 2008). It is a unique feature of transverse electric and magnetic waves. Polarization is an important feature of electromagnetic wave. Objects on the land and in the atmosphere can produce their unique polarized signals during reflection, scattering and transmission, which means, polarization can reveal abundant information about objects. In nature, natural polarisers exist: for instance, smooth leaves of a plant, soil, water surface, ice, snow, cloud, fog, etc. cause reflection of sunshine from such polarisers result in polarization. Based on this feature, polarized remote sensing provides new and potential information for objects of remote sensing: polarized remote sensing has become a new earth observation method which is receiving more attention.

Polarization is initially introduced to the study of optics by Newton between 1704 and 1706. The term "optic polarization" was coined by Malus in 1809, and Malus also identified polarization in the laboratory as a phenomenon. Maxwell created optical electromagnetic theory between 1865 and 1873, and this theory explained optical polarization in essence. According to electromagnetic theory, light is a transverse electric and magnetic wave and electric field vibrates vertically to the direction of light transmission. Based on the trajectory of electric field vibration, it has five polarization states: natural light (non-polarized light), linear polarized light, partial polarized light, circularly polarized light and elliptical polarized light (Ye et al. 2005).

Natural light has the same vibration amplitude in all directions. It may vibrate in each direction that is vertical to its spreading direction with the same amplitude. If we decompose the light in all directions to only two vertical directions, then we can find that the vibration energy and amplitude in the two directions are the same. Linearly polarized light means that, in the vertical plane to the direction of transmission, the electric field intensity vector only vibrates in a certain direction. Partially polarized light can be viewed as a mix of natural light and linearly polarized light, namely it has vibration range in a certain direction that is superior to other directions. Circularly polarized light and elliptical polarized light refer to light whose electric field intensity vector end has a round or elliptical trajectory on the plane vertical to the light transmission direction. To describe polarized light, we adopt the time average of the two components of light (for example, values along the x-axis and y-axis) that are vertical to the transmitting directions. Light spreading towards a certain direction can be viewed as a composition of two light waves along the x-axis (horizontal axis) and y-axis (vertical axis). The polarized electrical vector of polarized light can be measured by combination of x and y-values. At a stable vibration with a single frequency, the two components are also in a stable state and have a certain relationship with each other. So the final ends of the electrical vector will have a three-dimensional trajectory, and its location at a certain moment is called the "moment trail".

Plane electro-magnetic waves are transverse, where electric field and magnetic field are orthogonal, so when light travels along the z-direction, there are only x and y-components in the electric field. The electric field component of plane magnetic wave can be described as:

$$E = E_0 \cos(\tau + \delta_0) \tag{1.1}$$

Here, $\tau = \omega t - kz$, and this equation can be further broken down to:

$$\begin{cases} E_x = E_{0x}\cos(\tau + \delta_1) \\ E_y = E_{0y}\cos(\tau + \delta_2) \\ E_z = 0 \end{cases} \tag{1.2}$$

To obtain the curve comprised of end points of electric field vectors, we remove parameter τ, so we get:

$$\left(\frac{1}{E_{0x}}\right)^2 {E_x}^2 + \left(\frac{1}{E_{0y}}\right)^2 {E_y}^2 - 2\frac{E_x}{E_{0x}}\frac{E_y}{E_{0y}}\cos\delta = \sin^2\delta \tag{1.3}$$

Herein, $\delta = \delta_2 - \delta_1$, the equation above is an elliptical equation: because the determinant of coefficients is above zero, the end-points of electric field vectors will compose an elliptical track. This means that, at any moment, in the spreading direction, vector ends of all points in the space compose an elliptical shape when projected to the *x*-*y* plane. Such an electro-magnetic wave is referred to as elliptical polarized light. Other polarization states are special cases of elliptical polarized lighting. Figure 1.1 shows elliptical polarization with different values of phase difference.

(1) $\delta = 0$ or integer multiples of $\pm\pi$

With $\delta = \delta_2 - \delta_1 = m\pi\,(m = 0, \pm 1, \pm 2, \ldots)$, the ellipse will degenerate to a straight line, and we have:

$$\frac{E_y}{E_x} = (-1)^m \frac{E_{0y}}{E_{0x}} \tag{1.4}$$

Electrical vector E has consistent direction and is called linear, or plane polarization.

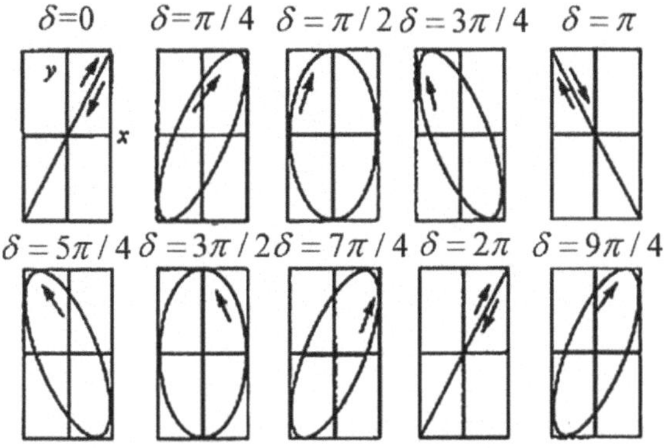

Fig. 1.1 Elliptical polarization with different values of phase difference

After linear polarization light goes through a polariser, when the polariser rotates around incident rays, the light may change in the strength. According to Malus's law, linear polarization of light of strength I_0, will change to:

$$I = I_0 \cos^2 \theta \tag{1.5}$$

Here, θ is the angle between direction of polarization after processing and that of incident light (Fig. 1.2).

(2) $\delta = \pm \frac{1}{2}\pi$

If vibration ranges of E_x and E_y are the same, and their phase difference is an odd multiple of $\pi/2$, namely $E_{0x} = E_{0y} = E_0$, $\delta = \delta_2 - \delta_1 = m\pi/2(m = \pm 1, \pm 3, \pm 5, \ldots)$, then the elliptical equation degenerates to a circular equation:

$$E_x{}^2 + E_y{}^2 = E_0{}^2 \tag{1.6}$$

Now all ends of the electrical vector E compose a circular track, and electrical vector E has circular polarization. Based on the rotating direction of the electrical field vector, we have left-turning polarized light and right-turning polarized light.

According to the theory of vertical vibration combination, circularly polarized light can be resolved into linear polarized light in two mutually vertical directions, light in one of which cannot go through the polariser, while the other can. The change in strength after the light goes through the polariser will be the same as non-polarized light, namely light strength is halved after polarization.

(3) δ as other values

When δ takes another value, based on the composition equation, the composed light will be elliptically polarized in any direction and elliptical polarization light can be levorotated or dextrorotated (Fig. 1.3).

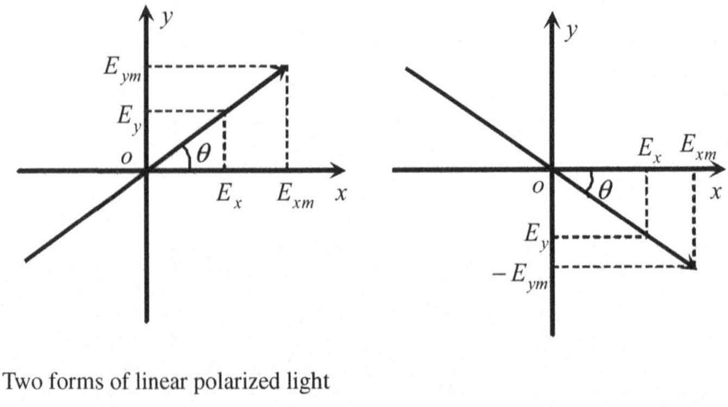

Fig. 1.2 Two forms of linear polarized light

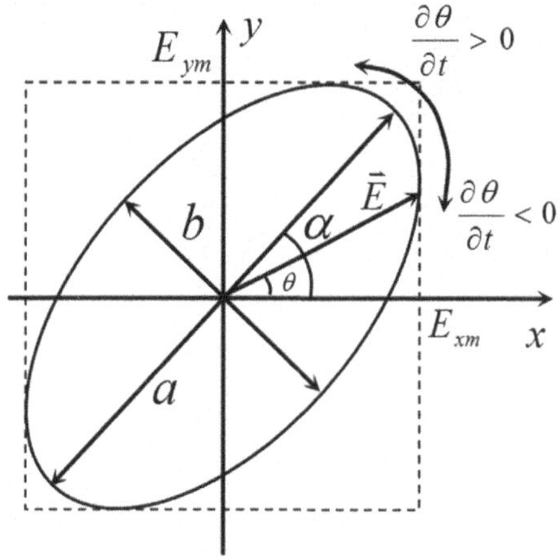

Fig. 1.3 Polarization of light: levorotation or dextrorotation

Based on the relationship between $\vec{E}(z_0, t)$ and $\hat{x}$, the angle between the two is:

$$\theta = \tan^{-1} \frac{E_{ym} \cos(\omega t - kz_0 + \delta_y)}{E_{xm} \cos(\omega t - kz_0 + \delta_x)} \quad (1.7)$$

The rotating speed of $\vec{E}(z_0, t)$'s end is:

$$\frac{d\theta}{dt} = \frac{\omega E_{xm} E_{ym} \sin(\delta_x - \delta_y)}{E_{xm}^2 \cos^2(\omega t - kz_0 + \delta_x) + E_{ym}^2 \cos^2(\omega t - kz_0 + \delta_y)} \quad (1.8)$$

With $0 < \delta_x - \delta_y < \pi$, $d\theta/dt > 0$, the electro-magnetic wave is right-turning elliptical polarization light. With $-\pi < \delta_x - \delta_y < 0$, $d\theta/dt < 0$, the electro-magnetic wave is left-turning elliptical polarization light.

(4) Non-polarized (natural) light

The internal movement status inside each molecule and atom in normal light source is random. Based on statistics, relevant light vibrations will cover all possible directions in the plane that is vertical to that of light spreading; and vibration ranges (light strength) of relevant light vectors in all possible directions are equal. Such light whose light vector vibrates in all directions vertical to light spreading direction is symmetrically distributed and is called natural light.

(a) Embodies distribution of natural light in the plane that is vertical to spreading direction. (b) shows natural light can be represented as two mutually-vertical light vectors. (c) demonstrates natural light is equivalent to a combination of two-plane polarized light that has no specific phase relationships but have vertical vibration directions (Fig. 1.4).

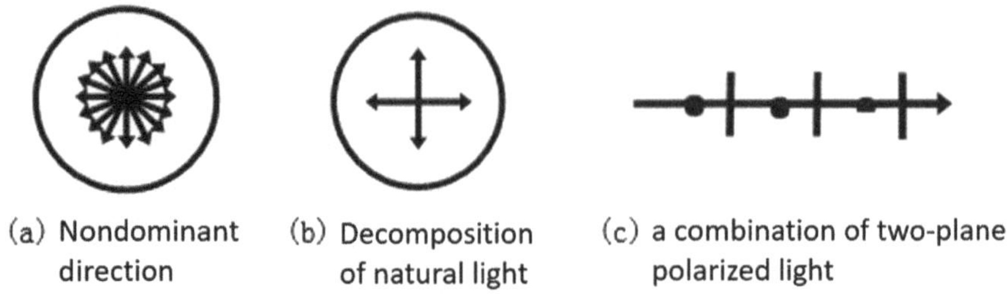

Fig. 1.4 Diagrammatic representation of natural light

(5) Partially polarized light

When non-polarized light goes through a polariser, linear polarized light along the transmission axis can pass through the polariser; while that normal thereto cannot. The polaroid serves as a polariser in this situation. So the light strength of non-polarized light will be halved after it goes through a polaroid and the vibration direction of its light vector remains the same as that of the transmission axis.

Linear polarization, elliptical polarization, and round polarization are all complete polarizations of light. Normally, light is a combination of complete polarized light and natural light, and such combined light is called partially polarized light. Partially polarized light has light vectors along all vibration directions that are in the plate vertical to the light spreading direction, but the vibration is not symmetric with stronger vibration in a certain direction, while vertical to which, the vibration is weaker. Light vectors of partially polarized light can be represented by two independent sub-vibrations that have different vibration ranges and vertical vibrating directions. Most light found nature is partially polarized.

1.2 Light Polarization on the Surface of Media

Light from the sun and common light sources are mostly natural light. When natural light undergoes reflection or refraction on the interface between any two kinds of similar media, reflected, and refracted, light are both partially polarized, except when detecting at zeniths of 90° or 0°. The relationship between spreading directions of reflected waves, transmitted waves, and incident waves can be described by reflection and refraction laws (the Fresnel formula):

Reflection law: $\theta_r = \theta_i$ (reflection angle is equal to incidence angle)
Refraction law: $n_1 \sin\theta_i = n_2 \sin\theta_t$, ($\varepsilon_r$, relative dielectric constant)

Fresnel formula:
As shown below, we present electric vector from any direction with the sub component (s Component) vertical to the incidence plane and that parallel to the incidence plane (p Component) (Fig. 1.5).

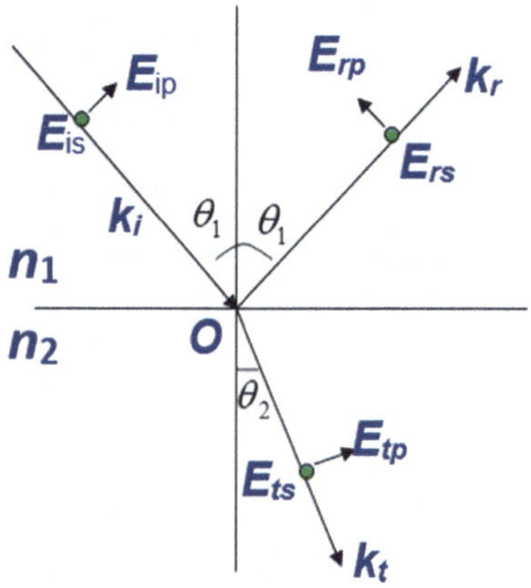

Fig. 1.5 Positive directions of s and p components

Reflection parameter of s component (electric vector vertical to incidence plane):

$$r_s = \frac{E_{0rs}}{E_{0is}} = -\frac{\sin(\theta_1 - \theta_2)}{\sin(\theta_1 + \theta_2)} = \frac{n_1 \cos\theta_1 - n_2 \cos\theta_2}{n_1 \cos\theta_1 + n_2 \cos\theta_2} \tag{1.9}$$

The transmission parameter is:

$$t_s = \frac{E_{0ts}}{E_{0is}} = \frac{2\cos\theta_1 \sin\theta_2}{\sin(\theta_1 + \theta_2)} = \frac{2n_1 \cos\theta_1}{n_1 \cos\theta_1 + n_2 \cos\theta_2} \tag{1.10}$$

Reflection parameter of p component (electric vector parallel to incidence plane):

$$r_p = \frac{E_{0rp}}{E_{0ip}} = \frac{\tan(\theta_1 - \theta_2)}{\tan(\theta_1 + \theta_2)} = \frac{n_2 \cos\theta_1 - n_1 \cos\theta_2}{n_2 \cos\theta_1 + n_1 \cos\theta_2} \tag{1.11}$$

The transmission parameter is:

$$t_p = \frac{E_{0tp}}{E_{0ip}} = \frac{2\cos\theta_1 \sin\theta_2}{\sin(\theta_1 + \theta_2)\cos(\theta_1 - \theta_2)} = \frac{2n_1 \cos\theta_1}{n_2 \cos\theta_1 + n_1 \cos\theta_2} \tag{1.12}$$

If refractive indices and incidence angels at both sides of an interface are known, then we can find the reflection and incidence parameters from the Fresnel formula.

The incidence reflection index and transmission index can be expressed as:

$$R_s = r_s^2 = \frac{\sin^2(\theta_1 - \theta_2)}{\sin^2(\theta_1 + \theta_2)} \tag{1.13}$$

$$R_p = r_p^2 = \frac{\tan^2(\theta_1 - \theta_2)}{\tan^2(\theta_1 + \theta_2)} \tag{1.14}$$

$$T_s = \frac{n_2 \cos\theta_2}{n_1 \cos\theta_1} t_s^2 = \frac{\sin 2\theta_1 \sin 2\theta_2}{\sin^2(\theta_1 + \theta_2)} \tag{1.15}$$

$$T_p = \frac{n_2 \cos\theta_2}{n_1 \cos\theta_1} t_p^2 = \frac{\sin 2\theta_1 \sin 2\theta_2}{\sin^2(\theta_1 + \theta_2)\cos^2(\theta_1 - \theta_2)} \tag{1.16}$$

When incident light is linear polarized light (R refers to reflectivity, T refers to transmission):

$$R = R_p \cos^2\alpha + R_s \sin^2\alpha \tag{1.17}$$

$$T = T_p \cos^2\alpha + T_s \sin^2\alpha \tag{1.18}$$

When incident light is natural light:

$$R = \frac{1}{2}(R_s + R_p) \tag{1.19}$$

$$T = \frac{1}{2}(T_s + T_p) \tag{1.20}$$

So according to the definition of the degree of polarization, for reflected light:

$$P_r = \left|\frac{I_{rp} - I_{rs}}{I_{rp} + I_{rs}}\right| = \left|\frac{R_p - R_s}{R_p + R_s}\right| = \left|\frac{\cos^2(\theta_1 + \theta_2) - \cos^2(\theta_1 - \theta_2)}{\cos^2(\theta_1 + \theta_2) + \cos^2(\theta_1 - \theta_2)}\right| \tag{1.21}$$

For refracted light:

$$P_t = \left|\frac{I_{tp} - I_{ts}}{I_{tp} + I_{ts}}\right| = \left|\frac{T_p - T_s}{T_p + T_s}\right| = \left|\frac{\sin^2(\theta_1 - \theta_2)}{1 + \cos^2(\theta_1 + \theta_2)}\right| \tag{1.22}$$

1.3 Characteristics of Polarization

The degree of polarization (DOP) is the most important parameter for measuring polarization. DOP is defined as: ratio of complete polarization light to total strength of partially polarized light.

$$DOP = \frac{I_P}{I_{\text{sum}}} \tag{1.23}$$

It can also be expressed as: $DOP = \frac{I_M - I_m}{I_M + I_m}$, wherein I_M and I_m are the maximum and minimum light strength in the two special directions whose phases are not relevant but with mutual orthogonality. The better to describe polarized light, we use the Electric Component Method, Jones Vector Method, Bangka Method, and Stokes Vector Method.

1.3.1 Electric Component Method

This method is that used in Sect. 1.1 when describing the status of light polarization. Electric Component Method is about describing polarized light based on classic vibration theories. In a Cartesian coordinate system, if we assume a beam of single-colour polarized light spreads along the z-axis (Fig. 1.6), then we can describe the three electric field vectors as:

$$\begin{aligned} E_x &= A\cos(\omega t - kz + \delta_x) \\ E_y &= B\cos(\omega t - kz + \delta_y) \\ E_z &= 0 \end{aligned} \tag{1.24}$$

where A and B are vibration ranges. ω is the time circular frequency. δ_x, δ_y are initial phases of E_x, E_y, and k is the value of a wave vector, (also called the special circular frequency).

E_x, E_y are shown below as a composite index:

$$\begin{aligned} E_x &= Ae^{i(\omega t - tz)}e^{i\delta_x} \\ E_y &= Ae^{i(\omega t - tz)}e^{i\delta_y} \end{aligned} \tag{1.25}$$

We combine the two vectors so that the combined electric vectors have their ends as elliptical tracks:

$$\left(\frac{E_x}{A}\right)^2 + \left(\frac{E_y}{B}\right)^2 - \frac{2E_xE_y}{AB}\cos(\delta_y - \delta_x) = \sin^2(\delta_y - \delta_x) = \sin^2\delta \tag{1.26}$$

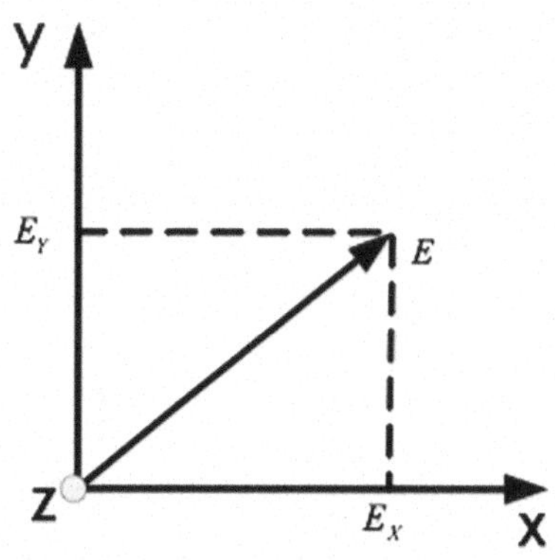

Fig. 1.6 Electric vector method diagram

When δ takes different values, changes in the combined electric vector can be different, which means we can get light of different polarization status. Based on the analysis above, the shape and directions of elliptical light will be decided by vibration ranges and their phase difference. The Electric Component Method is mostly used for analysing and calculating transmission of polarized light in a single optical device.

1.3.2 Stokes Vector Method

After analysis, polarized light is normally embodied as having been subjected to elliptical polarization. We need three mutually independent parameters to describe elliptical polarization, for example, amplitude E_x, E_y and phase difference δ (or elliptical length, short axis a, b and angle of orientation ψ). Stokes proposed use of the "Stokes Parameter" in 1852 to describe complete polarized light, as well as partially polarized light.

The Stokes vector has four parameters (three of them are mutually independent):

$$S = \left[\, S_0\ S_1\ S_2\ S_3 \,\right]^T \tag{1.27}$$

The four Stokes parameters are defined as:

$$\begin{aligned} S_0 &= \langle \tilde{E}_x^2(t) \rangle + \langle \tilde{E}_y^2(t) \rangle \\ S_1 &= \langle \tilde{E}_x^2(t) \rangle - \langle \tilde{E}_y^2(t) \rangle \\ S_2 &= \langle 2\tilde{E}_x(t)\tilde{E}_y(t)\cos\delta \rangle \\ S_3 &= \langle 2\tilde{E}_x(t)\tilde{E}_y(t)\sin\delta \rangle \end{aligned} \tag{1.28}$$

$\tilde{E}_x$, $\tilde{E}_y$ are components of the electrical field vector along x, y in the selected coordinates; δ is the phase difference between two vibration components; and $\langle\ \rangle$ denotes time-averaging. This four-dimensional vector can embody the status of any polarized light including polarization degree. S_0 in the above equation represents polarization light intensity; S_1 represents linear polarization light component; S_2 for 45° linear polarization light component directions; and S_3 for dextrorotation circular polarization light component.

Based on the Stokes vector, we can get polarization status information of any light as below:

$$\psi = \frac{1}{2}\arctan\frac{S_1}{S_2} \tag{1.29}$$

$$P = \sqrt{S_1^2 + S_2^2 + S_3^2}/S_0 \tag{1.30}$$

Herein, ψ is the azimuth angle of an ellipse, which is also its orientation, P describes the polarization degree of partially polarized light, and its value ranges from 0 under non-polarized conditions to 1 under complete polarization. For partially polarized light, P will be the mid-value. Sometimes, I, Q, U, V are used to represent S_0, S_1, S_2 and S_3.

1.3.3 Jones Vector Method

In 1941, Jones describe polarized light with a 2 × 1 matrix (the Jones vector). This describes the polarization status at a point along the direction of light spreading and is as expressed below with complex amplitude:

$$\begin{bmatrix} E_x \\ E_y \end{bmatrix} = e^{i(\omega t - kz)} \begin{bmatrix} A e^{i\delta_x} \\ B e^{i\delta_y} \end{bmatrix} = e^{i(\omega t - kz)} e^{i\delta_x} \begin{bmatrix} A \\ B e^{i\delta} \end{bmatrix} \tag{1.31}$$

Here, $\delta = \delta_y - \delta_x$, and δ has range $-\pi \le \delta \le \pi$. Accordingly, with $0 < \delta < \pi$, sub-vector x is the reverse of sub-vector y; with $-\pi < \delta < 0$, sub-vector y is the reverse of sub-vector x; with $\delta = 0$, sub-vector x and sub-vector y have the same phase; with $\delta = \pm\pi$, sub-vector x and sub-vector y have converse phases.

The coefficient $e^{i(\omega t - kz)}$ is the shared phase of sub-vector x, y. For certain polarized light in polarization optical systems, we are only concerned with the difference between two sub-vectors and so neglect this, but when we compare the phase of two sub-vectors, the two sub-vectors must be at the same location z, at the same time t. At location z, the phase of x, y both change in the form of ωt in terms of time, and change in the form of $-kz$ in space.

Without coefficient $e^{i(\omega t - kz)} e^{i\delta_x}$, a beam of polarized light can be described with the Jones vector as:

$$J = \begin{bmatrix} A \\ B e^{i\delta} \end{bmatrix} \tag{1.32}$$

Light strength I is:

$$I = A^2 + B^2 \tag{1.33}$$

Considering in many real cases of polarization measurement, we are not concerned with total light strength, we make some changes to the form of light strength as a shared factor:

$$J = \sqrt{A^2 + B^2} \begin{bmatrix} \cos\gamma \\ \sin\gamma \, e^{i\delta} \end{bmatrix} \tag{1.34}$$

In the equation above, $\cos\gamma = A/\sqrt{A^2+B^2}$, $\sin\gamma = B/\sqrt{A^2+B^2}$, and $\tan\gamma = B/A$ is the amplitude proportion. The domain of definition of γ is $(0,\ \pi/2)$.

When normalised by light strength, the final Jones vector is:

$$J_n = \begin{bmatrix} \cos\gamma \\ \sin\gamma\, e^{i\delta} \end{bmatrix} \tag{1.35}$$

With $\delta = \pm n\pi$, we have linear polarization light. With $\delta = \pm(2n+1)\pi/2$, we have circular polarization light. With $A = B$ and $n = 0, 1, 2, \ldots$, the normalised vector of related linear polarization light and circular polarization light are: $\begin{bmatrix} \cos\gamma \\ \pm\sin\gamma\, e^{i\delta} \end{bmatrix}$ and $\frac{1}{\sqrt{2}}\begin{bmatrix} \mu i \\ 1 \end{bmatrix}$. Here, −i represents dextrorotation circular polarization light, and +i the levorotation thereof.

1.3.4 Poincaré Sphere Method

In 1892, Poincaré introduce a sphere ($S_0 = 1$) into Stokes space. All complete light polarization has a one-to-one relationship with a point on the Poincaré sphere.

As indicated in Fig. 1.7, any point (P) has its longitude and latitude as 2χ, 2ϕ, then S_0, S_1, S_2, S_3 can be described as:

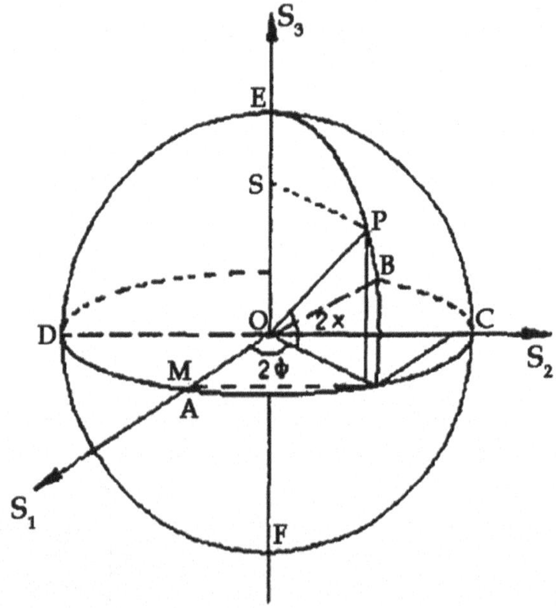

Fig. 1.7 Polarization of light described by Poincaré sphere

$$
\begin{aligned}
S_0 &= 1 \\
S_1 &= \cos 2\phi \cos 2\chi \\
S_2 &= \cos 2\phi \sin 2\chi \\
S_3 &= \sin 2\phi
\end{aligned} \tag{1.36}
$$

From the equation above, any point on the sphere is related to a form of unit intensity complete polarization. Each point on the Poincaré sphere represents for a polarization status. The azimuth angle denotes longitude on the Poincaré sphere and ellipticity denotes latitude: points on the equator of the Poincaré sphere denote linear polarization light; the top and bottom poles stand for dextrorotation polarization light and levorotation polarization light; all other points on the sphere denote elliptical polarization light. The upper hemisphere denotes dextrorotation elliptical polarized light, and the bottom hemisphere denotes levorotation polarized light. The centre of the sphere represents natural light, and all other points distributed within the sphere represent partially polarized light.

1.4 Physical Rationale for Polarized Remote Sensing and Its Three Parts

1.4.1 Two Bottlenecks in Remote Sensing and How Polarized Light Remote Sensing Was Proposed

Regular remote sensing is mostly passive, namely, the sensor passively receives the sunlight that is reflected from the land surface. Although solar radiation covers a large spectral range, visible light and near-infrared only account for a tiny part of the whole electromagnetic spectrum; however, visible light and near-infrared bands contain 80% of received solar energy. So the working spectra of most remote sensors are concentrated on this amplitude, but, when the near-infrared spectrum of visible light goes from the sun to a land surface, and is then reflected to a sensor, atmospheric attenuation occurs so two bottlenecks arise: an optical weakening effect and an atmospheric attenuation effect.

Bottleneck I: optical weakening

High-resolution remote sensing contains four definition dimensions: radiometric, spatial, spectral, and temporal resolutions. Among them, radiometric resolution represents the sensitivity of electromagnetic wave received by a sensor. Enough energy can guarantee sensitivity for object detection and can directly affect the sensitivity of high-resolution spectra as well as pixel sensitivity at high spatial resolutions. Currently, regular remote sensing can only catch good images within one third of the proper range of the entire solar spectrum. For other situations which are too bright (for example, solar flares, star or water flares) or too dark, it is quite difficult to capture a good image. This represents a bottleneck in remote sensing when used for advanced spatial detection and major geological disaster detection. Based on physical and chemical characteristics of different objects, polarized remote sensing has a strong contrast ratio, namely physical mechanisms including bright light "attenuation" and weak light "intensification". This brings about the possibility of resolving this bottleneck issue. This book is based on data that are collected over more than 30 years with support from ***The National Science Foundation***, from more than 300,000 sets of field experiments and over 100,000 sets of space mission experiments. Therein we proved the physical nature of "bright light attenuation, and weak light intensification". We generally confirmed multi-angle physical features, multi-spectral chemical features, roughness and density features, filter features and radiation energy features. This helps provide physical exploration methods able to expand the two ends (bright-dark) of the dynamic sensitive scope of earth detection systems, and achieve remote sensing of land objects by polarization inversion.

Bottleneck II: atmospheric attenuation

Atmospheric attenuation is the major source of error in remote sensing. Normally we measure this effect through experiments, but we do not have a theoretical method to accurately describe it. We take polarization as an atmospheric observation method, explore atmospheric polarized remote sensing theory, and lay a foundation for establishing and improving new atmospheric window theory. By introducing polarization into atmospheric window studies with a focus on the important attributes of light wave polarization (namely complementing the reaction between light on one hand and atmosphere and land objects on the other), the establishment of remote sensing technology helps improve atmospheric window theory and enable studies of this theory to make major progress. In our analysis, we found that the polarization from reactions between light and land objects and the atmosphere has followed atmospheric polarization rules. When we introduce polarization as a solution of atmosphere attenuation, we improve atmospheric polarization figures and the theory of the neutral point of atmospheric polarization. We further studied and summarised the meaning of polarized remote sensing, namely the rules and physical features of full-sky polarization, rules of atmospheric polarization neutral point distribution, multi-angle observations, and three-dimensional atmospheric analysis under full-sky polarization vector field rules, separation between objects and atmosphere, and bionic polarization navigation (Wu 2010; Guan 2010).

Solution: polarized remote sensing

Intensity, frequency, phase, and polarization are four major properties of electromagnetic wave. Traditional remote sensing mainly focused on light intensity of land objects. Polarization, as a new source of information, has allowed significant progress in recent years. Polarization is an important feature of an electromagnetic wave. Sunlight is a transverse electric and magnetic wave with polarization features, but the DOP in different directions is the same: however, with absorption, reflectance, and refraction by different land objects, its vertical and horizontal vectors vary, so it has different vibration intensities in different directions, which leads to partial, and linear, polarization. In nature, many items serve as natural polarisers, including smooth leaves, surfaces of rivers, lakes, seas, snow, ice, soil, desert, clouds, fish scales, and feathers. The polarization status of reflected light might be affected by the surface structure, texture, chemical composition, water content, and angle of incidence, so polarization features can indicate different information about objects. In addition, heat radiation from objects also causes polarization effects and contains polarization features with much information about the targeted objects. This can provide lot of new information for identifying new objects and is thus quite valuable.

Polarization measurement of natural objects started in 1964 (Talmage and Curran 1986). Since then, many scholars have studied polarization reflectance features of land objects. This marks a new remote sensing area to be further studied. In 1996, POLDER (polarization and directionality of the earth's reflectance), a sensor developed in France, was carried on ADEOS and launched into space. POLDER-3 is the only on-track working polarized remote sensor available. The International Society for Optical Engineering (SPIE) hosts professional academic sessions about polarization measurement, analysis, and remote sensing to date they have published 19 series of conference papers. Generally, remote sensing is shifting from qualitative studies to quantitative studies. Polarized light detection is presenting a new trend and flourishing in all dimensions, shifting from single angle to multiple angles, single wave amplitude to full spectrum, from linear polarization measurement to full polarization detection, from single parameter to multiple parameters, from measurement in the laboratory to field experimentation, from ground measurement to aerial, and even spatial, detection. In this process, calibration of instruments is key.

China has come late to polarized remote sensing research which started in the mid-1970s. Three institutes of The Chinese Academy of Sciences, including Changchun Institute of Optics, Fine Mechanics and Physics; Shanghai Institute of Technical Physics; and Anhui Institute of Optics and Fine Mechanics, have developed related polarized detection devices based on scientific need. Northeast Normal University, Peking University and Beijing Normal University mainly investigate the polarization characteristics of land objects. In recent years, polarized remote sensing has developed rapidly in China with more researchers paying close attention to this area. Peking University, Institute of Remote Sensing and Digital Earth of Chinese Academy of Sciences, and Yunnan Observatories of Chinese Academy of Sciences as well as other institutes have carried out sky polarization detection especially atmospheric polarization research. Land surface and sky (or atmosphere) research has helped to study remote sensing land object data inversion. Sky polarization observation stems from atmospheric studies, while land surface polarization observation is a new area of remote sensing. In terms of polarization instrumentation platforms, high-accuracy devices such as polarization BRDF instrument, aviation polarization imagers, and complete Stokes parameter measuring devices have been developed. Polarization calibration has also been top of the agenda. As for polarization characteristics of land objects, polarization reflectance spectrum measurement and characteristic analysis

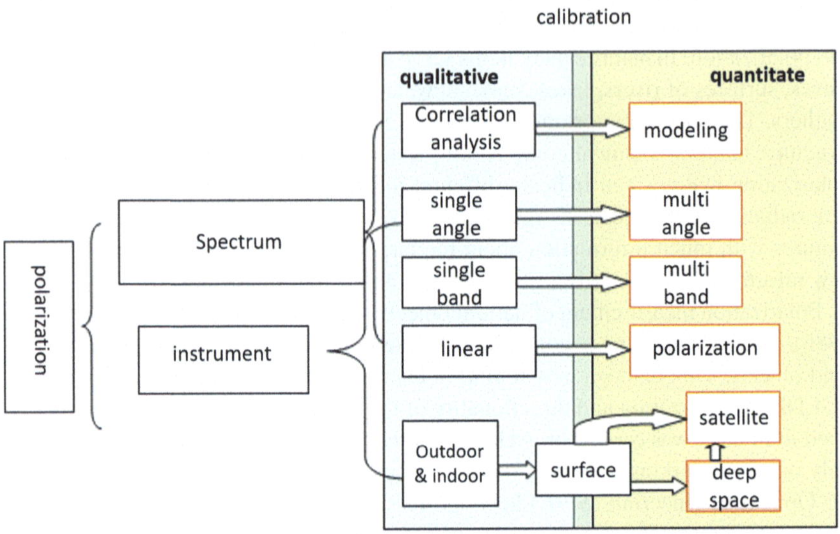

Fig. 1.8 Trend of polarization remote sensing detection

have been conducted on soil, rock, waters, and plant leaves. Quantitative model studies have also made good progress. Studies of the neutral point of atmospheric polarization help to lay the foundation of land remote sensing through aviation and space platforms. Studies of polarized navigation present a blueprint for bionic polarization-based navigation. Driven by improvements in polarization detection instruments and polarized remote sensing quantitative models, polarized remote sensing is entering a flourishing era (Yan et al. 2010) (Fig. 1.8).

Assuming incident radiation is of unit intensity, multiple scattering interaction exists on the ground surface, underground, and at intermediate depths. The final radiation received by sensors would be the total of all scattering components. Normally, radiation on the land surface will be scattered back to atmosphere, be absorbed, or transmitted to depth. Radiation transmitted under the land surface will cause a biochemical reaction with substances underground, such as soil water. Part of the radiation will also be scattered or absorbed: land surface scattering is related to substance composition beneath the land surface. Its modulating effect on radiation is mainly reflectance-based and this part of the total radiation is non-polarized (Knyazikhin et al. 2013a, b) as shown in Fig. 1.9.

Fig. 1.9 Reaction between radiation and the land surface

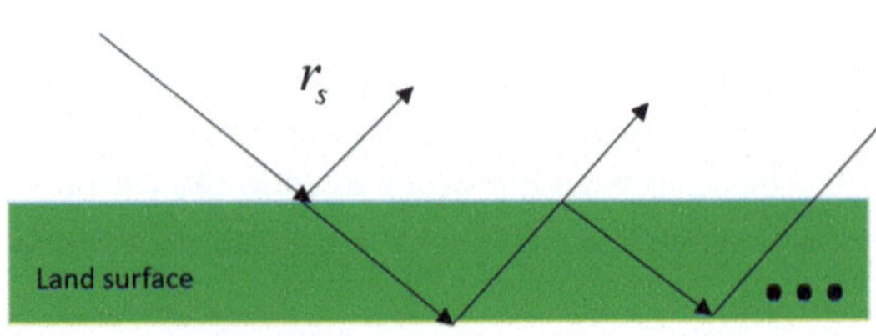

Assuming that scattering albedo is ω, direct reflectance from the land surface is r_s, the probability of upward scattering from within the ground is ρ, and the probability of downward scattering is τ, according to Fresnel theory, the radiation directly reflected from the land surface is partially polarized, so r_s can be further broken down as:

$$r_s = r_{sp} + r_{snp} \tag{1.37}$$

Here, r_{sp} stands for the part of linear polarization, and r_{snp} stands for non-polarized part.

After entering inside of land surface and mutual interaction, the total probability of radiation upward scattering is:

$$r_i = (1 - r_s)\omega\rho \tag{1.38}$$

The probability of radiation absorption and transmission are:

$$a_i = (1 - r_s)(1 - \omega) \tag{1.39}$$

$$t_i = (1 - r_s)\omega\tau \tag{1.40}$$

The total energy of radiation received by a sensor after interacting on the land surface and within can be described as;

$$r = r_s + (1 - r_s)\omega\rho \tag{1.41}$$

The DOP detected by a sensor is:

$$p = \frac{r_{sp}}{r_s + (1 - r_s)\omega\rho} \tag{1.42}$$

Normally, r_{sp} can be viewed as a constant, namely it does not change with wavelength. When land surface reflectance is low, which means when $\omega\rho$ has a small value, the target is quite dark visually, then DOP p can be relatively large. This means that "weak light intensification" is realised. On the contrary, when the reflectance of land surface is high, namely when $\omega\rho$ is large, the target is quite bright visually, then DOP p can be relatively small, and this means that "bright light attenuation" is realised.

1.4.2 The Four Study Targets and Five Characteristics of Land Object Remote Sensing

The most typical land objects are water, soil, rock, and vegetation. These are also the four targets for land polarization studies from which five features of land objects can be summarised: multi-angle physical characteristics, multi-spectrum chemical characteristics, roughness and density characteristics, filtering characteristics, and radiation characteristics.

1. **Multi-angle physical characteristics**

During reflection, scattering, and transmission, multi-angle spectral feature and polarized feature based on intrinsic characteristics of land objects arise. By studying the multi-angle spectral features and polarized features we can identify multi-angle reflectance, and polarization reflectance, laws in 2π-space. Those potential laws, together with abundant information about angle and polarization bring new methods to remote sensing data analysis.

When the angle of incidence is small, there is no mirror effect, and mostly objects are expressed as Lambert features: at greater angles of incidence, obvious mirror effects arise as a result of the combination of mirror reflectance and diffuse reflectance (Zhao 2004). In the mirror reflectance direction, there is no obvious discontinuity, only a continuous, smooth rise.

The following conclusions can be drawn from studies of angle polarization reflectance and bidirectional reflectance of multiple rock types, including: peridotite, granite, limestone, gabbro, and basalt (Zhao et al. 2003, 2004a, b).

(1) Reflection light is partially polarized light where the vibration vertical to the incident surface is at an advantage. Refraction light is the partially polarized light where the vibration horizontal to the incident surface is at an advantage. With increasing angle of incidence, vibration vertical to the incident surface accounts for a larger proportion of the reflected light, which means that the polarization of reflected light increases. This shows that the amplitude in the incident direction decreases and that vertical to the incident direction increases: polarization always takes place in the direction towards the normal to the incident surface.
(2) Polarization of light reflected by land object surfaces is theoretically affected by two factors: the first being the angle of incidence, the second being the refractive index. The different compositions and structures of land objects give rise to different refractive indices, so polarization of reflected light varies. Physical features of land objects can be inverted from polarization signal data and vice versa.
(3) The DOP of the reflection spectrum does not change much in 2π-space with azimuth or viewing angle and it fluctuates by around 5%, and curves of different viewing angles exhibit a certain twist. This means that DOP, as a physical indicator, is relatively stable throughout 2π-space and does not change to any significant extent (Zhao et al. 2008).

2. **Multi-spectral chemical characteristics**

To compare reflectance and DOP spectra when reflectance is at its maximum, the DOP is normally at its minimum and vice versa. When incident on a dielectric surface, after reflection and refraction, natural light will normally become partially polarized light: the definition of DOP is the based on the ratio of complete polarization light to the total intensity of partially polarized light. Since complete polarization of light varies according to wavelength and measurement angle, there may be a loss of light after absorption, reflection, and refraction in any polariser, so the inverse relationship between spectra is not constant, namely, there is no constant inversion coefficient. It is only demonstrated as a qualitative relation with inverse trough-crest data.

(1) When the angle between polarization direction of incident light and transmission axis is 45°, the intensity after passing through a polariser is halved and the arithmetic average of light intensity in the transmission, and extinction, directions is equal to the intensity when the angle between the polarization direction of incident light and its transmission axis is 45°.
(2) The 45° polarization reflectance, bi-directional reflectance, arithmetic average of reflectance of light intensity in the transmission, and extinction, directions are the same at a given azimuth, zenith, viewing angle, and pathway (Wu et al. 2010).

These conclusions summarise the quantitative relationship between multi-angle polarization reflectance and bidirectional reflectance: they can help to improve quantitative accuracy of current remote sensing methods. Quantitative relationship between reflectance and DOP can be used as a benchmark against which to measure spectral quality (Xiang 2010).

3. **Filtering characteristics**

Regular remote sensing methods cannot capture the information needed in too dark, or too bright, an environment. Polarized remote sensing is characterised by "bright light attenuation" and "weak light intensification", and can help analyse characteristics of land surfaces with strong reflection. Studies of the characteristics of land surfaces with strong reflection are carried out from two standpoints: to study the direction and polarized spectral characteristics of polarization information about the land surface based on the relationship between radiative transfer theory, Stokes coefficient, strong-refraction land surface refraction, and emission polarization theories; secondly, to build a relationship between polarization characteristics and snow features after calculating background feature results using polarization methods. The two images below are related to plane and water surface experiments with a high signal-background ratio (Figs. 1.10 and 1.11).

(a) Intensity image (b) DOP image (calculated)

Fig. 1.10 Intensity image and DOP image of aircraft

(a) Before rotating polarization film (b) When polarization film is rotated to certain angle

Fig. 1.11 Effects of different angles of rotation of a polarising film on a water surface

This study shows that:

(1) Experiments show strong intensity of background and weak intensity of the plane. Planes which showed less obvious details in intensity images can be very obvious in a DOP image. This means that, with strong background intensity and weak object intensity, polarization can help detect detailed information about objects.

(2) When a polarising film is rotated on a water surface, information about underwater objects cannot be detected at certain angles because the reflected light intensity is too high, but at certain other angles, such reflected light can be filtered with a polarization film, and underwater object information can be detected: this means that polarization can detect details of an object even with a high-signal background ratio where there is a strong water reflection noise intensity and weak underwater object intensity.

4. **Radiation characteristics**

Knjazihhin's work, published at PNAS (Proceedings of the National Academy of Sciences of the United States of America) in 2013, demonstrated that the high-spectral performance of vegetation is not only related to the optical features of leaves and incidence/observation directions, but also to the canopy structure of vegetation (Knyazikhin et al. 2013c). Conclusions from earlier research, indicating that the BRF of vegetation is linearly related to its nitrogen content (Ollinger et al. 2008), might be wrong because the sample distribution of broad leaves and needle leaves is basically linear. Moreover, the BRF information of these studies has to consider surface and internal scattering. To understand the internal chemical composition of leaves, we must adopt polarization methods to eradicate the effects of surface mirror scattering.

Remote sensing of radiology and phonology has been conducted for many years, and polarized remote sensing represents the new trend in space remote sensing technology development. Compared to traditional remote sensing, polarized remote sensing is unique in that it can resolve some issues that cannot be solved using ordinary methods, for example, particular diameter distributions of clouds and aerosol particles. Polarization measurement accuracy can be high without accurate radiation calibration, and radiation information can be obtained together with polarization data.

1.4.3 Two Theoretical Findings and Chromatographic Observation of Atmospheric Polarized Remote Sensing

The development of remote sensing science and technology is now at a new stage: to be specific, theoretically it is shifting from simple qualitative research to quantitative research, from just describing the relationship between measured radiation value and land surface phenomena to quantitatively describing the relationship around bidirectional reflectance and radiation with the radiative transfer model, from forward radiative transfer model to quantitative inversion of radiative transfer model, and from focusing on separate amplitudes (for example) to integrating wave spectrum amplitudes. From the view of technical development, it has developed from single band to multi-spectral, multi-angle, multi-polarized, multi-temporal, and multi-modal measurement, as wells as from single sensor to integrating multi-sensor detection (Chen 2013).

The bottleneck issue facing remote sensing observation, namely remote sensing atmospheric window attenuation, is the most important source of remote sensing observation error. Regular remote sensing observation is inevitably affected by atmosphere attenuation through the atmospheric window. Although atmospheric window studies help remote sensing observations, attenuation of electromagnetic wave in the atmosphere is significant at all bands. We urgently need to improve atmospheric window theory. Moreover, current studies on attenuation effects of electromagnetic waves by introducing polarization into atmospheric window studies with a focus on the important attributes of light wave polarization (namely complementing the reaction between light on one hand and atmosphere and terrestrial objects on the other hand), the establishment of remote sensing technology helps to improve atmospheric window theory.

Polarization studies on the interaction among light, land objects, and atmosphere have found a suitable atmospheric polarization rule. When introducing polarization as a solution to atmospheric attenuation, studies implemented atmospheric polarization modes and atmospheric neutral point theory. With further research and application, we may investigate the physical features of aerosol and air pollution particles, ground-atmosphere separation, and bionic polarized navigation (Wu 2010; Guan 2010).

1. **Exploration on atmospheric polarization modes and physical features of atmospheric polarized remote sensing**

Many plants and creatures navigate by taking advantage of polarized light because some plants and creatures have very sensitive vision systems that help them detects the direction of polarization of sky light. Secondly, there is a relatively stable atmospheric polarization pattern in the sky (Liang 2004; Schott 2009; Nnen 1985; Chen and Rao 1968).

(1) Observation analysis of atmospheric polarization mode under ideal weather conditions

In the real world, the distribution of model figures of atmosphere atmospheric polarization is affected by many factors, especially climate. We analysed its distribution under different climate conditions, Fig. 1.12 shows the spatial distribution of DOP and polarization angles under different climate conditions.

a. **DOP measurements in different weather conditions**

Since the distribution of atmospheric polarization light is closely related to scattering of sunlight by atmospheric particles, we conducted comparative studies of DOP distributions under sunny and cloudy conditions (Fig. 1.12).

It can be found that atmospheric DOP distributions are annular and centred around polarization neutral points. From a neutral point to the outer area, DOP increases until reaching the areas shown in black where DOP is the highest, and then DOP decreases to another neutral point. From images recorded on cloudy days, it can be found that the DOP is much smaller than on cloudless days, because multiple scattering in the atmosphere on cloudy days creates depolarization effects which decrease the DOP.

b. **AOP measurements in different weather conditions**

Creatures can not only navigate using sky light, but also with related polarization vision to detect space polarized light, and moreover to obtain direction information based on the distribution of angle of polarization.

By analysing polarization angles (Fig. 1.12), it can be found that there is some similarity between images recorded in cloudy and sunny weather: the distribution of AOP is symmetrical around a polarized neutral point. From images recorded

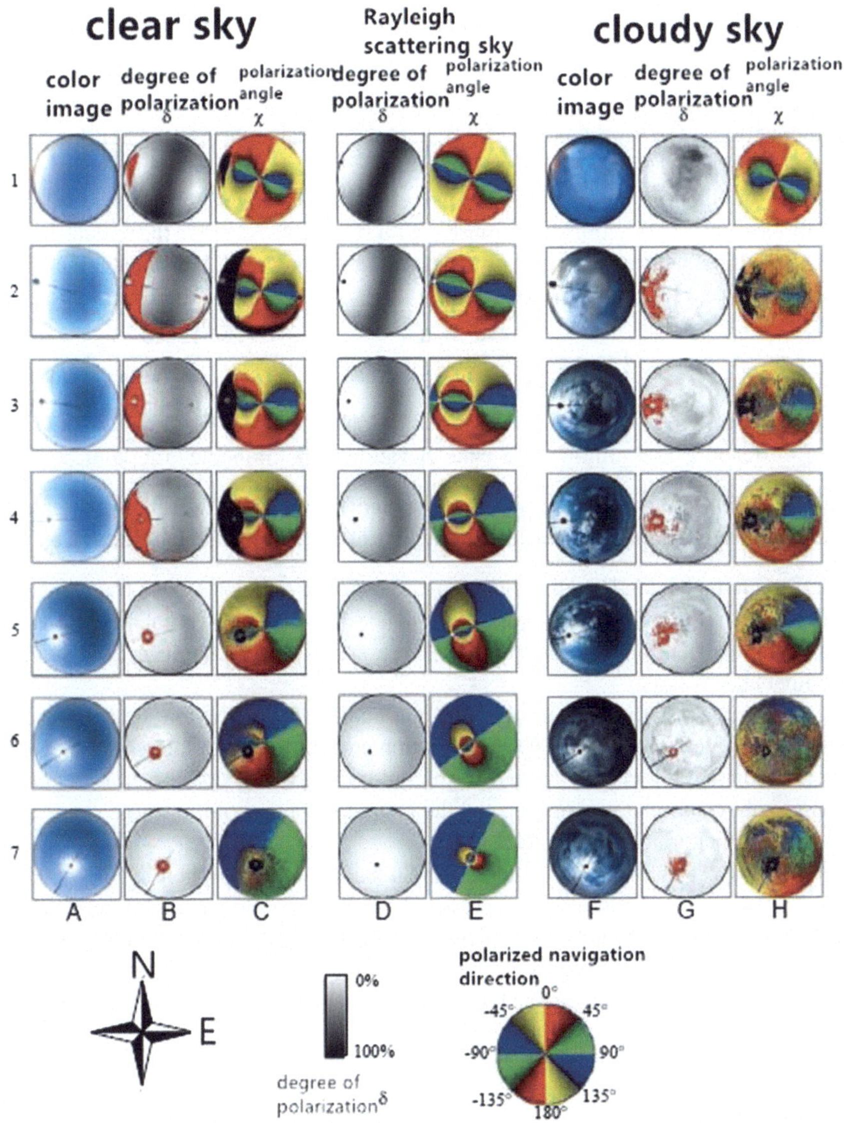

Fig. 1.12 Spatial distribution of DOP and polarization angles under different climate conditions

on cloudy days, it can be found that DOP is much smaller than on cloudless days, because multiple scattering by particles creates depolarization effects which decrease the DOP.

2. **Exploration of distributions around the neutral point area and discussion related to changes to atmospheric window theory**

(1) **Distribution of atmosphere polarization around neutral points**

Atmosphere polarization neutral points are the points whose DOP is zero in this area. At the beginning of the nineteenth century, three neutral points were discovered, including the Arago neutral point (1809), Babinet neutral point (1840), and Brewster neutral point (1842) (Fig. 1.13).

Here, the Babinet neutral point is the best choice for polarized ground observation, and its location is of functional relationship with solar altitude. When the optical depth is 0.10, Babinet neutral point curves bend a little. To understand this better, we split it into two linear portions:

$$\begin{cases} y = 0.9x + 18 & (0 < x < 30) \\ y = 0.75x + 22.5 & (30 <= x < 90) \end{cases} \tag{1.43}$$

where x denotes solar altitude, and y the elevation angle of the neutral point.

(2) **Detection of atmosphere polarization neutral point and discussion related to changes to the attenuation coefficient of the atmospheric window**

The better to understand the distribution of atmospheric polarization neutral points in real world conditions, we analysed time sequence data for neutral points. Figure 1.14 shows the distribution of full-sky DOP: there are two polarization neutral points (Babinet and Arago).

As in Fig. 1.14, from location of the sun and DOP distribution at all times, it can be seen that atmospheric DOP distribution and neutral point location vary with sun position and are closely related to solar altitude. DOP is distributed as a circle around polarization neutral points, and the DOP distribution of the whole sky is the overlapping pattern of two circles.

From Figs. 1.14 and 1.15, it can be concluded that an atmosphere DOP window does exist and DOP is distributed regularly around polarization neutral points. Since the linear DOP of polarization neutral points is zero, ground observation from neutral points can remove atmospheric polarization, and therefore, obtain information about targets directly. This serves as the basis for an atmosphere window attenuation law.

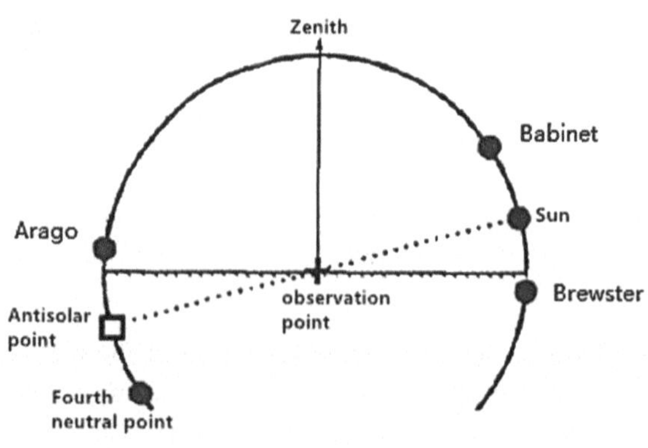

Fig. 1.13 Neutral points in space

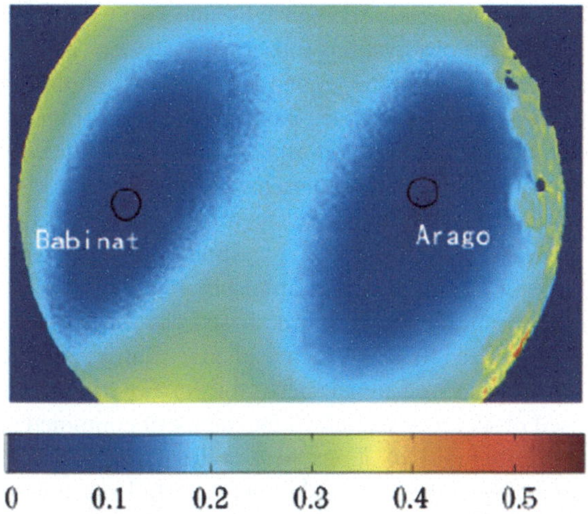

Fig. 1.14 Distribution of full-sky DOP

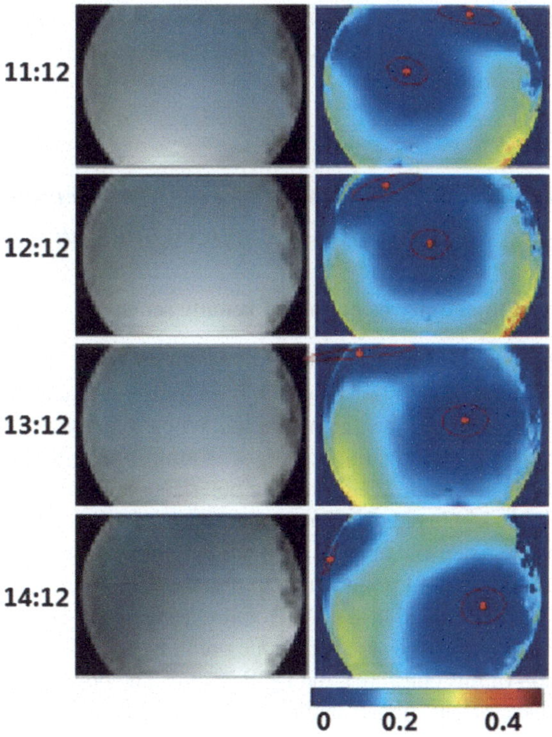

Fig. 1.15 Distribution of full-sky DOP based on continuous detection

3. **Atmospheric polarization pattern—interpretation of neutral point area and atmosphere combination and chromatography distribution of polluting particles**

Remote sensing observation falls into two classes: the first is airborne-based and space-based observations (observation devices are carried by a balloon, aircraft, or satellite platform); the second is ground-based observation (land photometry, etc.). Airborne-based and space-based observations conduct inversion of parameters by reflection signal from atmosphere and land surfaces and ground-based observation usually conducts atmosphere inversion by direct or scattered sunlight. Characteristics of different atmospheric layers require analysis of aerosol and other atmospheric physical features using dual measurement (Cheng et al. 2008; Sun and Zhao 2009). Based on relatively stable atmospheric polarization modes and atmosphere polarization neutral points, airborne-based, satellite-based, and ground-based observation are conducted to analyse atmosphere polluting particles and aerosol features.

Dual polarized signals derived from airborne-based, satellite-based, and ground-based observation explained the theory of combined atmospheric interpretation and chromatography of atmosphere polluting particle distributions. With further study and analysis, the relationship between full-sky polarization signals from airborne-based and satellite-based platform observations and optical depth aerosol as well as aerosol models will be discovered. We theoretically predicted that full-sky atmospheric polarization models derived from airborne-based and satellite-based observation can serve as an important basis for aerosol information separation inversion.

1.4.4 Physical Theory of Polarized Remote Sensing and Its Objectivity, Uniqueness, and Stable Repeatability

1. **Objectivity of polarized remote sensing**

The most typical application of remote sensing is in advanced space detection and global changes studies, which is an important method for making breakthroughs in major international science issues. When solar radiation reaches the earth's surface, it embodies polarization feature after atmospheric scattering: if atmosphere scattering is mostly single scattering, the polarization of this kind of solar radiation may compose a stable polarization distribution, and this is called the atmospheric polarization mode pattern. Atmospheric neutral point contains two meanings: first, in real situations, a neutral point is not an independent point due to multiple scattering effects. It is an area with a sky DOP of zero (or near zero). Secondly, a neutral point is not a fixed area in the sky, but a cone pointing towards a neutral point area. At any elevation of this cone area, the DOP is zero. The objectivity of astronomical polarization observations of stars and planets inspired the complementarity and convertibility of polarization observation methods in astronomy and remote sensing applications. All-field comparison and objectivity of whole sky polarized field, gravitational field, and geomagnetic fields indicate the possibility of breakthrough in observation methods of different astronomical phenomena.

2. **Uniqueness of polarized remote sensing**

The brightness ratio of the moon to regular sunlight is one to four million, and its radiant features are similar to the terrestrial surface. Changes in moon phase can make moon calibration more difficult, but the moon is the only feasible calibration source that has a certain brightness and its surface intensity is barely affected by atmospheric effects. Its irradiance at certain phases does not change to any significant extent and the annual change is 10^{-8} in magnitude. This is why the moon can be regarded as an ideal calibration source for satellite sensors. Moreover, in terms of the relationship between biochemical content of vegetation and BRF, the linear correlation between vegetation BRF and nitrogen content may be wrong: BRF information from these studies is based on surface and internal light scattering. The better to understand the internal chemical composition of leaves, we must adopt polarization methods to eradicate effects of surface mirror scattering. All of the above demonstrates the uniqueness of polarized remote sensing.

3. **The stable repeatability of polarized remote sensing**

Some reptiles (and for example, ants) can navigate using the angle between the atmospheric polarization vector line and the axis of their body. This is the best example the can be used to improve stable repeatability of polarized remote sensing applications. Bionic polarization navigation is an original navigation method. Foreign studies in this area have come a long way, but are mostly concerned with the polarization light navigation capability of some creatures and chemical rectification of their actions, seldom on navigation applications. Research into this area started late in China, remains in its infancy, and has not yet generated any applicable products. The slow progress in the application of polarization light navigation has arisen because, firstly, atmospheric polarization figures, which are the basis for such work, are severely affected by the environment. Secondly, sensors to acquire atmospheric polarization light direction information still need to be improved. So it is necessary to begin pioneering studies both in theoretical and practical contexts.

1.5 Organisation of Chapters

This book is made up of eleven chapters.

This chapter stands alone and mainly covers the physical basis of polarized remote sensing. This chapter discussed two bottlenecks in remote sensing studies and proposed polarization as a solution thereto.

The first part is about land object polarization physics (Chaps. 2–6). With the study of the four main land objects, five features can be identified as key: multi-angle physical characteristics, multi-spectral chemical characteristics, roughness and density characteristics, filtering characteristics, and radiation characteristics.

The second part covers atmospheric polarized remote sensing physics (Chaps. 7–9). In this part, we discuss the meaning of atmospheric polarized remote sensing from the perspectives of its governing laws and physical explanation of atmospheric polarization patterns (Chap. 7), rules governing atmospheric polarization neutral point areas and ground-atmosphere separation (Chap. 8), and multi-angle observations and three-dimensional chromatography of atmosphere particulates (Chap. 9).

The last part of this book covers the application of, and progress in, polarized remote sensing (Chaps. 10 and 11). By introducing bionic polarization self-navigation techniques based on terrestrial polarization vector fields (Chap. 10), and methods for advanced space detection and global changes studies (Chap. 11), it proves that polarized remote sensing is objective and unique and can be repeatedly applied on a stable basis, which can help China play a leading role in theoretical, methodological, and application-based work in polarized remote sensing physics.

This book gives the physical basis of polarized remote sensing, land object polarized remote sensing physics, atmospheric polarized remote sensing physics, and multiple applications thereof as its main thread. The book introduces the laws governing various polarized remote sensing methods and offers insights into related polarized remote sensing applications. This is the first systematic monograph about polarized remote sensing published anywhere.

References

Chen W (2013) Research on non-spherical aerosol properties and optical properties inversion method. PhD thesis, Peking University

Chen HS, Rao CRN (1968) Polarization of light on reflection by some natural surfaces. J Phys D Appl Phys 1(9):1191–1200

Cheng T, Gu X, Chen L et al (2008) Study on multi-angle polarization characteristics of cirrus. Acta Phys Sinica 57(008):5323–5332

Guan G (2010) Biomimetic polarization navigation orientation mechanism and experimental verification research. PhD thesis, Beijing Institute of Technology

Knyazikhin Y, Schull MA, Stenberg P et al (2013a) Hyperspectral remote sensing of foliar nitrogen content. Proc Natl Acad Sci USA 110:E185–E192

Knyazikhin Y, Lewis P, Disney MI et al (2013b) Reply to Ollinger et al.: remote sensing of leaf nitrogen and emergent ecosystem properties. Proc Natl Acad Sci U S A 110:2438

Knyazikhin Y, Lewis P, Disney MI et al (2013c) Reply to Townsend et al.: decoupling contributions from canopy structure and leaf optics is critical for remote sensing leaf biochemistry. Proc Natl Acad Sci U S A 110:E1075

Liang S (2004) Quantitative remote sensing of land surfaces. Wiley-IEEE, Washington

Nnen GP (1985) Polarized light in nature. Cambridge University Press, Cambridge

Ollinger SV, Richardson AD, Martin ME et al (2008) Canopy nitrogen, carbon assimilation, and albedo in temperate and boreal forests

Schott JR (2009) Fundamentals of polarimetric remote sensing. Society of Photo Optical, Bellingham, Washington

Sun X, Zhao H (2009) Inversion of optical properties of aerosols on land based on POLDER data. Acta Opt Sinica 007:1772–1777

Talmage DA, Curran PJ (1986) Review article: remote sensing using partially polarized light. Int J Remote Sens 7(1):47–64

Wu T (2010) Study on the properties of land objects and ground-gas separation methods in polarization remote sensing. PhD thesis, Peking University

Wu T, Yan L, Xiang Y et al (2010) Multi-angle polarization spectrum characteristics of water bodies and their applications in water color remote sensing. Spectro Spectral Anal 30(2):448–452

Xiang Y (2010) Preliminary study on multi-angle polarization spectrum measurement and related characteristics of rock. PhD thesis, Peking University

Yan L, Xiang Y, Li Y et al (2010) Progress in Polarization Remote Sensing. J Atmos Environ Opt 5(3):162–174

Yao Q (2008) Optical tutorial, 4th edn. Higher Education Press, Beijing

Ye Y, Rao J, Xiao J et al (2005) Optical tutorial. Tsinghua University Press, Beijing

Zhao H (2004) Multi-angle reflectance spectroscopy and polarization reflectance spectroscopy characteristics of rock. PhD thesis, Peking University

Zhao H, Zhai L, Zhao Y (2003) Multi-angle polarization reflectance spectroscopy study of basalt. Geogr Geo-Inf Sci 19(4):81–83

Zhao H, Yan L, Zhao Y (2004a) Multi-angle polarization reflectance spectroscopy of peridotite. Geol Explor 40(2):51–54

Zhao H, Yan L, Zhao Y (2004b) Multi-angle polarization reflectance spectroscopy study of granite. Mineral Rock 24(2):9–13

Zhao H, Xu L, Ma W (2008) Study on the influence of solar elevation angle on the 3D spatial reflectivity of ground surface. Remote Sens Inf 4:3–6

Part II
Physical Rationale of Polarization Remote Sensing of Ground Objects

As a new remote sensing observation method, and a supplement to conventional remote sensing, polarization remote sensing possesses the important physical characteristics of being able to "attenuate bright light" and "intensify weak light" so as to extend the application of the two ends (the "bright" and "dim" areas) of remote sensing detected light spectra. Such capabilities can be summarised into five characteristics of polarization remote sensing: multi-angle reflection physical features (Chap. 2), multispectral chemical properties (Chap. 3), structural properties of roughness and density (Chap. 4), characteristics of info-background high contrast ratio filter (Chap. 5), and the features of radiative energy transfer (Chap. 6).

Chapter 2
Remote Sensing of Ground Objects 1: Physical Characteristics of Multi-angle Polarized Reflectance

This chapter describes the first and most important feature of remote sensing of surface features—the physical characteristics multi-angle polarized reflectance. Specifically including: the physical detection geometry of polarized reflectance and the physical basis of multi-angle polarized reflectance to ensure the accuracy of quantitative research into polarization remote sensing of ground objects; Multi angle polarized reflectance analysis instrumentation and measurement methods are used to realise sample analysis of polarimetric remote sensing objects; analysing the angle of incidence and the law and mechanism of multi-angle beam of reflectance of ground object samples to explore natural light's effect of root cause on polarized reflectance; analysing the characteristics and laws of multi-angle non-polarized spectra, and the degree of polarization to explore the nature of physical characteristics of multi-angle polarized reflectance.

2.1 The Physical Basis of Physical Detection Geometry of Polarized Reflectance and Multi-angle Polarized Reflectance

This section explores the physical basis of the design of multi-angle polarization reflectance spectroscopy instruments based on formula used to calculate the degree of polarization.

2.1.1 Degree of Polarization Calculation

Degree of polarization (DOP) can be represented as (Ye et al. 2005)

L. Yan et al., *Polarization Remote Sensing Physics*, Springer Remote Sensing/Photogrammetry, https://doi.org/10.1007/978-981-15-2886-6_2

$$DOP = \frac{I_M - I_m}{I_M + I_m} = \frac{I_{90} - I_0}{I_{90} + I_0} = \frac{R_{90} - R_0}{R_{90} + R_0} \tag{2.1}$$

where I_M and I_m are the maximum and minimum light intensities corresponding to the two special directions that are phase-independent and mutually orthogonal. In this experiment, the two directions correspond respectively to 90° and 0° in the polarising prism. Calculations are derived from the ratios of 90° polarization with respect to the whiteboard reflectance (R_{90}) and 0° polarization with respect to the whiteboard reflectance (R_0).

2.1.2 Detection Geometry: Multi-angle Polarized Reflectance

When conducting multi-angle detection, the coordinate system has to be established. The detection geometry is specified as follows: with a viewing zenith angle probing vertically downwards at 0°, the forward direction of beam is positive while the backward direction is negative; where the azimuth viewing angle is 0° rotating clockwise around the light source (Fig. 2.1).

The area of the detected field of view will change during vertical and tilt observations as shown in Fig. 2.2: α is the detector half-angle field of view, β is the viewing zenith angle, y is the vertical probe, or the radius field when apex angle is 0°, while the current field of view is circular. During tilted observation, namely when detecting

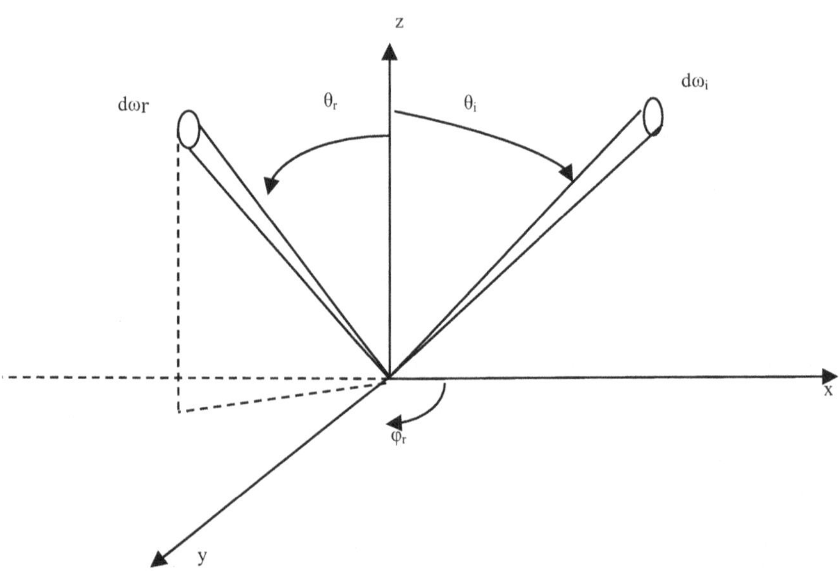

Fig. 2.1 Measurement geometry schematic (θ_i: incident zenith angle; θ_r: viewing zenith angle; φ_r: viewing azimuth angle)

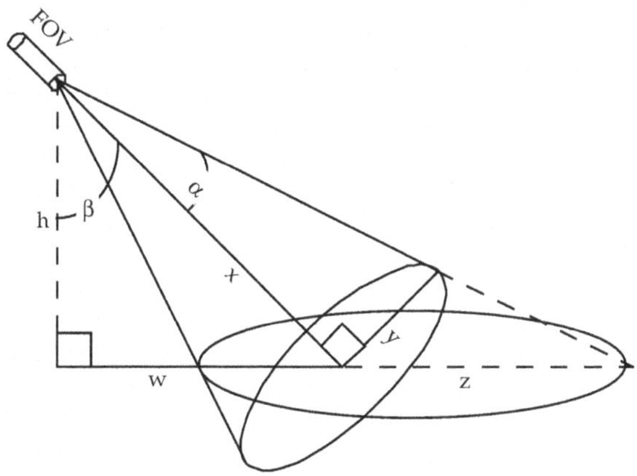

Fig. 2.2 Schematic diagram of tilt detection

the change of zenith angle, changes in the field of view are elliptical while there is corresponding elongation of y to z, thus:

$$y = x \tan \alpha \tag{2.2}$$

$$z = h[\tan(\beta + \alpha) - \tan(\beta)]. \tag{2.3}$$

2.1.3 Polarization Reflectance Spectrometer for Multi-angle Polarization Mechanism Verification

(A) Instrument

Multi-angle polarization reflectance spectroscopy was developed by the Changchun Institute of Optics and Fine Mechanics, Chinese Academy of Sciences. The instrument consists of three major components: a light source system, a bidirectional (polarized) reflection photometer and a control system. Its layout is shown in Fig. 2.3 and the various parts are discussed in detail below.

(1) Lighting system

In the light source system, a bromine tungsten lamp is used as a radiant light source, together with an optical imaging system that allows the light to form a uniform spot of 10–15 mm diameter on the illuminated object. Since the colour temperature of the bromine tungsten lamp is generally above 3000 K and the light emission is stable, it is suitable as a light source for measuring light intensity with a rated power of 15 W.

To change the incident angle of the light source, a light source holder is employed and provides an incidence azimuth every 10° ranging from the zenith angle and changing from 0° to 60°, so that the light source on the stent will be able to suit

Fig. 2.3 Multi angle polarized reflectance spectrometer

particular needs by adjusting the angle of incidence of the light source. At the same time, the rotatable polariser can be installed at the entrance of the light source to form different polarization states after light passes through the polariser, irradiating target objects.

(2) Bidirectional reflectance photometer system

The detection frame of the photometer is rotatable, therefore the azimuth is 0°–360°, and data can be collected every 10° such that 36 sets of data can be measured in the horizontal direction at one time. Simultaneously, while there are seven detectors on the detection frame and the angle interval is 10° so that the scope of viewing angle is 0°–60°. If the polariser is not positioned at the incident beam, the measured data form a multi-angle bidirectional reflectance spectrum of the ground object; when equipped with a polariser, the 0° axis of the polariser is the direction of the transmission axis, and the measured data is defined as the 0° polarization data of the feature: when aligned at 90°, which is the direction of the extinction axis, the data represent polarization data at 90° to the surface features.

(3) Band settings and control system

Two detection frames coexist because the instrument has both A- and B- bands. When measuring the A-band, the B-band lies horizontally on the stage, thus affecting the light irradiation. Also, when measuring the B-band, the A-band lies horizontally.

The wavelength of the A-band is 630–690 nm while the wavelength of the B-band is 760–1100 nm.

To collect data quickly and reduce human intervention, an automatic control system was designed: the detector can rotate and collect data automatically on the platform by computer control, thus realising the automatic measurement, acquisition, storage, and display of spectral data.

Through analysis of the instrument above, it can be found that when a single incident angle is fixed and a band is fixed, the system is able to collect data of a certain ground object from 36×7 directions within the 2π spatial orientation. In actual measurement, since the detection frame is inside the light source holder, the detection frame would block the incident beam at the 0° azimuth position, so that all the data collected at the 0° azimuth is invalid, causing the number of actual data collected to be $35 \times 7 = 245$. Finally, counting in the changes of the six incident angles, the changes in two bands, and the three states of the polariser (no polariser, 0° polarization, and 90° polarization), there would be $245 \times 6 \times 2 \times 3 = 8820$ sets of data collected for each surface feature.

The instrument has a revolving stage at a radius of 15 cm, an arm radius of 15 cm, and a revolving stage holder with a height of 12 cm. Its systematic error is 5% and its accuracy is 4%.

All experimental work was carried out under darkroom conditions at night to prevent external sunlight and indoor ambient light from interfering.

(B) Production and measurement of samples

To standardise rock samples and guarantee comparability between them, a flat surface will be cut on each of the 20 types of rock samples selected. Then the measured object is placed on the central stage of the Multi-angle Polarization Reflectance Spectroscopy system after adjusting the height and horizontal position. urn on the light source, we first measure in a situation where the polariser is not placed at the incident light source before rotating the polariser to the desired angle. Reflectance spectra are measured in the 2π spatial orientation in Band-A (690–760 nm) and Band-B (760–1100 nm) in three states: without a polariser, 0° polarization, and 90° polarization, while simultaneously varying the incident angle of the light source to measure spectral reflectance values at different zenith angles. In this way, the five factors (incident angle, viewing angle, azimuth angle, band, and characteristics of polarization) are all changed to form an independent variable factor, in which the viewing angle and the azimuth angle are combined to form the 2π-space of the ground object. As such, their influence can be analysed for their effect on multi-angle reflection spectra and polarization spectra of the samples in 2π-spatial orientation.

In the measurement, the incandescent light source is irradiated from different zenith angles, and the azimuth position of the light source is always fixed, so that the azimuth angle of the light source is $\varphi_0 \equiv 0°$. The viewing angle is rotated along different azimuth angles in the horizontal plane, divided into an observation azimuth intervals of every 10° within 360°, the vertical plane is divided into seven channels each having a phase difference of 10°. Consequently, data can be sketched to show a retro-reflective diagram of the reflective spectrum in 2π-spatial orientation (since the maximum angle of detection is 60°, strictly speaking, it is not a complete hemisphere). To study the characteristics of the polarization reflectance of the rocks, a rotatable polariser is installed at the exit of the incident beam. At the same time, it is prescribed that the direction of the transmission axis of the polarising plate is 0°, and the direction of the orthogonal direction is defined as 90° (hereinafter referred to as 0° polarization and 90° polarization). As a result, another retro-reflective diagram of the rock sample reflective spectrum at 0° and 90° polarizations in the 2π-spatial orientation could be obtained.

2.2 Multi-angle Polarization Reflectance Analytical Instruments and the Methods of Measurement

Multi-angle polarization reflectance spectroscopy can be sorted into indoor, and outdoor, types, and is based on the principle of measuring the polarization intensity distribution in the 2π-space from different incident detection orientations. The introduction of indoor and outdoor multi-angle polarization reflectance spectroscopy will be mentioned in the following section. Also, experiments on the multi-angle polarization of outdoor samples will be conducted outdoors first (such as vegetation observation); after an in-depth analysis of the physical characteristics of the measurements made outdoor, further collection of specimens for indoor fine measurement to eliminate external disturbances, simplification of experimental conditions, and the fine multi-angle polarization reflection law of ground objects. Rock or soil sample analysis often uses this method. Without the loss of generality, this section contains a multi-perspective analysis of the laws using rock samples.

2.2.1 Outdoor Multi-angle Polarization Measuring Device

The outdoor polarization multi-angle measuring device is generally equipped with a spectrometer by a device capable of multi-angle measurement, and a polarising prism is installed in front of the spectrometer probe to allow polarization multi-angle measurement. Staff at The Institute of Geographical Sciences and Natural Resources Research, China Academy of Science invented an omnidirectional multi-angle ground-based measurement platform. Although it can achieve multi-perspective, multi-angle measurements of the target, the shadow of the observation frame itself affects the quality of the observed and this is not only limited to the specific remote sensing observation field, but also cannot carry multiple instruments to achieve simultaneous observation. Beijing Normal University has a manually operated portable multi-angle observation device, which solves the problem of high uniqueness, poor portability, and large shadow interference in remote sensing observation, but suffers the disadvantage of having a small observation platform and being incapable of bearing heavy load. Anhui Institute of Optics and Fine Mechanics of the China Academy of Science designed a multi-angle measurement system consisting of an automatic measuring stand and two field spectroscopies produced by the ASD (Analytical Spectral Devices) Company (Li et al. 2008). One spectrometer is fixed on the measuring platform to measure the reflection of the target in each direction, while the other spectrometer is placed on the ground measuring the reflection of the diffuse reflector. The system was used to conduct an experiment measuring an outdoor lawn, with the entire period (66 positions) taking 10 min and the measurement of the principal plane (with an interval of 5°, and a total of 31 positions) taking 2.5 min in total.

Staff at Beijing Agriculture Information Technology Research Centre designed a ground multi-angle remote sensing device (GAMOD) for quick access to information about agricultural crops and an allied information acquisition and control system (Zhang et al. 2013), and used the device to synchronise a mounted HD camera and the visible and near infrared imaging spectroscopy (VNIHS) system to conduct field observations, which took 10 min (the time required is determined by the viewing angle) and was able to access the image data of soybean vegetation at four zenith angles (0°, 20°, 40°, and 60°) under main plane of the sun (at 0° and 180°) and the main vertical plane of the sun (at 90° and 270°).

2.2.2 Large Indoor Multi-angle Measuring Device

Figure 2.4 shows the multi-angle polarization measurement platform, a large indoor BRDF measuring device developed by a team of Changchun Institute of Optics and Fine Mechanics. The instrument consists of three major components: a light source system, detection frame, and control system, which are all able to rotate with the centre of the load platform as the centre of the ball.

Light source
A visible light near-infrared light source is commonly used in typical experiments; here, a 100 W tungsten halogen lamp, is used as the illumination source. Figure 2.5 shows the radiance of the light source on the whiteboard. A wall-mounted light source holder (radius 1.5 m) ensures incidence angle of light changes from zenith angle 0°–90° at any angle, with a scale accuracy of 0.5°, to obtain a light spot with a uniform diameter of around 90 mm on the illuminated object through the lens.

(1) Stage and detection frame

The stage is circular, fixed, and has a diameter of 200 mm. Samples can be placed on top, while the stage can be adjusted up and down so that the surface height of the sample is at the level of the centre of the entire detection system.

About 100 mm from the stage, a rotatable table (diameter, 1 m) of considerable mass to ensure that azimuth angle and zenith angle change gradually during the

Fig. 2.4 Indoor multi-angle measuring device

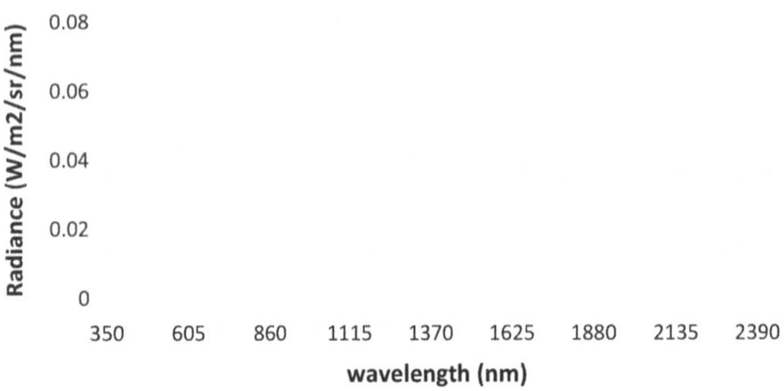

Fig. 2.5 The radiance of the light source on the whiteboard (incident zenith angle 50°, zenith angle measurement 50°, relative azimuth detection 180°)

process of rotation and to minimise vibration of the device. The rotating platform enables azimuth detection equipment to be rotated at any angle within the detection range of 0°–360°, with a scale accuracy of 0.5°. At the same time, on the rotating platform there is an oscillatory detection frame which can move from 0° to 90° within the range of angle of elevation thereof. The detection frame can be rotated in a plane perpendicular to the stage, and carries a detecting arm extending towards the centre of the stage, which can be used to fix ASD spectroscopes and the measurement fibre of the Ocean Optics spectroscope and the like. A polarising plate, a polarising prism, etc. can be loaded in front of the device. The detection device can theoretically move within a variable radius from 100 mm to 1.2 m from the centre of the ground object, but the radial distance depends on the viewing angle of the spectroscope and the surface area of the ground object to ensure the surface of the ground object fills the field of view of the spectrometer.

Theoretically, the detection frame can probe within −90° to 90° of the zenith angle, yet due to the instrument itself being occluded on the same side as the light source, some angles cannot be detected. Similarly, except for the angles of overlapping shadows caused by azimuth angle and the light source, all other angles within the range of 360° can be detected.

(2) Control system

The control system uses a motor to control the azimuth and zenith angle of the detector. Controlling the zenith angle relies entirely on the motor having the angle displayed on the control screen, with an angular resolution of up to 0.01°, while changing the azimuth angle requires visually alignment of the scale on the dial with an accuracy of 0.5°; however with a wide scale spacing, the error can be controlled to within 0.1°.

2.2.3 Process, and Error Analysis, of Experimental Measurements

(1) Process

The process of measurement and analysis is summarised in Fig. 2.6, wherein due to the need for frequent replacement of rock samples and whiteboard, thus it is important that the rock samples are placed consistently. In addition, because of the limitations of rock sample processing conditions, it is difficult to find a uniform component structure for larger rock samples while spectral measurements require a certain surface area: rock samples of the right size that were finally found were difficult to cut, moreover errors occur often as the standard of cutting surface is usually uneven, which made the placement of the sample more difficult. During the experiment, a lot of time is spent to ensure the standardisation of samples, and the spectral quality test was carried out after experimental analysis of the qualitative and quantitative relationship between the nonpolarized, and polarized, spectra over several measurements. The results finally obtained meet the requirements of having the smallest error.

(2) Error analysis

Error analysis, during measurement, is conducted as shown in Fig. 2.7. In the actual measurement process, almost all factors will cause measurement errors occur (Zhao et al. 2007; Yuan et al. 2010). While analysing and calculating errors, it is impossible

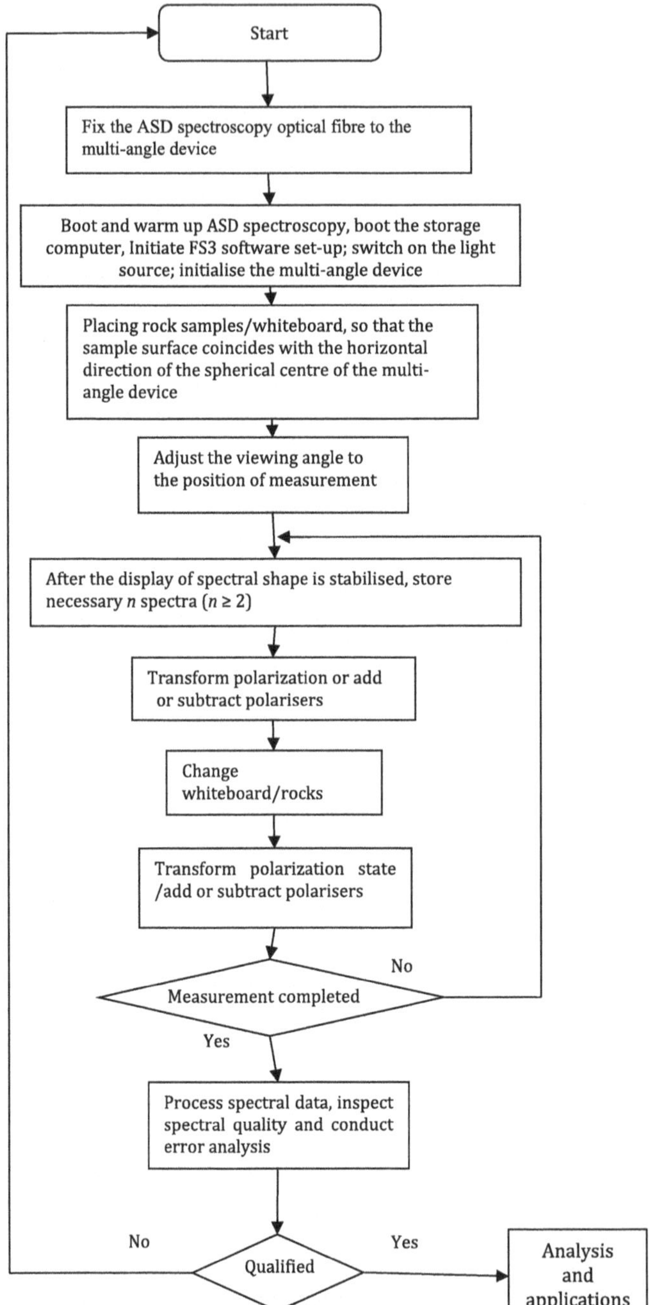

Fig. 2.6 Measurement flowchart

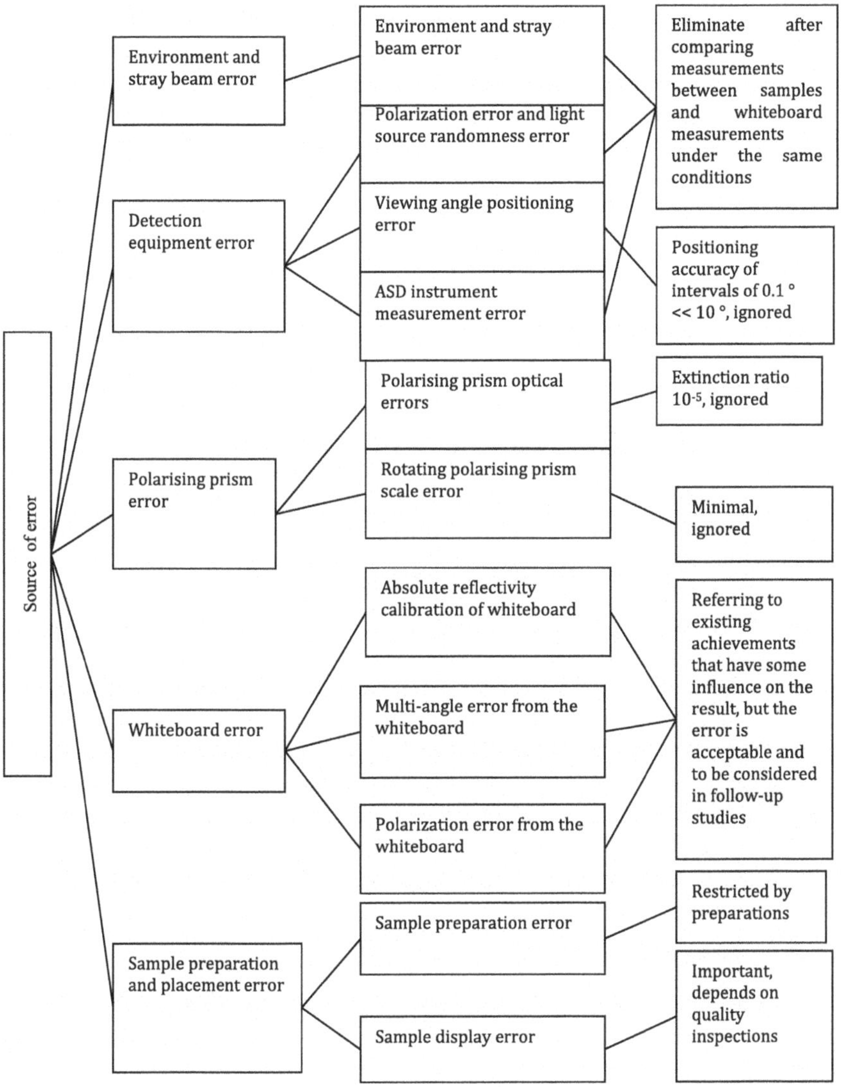

Fig. 2.7 Sources of measurement errors and their analysis

and unnecessary to consider all factors and calculate all errors of measurement, thus only the main factors causing measurement error will be analysed.

(3) Measurement of multi-angle spectra

Multi-angle measurement platform can be collaborated with ASD FieldSpec3 Spectroradiometer to measure reflectance spectra. The effective wavelength of the spectroscopy ranges from 350 to 2500 nm, with a resolution of 3 nm at 350 to 1000 nm and 10 nm at 1000 to 2500 nm (equivalent noise: UV/VNIR 1.1 × 10^{-9} W cm^{-2} nm^{-1} sr^{-1} @ 700 nm, NIR 2.4 × 10^{-9} W cm^{-2} nm^{-1} sr^{-1} @ 1400 nm, NIR 4.7 × 10^{-9} W cm^{-2} nm^{-1} sr^{-1} @ 2100 nm). The polarising prism is equipped in front of the fibre optic lens of the spectroscope. If a polariser is not installed, the bidirectional multi-angle reflectance spectra are measured; when equipped with a polariser, spectra from different polarization states can be measured. To reduce light interference, measurements should be done in a darkroom.

Since the ASD spectrometer needs to use the whiteboard for initial optimisation and as a reference when measuring the reflectance parameter, we measure DN values of the ground object, which are then divided by the whiteboard DN value automatically to deduce the reflectivity. Therefore, when specular reflection occurs, if the reflected light of the object is stronger than the reflection from the whiteboard, the instrument will issue a saturation alarm and stop measurement. As for the measurement of radiance parameters, the light source can be optimised directly without triggering this saturation alarm; therefore, spectroscopy with factory-calibrated radiance is employed in this experiment to measure radiance parameters.

The maximum radiation value that ASD FS3 spectroscopy can measure is more than twice the radiation reflected by a 100% whiteboard at a zenith angle of 0°, therefore, the linear response of the strong specular reflection instrument generated on the smooth surface is poor, and the measurement error is large: however, the smoother the surface of the ground object, the reflection value of the specular reflection orientation and its vicinity angle will be far more than several times that of other angles, so instrument detection error has little effect on this analysis.

During spectral measurements, the sample under test is placed in a multi-angle measurement platform with black insulating tape will be placed beneath. After adjusting the horizontal position and height of the sample surface, set the ASD spectroscope sampling count to 25 and duplicate records at least twice for each spectral measurement, then measure two to four further times for noisier curves and finally take the average spectrum as the final result.

(4) Experimental analysis of multi-angle spectra combination

It should be noted that the effective range of the ASD spectrometer from 350 to 2500 nm is composed of three sub-spectral detectors with its range of wavelength being VNIR 350–1000 nm, SWIR1 1001–1830 nm and SWIR2 1831–2500 nm respectively, while 1000 nm and 1830 nm demarcate the junction between the three sections of the spectrum. Sometimes due to detection performance of the instrument, spectrum-jumping occurs at the junction. SWIR1 band data will be affected by errors in calibrating the connection during the measurement by taking the results from 350 to 1000 nm and 1831 to 2500 nm to compare and calibrate them against those from 1001 to 1830 nm to connect the three sections of the spectral curves smoothly. Figure 2.8 shows the comparison before, and after, correction of the radiance spectrum on the whiteboard.

After completing spectral measurement, a comparison is firstly conducted within the result of recordings for repeated measurements at least twice. If there are significant differences between results, it means that the environment is unstable when the

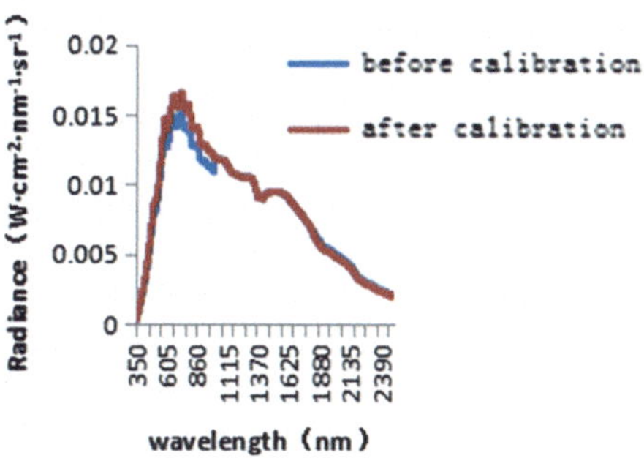

Fig. 2.8 Connection matching of whiteboard curves

measurement is performed, necessitating re-measurement. Secondly, quality inspection is conducted on all spectra measured. After several trials relating to the standards of inspections, a summary is provided in Chap. 4. If the standard is adhered to, then there is little error in the positioning of the samples and the quality of the spectrum is thus affected; however, if the standard is not adhered to, this means that the measurement error is large. In this case, the cause should be analysed, and the error should be avoided or minimised when repeating the measurement.

Spectrometric process will inevitably produce errors: mathematical methods can then be used to pre-process the error portion of the subtractive spectrum that is affected by random factors. The mathematical pre-processing of the spectrum is an important part of the spectrum analysis test process.

Spectral data pre-processing mainly includes three aspects (Li 2006): firstly, eliminating abnormal results which have greater spectral data errors therein; the second is to eliminate the influence of noise and other irregular factors, such as the elimination of random interference, background interference, changes in sample optical path, and the influence of factors such as spectral differences caused by the sample device on the calibration results; thirdly, we optimise the spectral range and purify mapped information, or to select prominent spectral regions which reflect sample information, therefore selecting the most effective spectral region to improve operational efficiency.

Due to the diversification of specific measurement conditions and the complexity of noise and errors, a single mathematical pre-treatment method cannot be applied to all substances and conditions. Representative spectral pre-treatment methods include K-M conversion, smoothing, differentiation, normalisation, and the use of multiplicative scatter functions (MSF). More complex spectral transformation methods include principal component analysis, Fourier transform analysis, and the wavelet transform (Li 2006).

In this study, due to the use of a polarising prism in the measurement of the spectrum, the intensity of light is reduced and therefore there are some random errors in the spectrum. In particular, the 0° polarization spectrum is generally low due to the corresponding extinction conditions. For the ideal state of complete linearly polarized light, the 0° polarization should be zero, but the ideal state is difficult to achieve under actual experimental conditions. Factors such as the influence of a

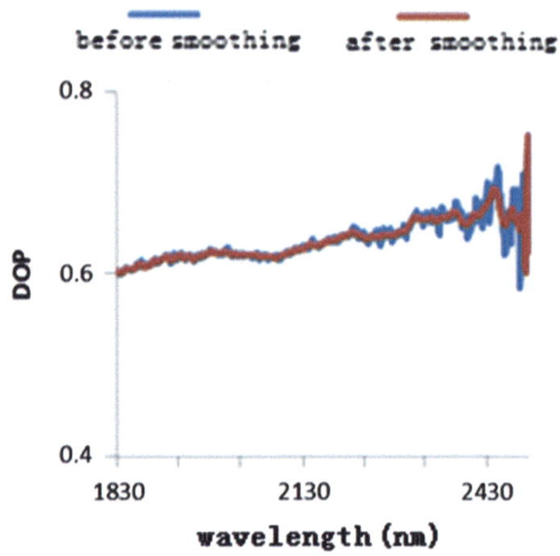

Fig. 2.9 Contrast before, and after, smoothing

straying beam, the dark current of the instrument, and light source stability errors will cause the entire system to be prone to error, therefore, the use of certain spectral smoothing algorithms to reduce or eliminate the effect of noise is essential.

The easiest smoothing method is the moving average method, which substitutes the current coordinate K-point (Y_k) on the spectrum with the average value of total $2r + 1$ numbers (r numbers before, Y_k and r numbers after).

$$y_k = \sum_{i=k-r}^{i=k+r} y_i \tag{2.4}$$

In the later stage, the moving average algorithm was used to deal with polarization data containing higher noise. Figure 2.9 shows the example of cases with, and without, smoothing (seven averaged numbers where $r = 3$).

2.3 The Angle of Incidence and the Mechanism Underpinning Multi-angle Reflectance of the Ground Object

Here, the incident angle is taken as the main influencing factor, and the influence of the incident angle on the multi-angle reflection spectrum of the ground object sample, the law governing the degree of polarization with the angle of incidence and the influence thereof on the spectra at different detection angles are discussed.

2.3.1 Qualitative Analysis of How the Angle of Incidence Affects the Multi-angle Reflection Spectrum of Rocks

When measuring the surface (flat) of 20 rock samples, a surface roughness of approximately 0.01 mm is invisible to the naked eye but is evident under near-infrared bands. From the experimental data, most of the rock experimental data shows: when the angle of incidence is small (generally less than, or equal, to 20°), specular reflection do not seem to exist, which coincide with the views of many American scholars; but when the incident angle is wide (generally greater than, or equal to, 30°), specular reflection from the rock surface occurs, which is consistent with the view of Academician Tong Qingxi (Tong 1990), thus it cannot be generalised.

Taking the most common basalt sample as an example, Figs. 2.10, 2.11, 2.12, 2.13, 2.14 and 2.15 show the spectra of basalt in 2π-spatial orientation from different angles of incidence (10°–60°) in the B-band, wherein the abscissa represents azimuth angle θ changing from 0° to 360°, and the ordinate represents the strength of the reflection spectrum of the rock. Each curve represents a viewing angle at a different height, from 10° to 60°.

As shown in Fig. 2.10, according to the view of academician Tong Qingxi (Tong 1990), the increase in reflectivity should be seen in the mirror direction (i.e., the detection angle is 10° and the azimuth angle is 180°), when the angle of incidence is 10°. In fact, the curve for a viewing angle of 10° was very smooth, but when the incident angle is large as shown in Figs. 2.13, 2.14 and 2.15, reflectivity increases can be seen in the corresponding mirror direction, which is consistent with the view

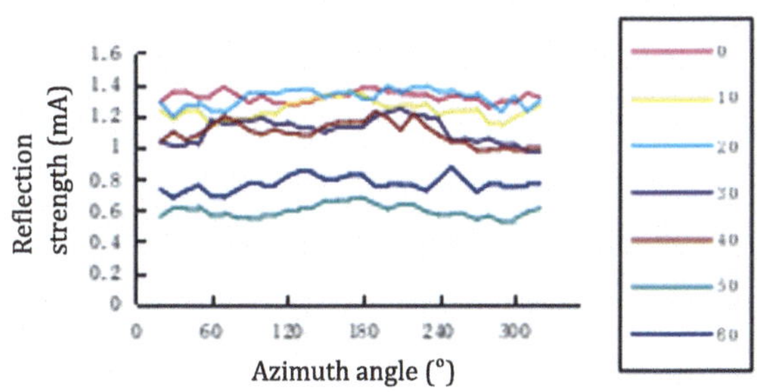

Fig. 2.10 Spectra of basalt with incident angle at 10°, B-band, without polariser

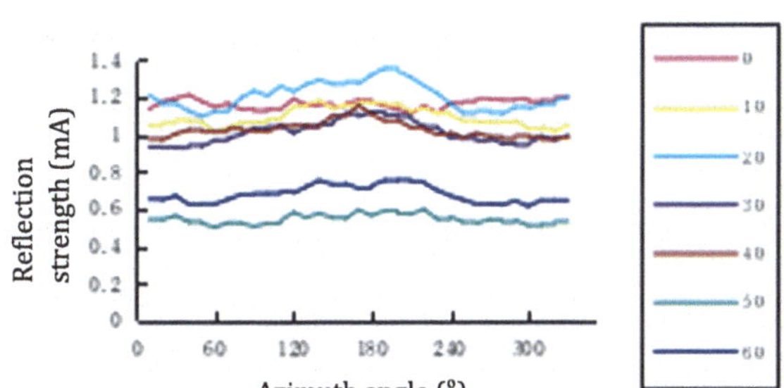

Fig. 2.11 Spectra of basalt with incident angle at 20°, B-band, without polariser

Fig. 2.12 Spectra of basalt with incident angle at 30°, B-band, without polariser

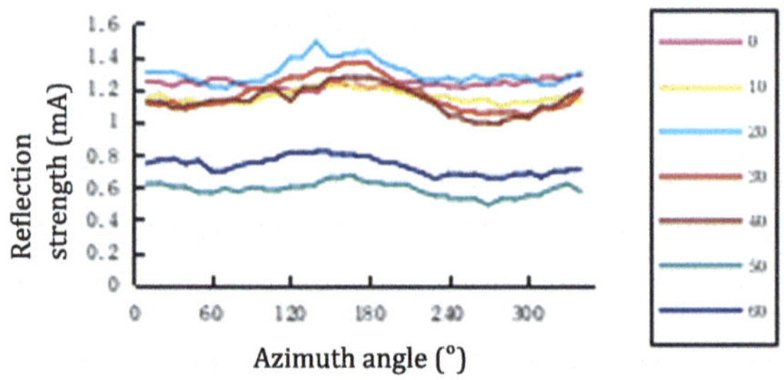

Fig. 2.13 Spectra of basalt with incident angle at 40°, B-band, without polariser

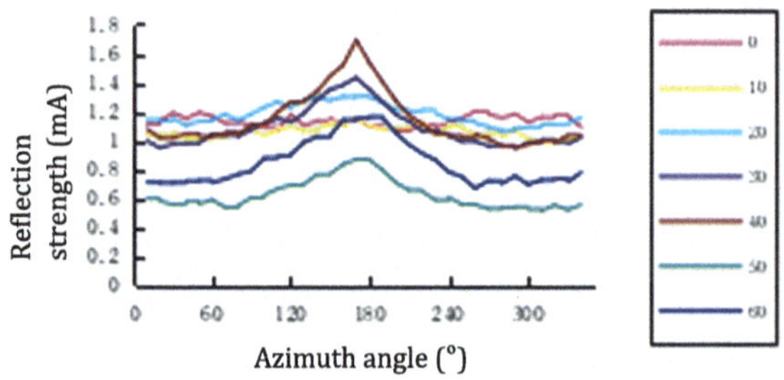

Fig. 2.14 Spectra of basalt with incident angle at 50°, B-band, without polariser

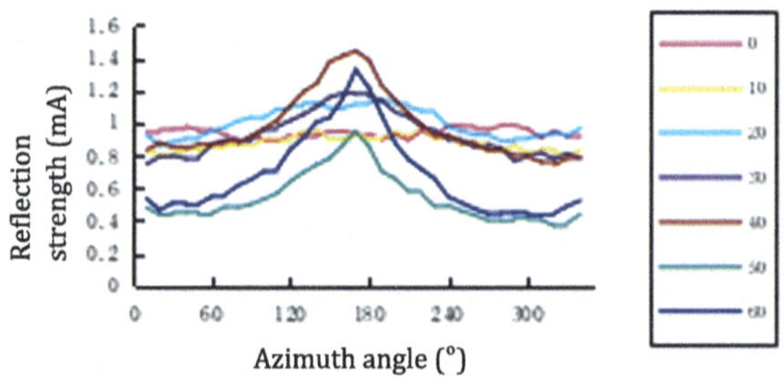

Fig. 2.15 Spectra of basalt with incident angle at 60°, B-band, without polariser

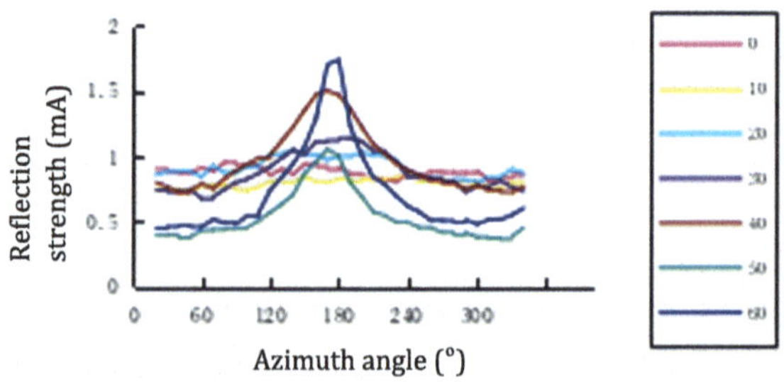

of academician Tong Qingxi. These phenomena show whether the reflectivity of the rock surface is a mixture of specular reflection and diffuse reflection is largely dependent on the angle of incidence.

Meanwhile the composition of the rocks also has a strong impact on whether the reflectivity of the rock surface is a mixture of specular reflection and diffuse reflection. Take two extreme but characteristic examples: firstly, oil shale is taken as an example: Figs. 2.16, 2.17, 2.18, 2.19, 2.20 and 2.21 show spectra of oil shale in a 2π-spatial orientation at different angles of incidence (10°–60°) in the B-band (the azimuth angle θ changes within the range of 0°–360° while the detecting elevation angle changes from 10° to 60°).

From Figs. 2.16, 2.17, 2.18, 2.19, 2.20 and 2.21, in spite of the size of the incident angles, mirror effects occur at different angles of incidence, matching the views of Academician Tong Qingxi (Tong 1990). For example, in Figs. 2.16 and 2.17, the curves for detection angles of 10° and 20° contain peaks, while in Fig. 2.17, the 10° curve becomes smooth. It is worth noting that this mirror effect is not strictly coherent with theory. From Figs. 2.16, 2.17 and 2.18, there is a double peak seen, which implies that a mirror effect appears on the curves when the incident angle is

Fig. 2.16 Spectra of oil shale with incident angle at 10°, B-band, without polariser

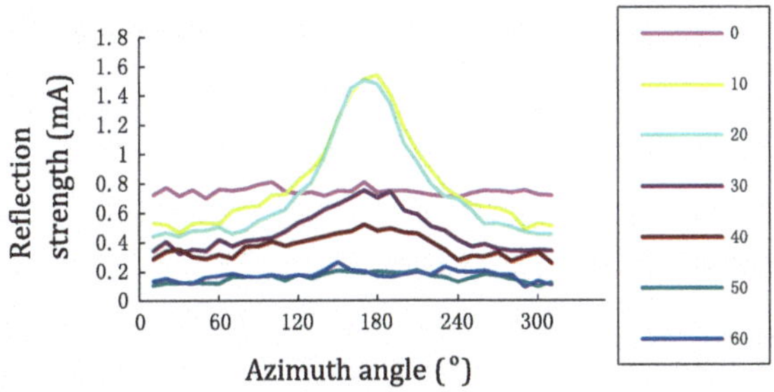

Fig. 2.17 Spectra of oil shale with incident angle at 20°, B-band, without polariser

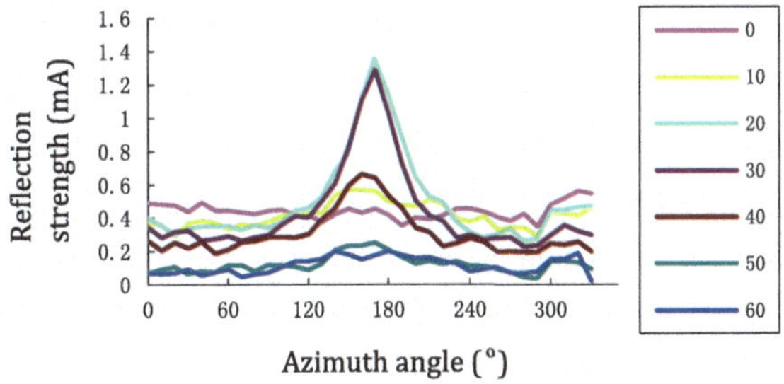

Fig. 2.18 Spectra of oil shale with incident angle at 30°, B-band, without polariser

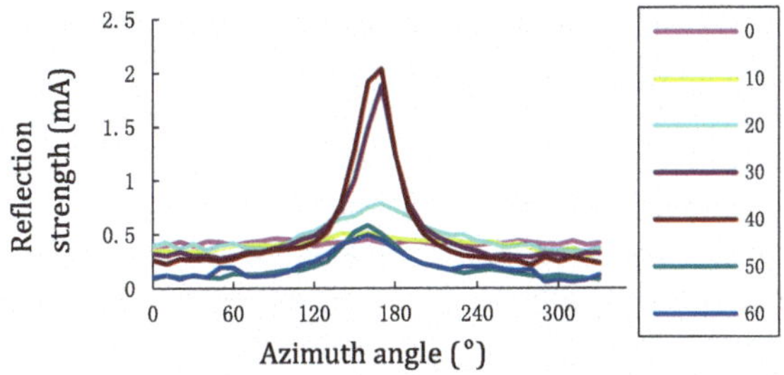

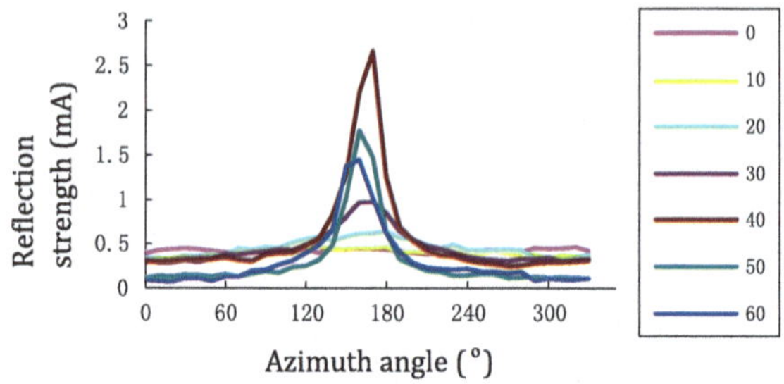

Fig. 2.19 Spectra of oil shale with incident angle at 40°, B-band, without polariser

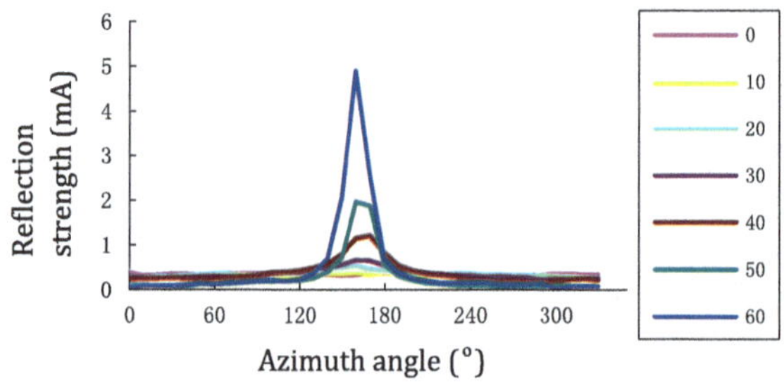

Fig. 2.20 Spectra of oil shale with incident angle at 50°, B-band, without polariser

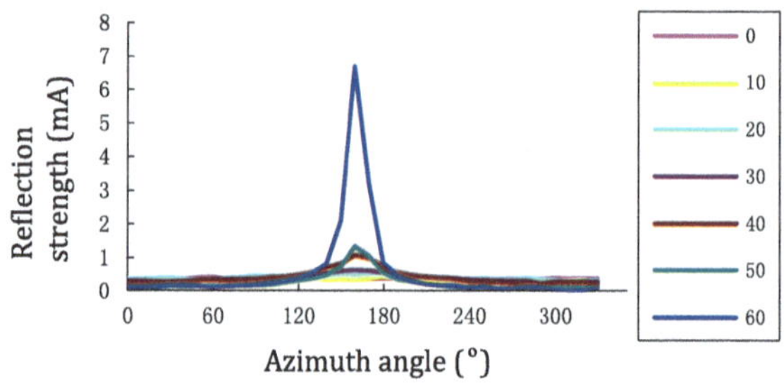

Fig. 2.21 Spectra of oil shale with incident angle at 60°, B-band, without polariser

10° and detection angles are 10° and 20°. In theory, only in the curve for a detection angle of 10° would we see significant changes in the direction of the mirror, which also applies to incidence angles of 20° and 30°: such a case is not fully consistent with prevailing theory.

Figures 2.22, 2.23, 2.24, 2.25, 2.26 and 2.27 show spectral graphs of quartz porphyry in the 2π-spatial orientation under different angles of incidence (10°–60°) in the B-band, wherein the azimuth angle θ changes from 0° to 360° and the elevation angle of detection ranges from 10° to 60°.

From Figs. 2.22, 2.23, 2.24 and 2.25, all waveforms for all viewing angles are flat and did not show any evidence of the mirror effect, which is coherent with the claims of American scholars and distinct from the views of Academician Tong Qingxi. The mirror effect occurs only at a large angle of incidence (greater than or equal to 50°).

The aforementioned experiments demonstrate whether, or not, the reflectivity of the rock surface is a mixture of specular reflection and diffuse reflection has a strong

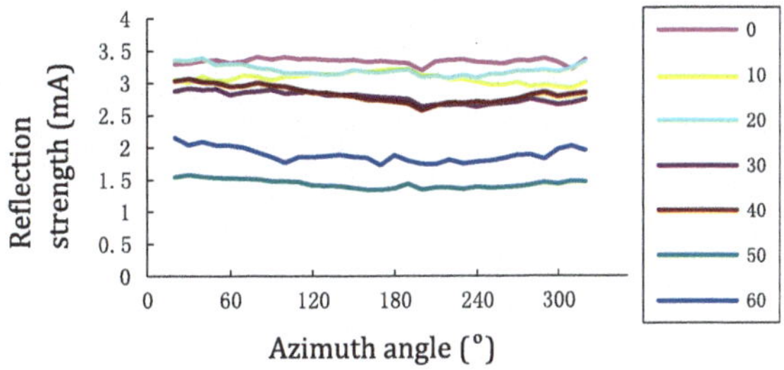

Fig. 2.22 Spectra of quartz porphyry with incident angle 10°, B-band, without polariser

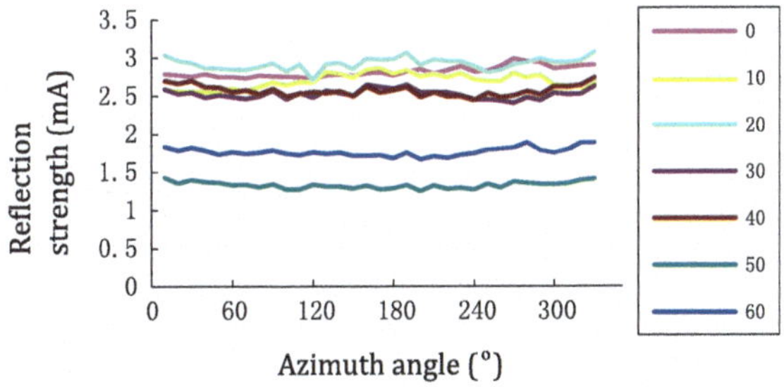

Fig. 2.23 Spectra of quartz porphyry with incident angle 20°, B-band, without polariser

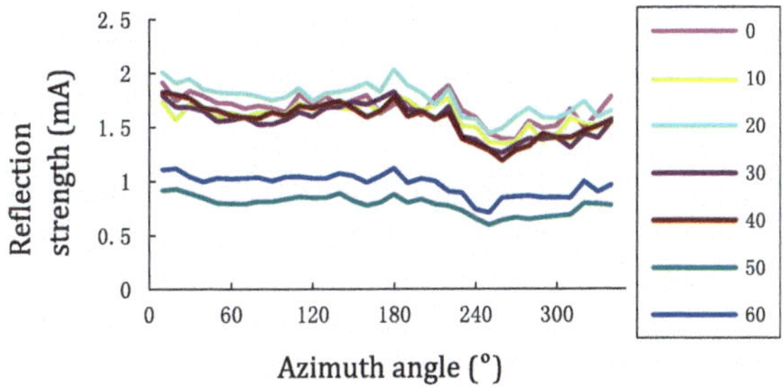

Fig. 2.24 Spectra of quartz porphyry with incident angle 30°, B-band, without polariser

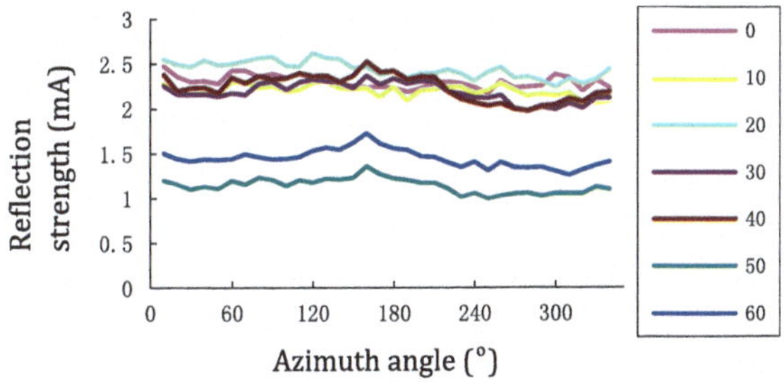

Fig. 2.25 Spectra of quartz porphyry with incident angle 40°, B-band, without polariser

dependency on the size of the angle of incidence and the composition of the rock, thus it should not be generalised; however, there is generally no mirror effect at small incident angles.

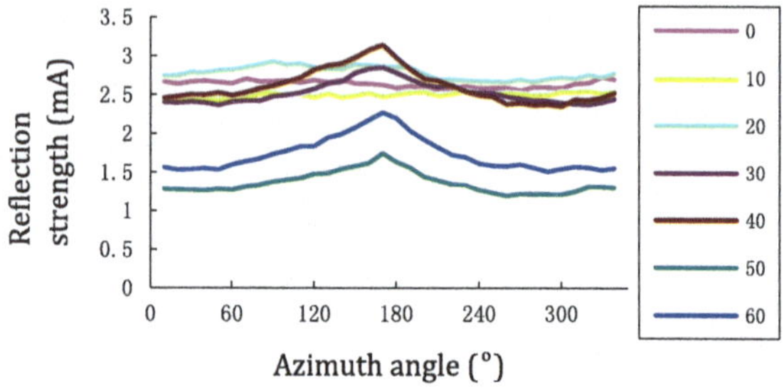

Fig. 2.26 Spectra of quartz porphyry with incident angle 50°, B-band, without polariser

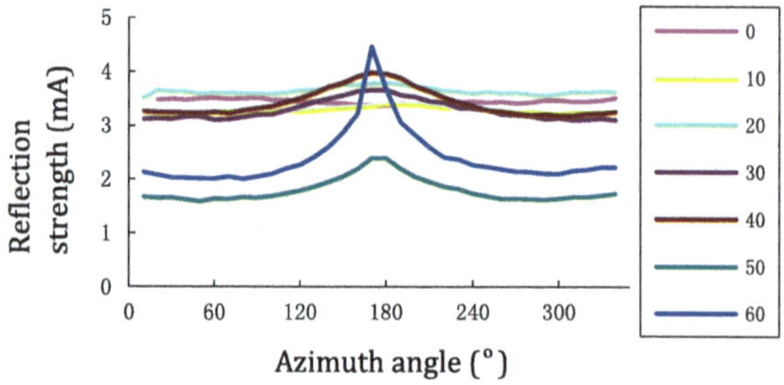

Fig. 2.27 Spectra of quartz porphyry with incident angle 60°, B-band, without polariser

2.3.2 *Mechanism of Variation of Degree of Polarization with Incident Angle in Rock Polarization Detection*

Table 2.1 lists the degrees of polarization of quartz porphyry at different angles of incidence: the degree of polarization of quartz porphyry rises with increasing incident angle until it reaches 60°, the degree of polarization then begins to decrease.

To explain this phenomenon, further analysis on the degree of polarization is needed. Since light intensity is the square of the amplitude of the electrical field vector, with the use of Fresnel's formula, it can be found that:

Table 2.1 Degree of polarization of quartz porphyry at different angles of incidence

Incident angle	Viewing angle					
	10°	10°	10°	10°	10°	10°
10°	3.47%	–	–	–	–	–
20°	–	5.53%	–	–	–	–
30°	–	–	7.54%		–	–
40°	–	–	–	10.3%	–	–
50°	–	–	–	–	11.61%	–
60°	–	–	–	–	–	9.37%

$$P = \frac{I_\perp - I_=}{I_\perp + I_=} = \frac{E'^{2}_{10\perp} - E'^{2}_{10=}}{E'^{2}_{10\perp} + E'^{2}_{10=}} = \frac{E^{2}_{10\perp}\frac{\sin^2(\alpha-\beta)}{\sin^2(\alpha+\beta)} - E^2_{10=}\frac{tg^2(\alpha-\beta)}{tg^2(\alpha+\beta)}}{E^2_{10\perp}\frac{\sin^2(\alpha-\beta)}{\sin^2(\alpha+\beta)} + E^2_{10=}\frac{tg^2(\alpha-\beta)}{tg^2(\alpha+\beta)}}$$

When the incident beam contains only natural light, vibration of the vertical incident plane and the parallel incident plane account for half of the other because the electric field vector direction of the light is varying. Therefore, $E^2_{10\perp} = E^2_{10=}$, so the degree of polarization can be further simplified to:

$$P = \frac{\frac{\sin^2(\alpha-\beta)}{\sin^2(\alpha+\beta)} - \frac{tg^2(\alpha-\beta)}{tg^2(\alpha+\beta)}}{\frac{\sin^2(\alpha-\beta)}{\sin^2(\alpha+\beta)} + \frac{tg^2(\alpha-\beta)}{tg^2(\alpha+\beta)}} = \frac{1 - \frac{\cos^2(\alpha+\beta)}{\cos^2(\alpha-\beta)}}{1 + \frac{\cos^2(\alpha+\beta)}{\cos^2(\alpha-\beta)}} = \frac{\cos^2(\alpha-\beta) - \cos^2(\alpha+\beta)}{\cos^2(\alpha-\beta) + \cos^2(\alpha+\beta)}$$

Expanding the above formula:

$$P = \frac{2\cos\alpha\cos\beta\sin\alpha\sin\beta}{\cos^2\alpha\cos^2\beta + \sin^2\alpha\sin^2\beta} = \frac{2}{\frac{1}{tg\alpha tg\beta} + tg\alpha tg\beta} \tag{2.5}$$

At this point, the maxima and minima of the functional (2.5) can be deduced:

When $tg\alpha tg\beta$ is in the interval (0, 1), the derivative of function P is greater than 0, and the function is a monotonically increasing function;

When $tg\alpha tg\beta = 1$, the derivative of function P is 0, then the function has a maximum value of 1;

When $tg\alpha tg\beta > 1$, the derivative of function P is less than 0, and the function is a monotonically decreasing function;

The above analysis shows that the DOP function monotonically increases first in the range of incident angles on (0°, 90°), then reaches its extreme value, and finally begins to decrease. This reflects Brewster's law from another perspective which states that the tangent of the Brewster angle is equal to the refractive index N of the dielectric. According to the law of refraction, when light waves are incident to the interface at Brewster angles:

$$\frac{\sin\theta_b}{\sin\beta} = \frac{N}{N_s} \tag{2.6}$$

Here $\beta = 90° - \theta_b$, then formula (2.6) becomes

$$\frac{\sin\theta_b}{\sin\beta} = \frac{\sin\theta_b}{\sin(90° - \theta_b)} = \frac{\sin\theta_b}{\cos\theta_b} = tg\theta_b = \frac{N}{N_s} \tag{2.7}$$

Meanwhile for air, $N_s = 1$, so:

$$tg\theta_b = N \tag{2.8}$$

Formula (2.8) (Brewster's law) shows that the tangent of the Brewster angle is equal to the refractive index N of the dielectric irradiated by light waves. Due to the dispersivity of N (varying with wavelength), the Brewster angle also changes with the wavelength of the incident beam. When the angle of incidence is equal to the Brewster angle, the degree of polarization is equal to 1, indicating that at this time the reflected light is solely vibrating on the vertical plane of the vibration surface, or the linearly polarized light. At other incident angles, it is partially polarized.

In conclusion, the degree of polarization in the reflection spectrum of rock starts to increases with the increase of incident angle and reaches the maximum at the Brewster angle and then begins to decrease.

2.3.3 The Influence of Incident Angle on Different Viewing Angle Spectra

Experimental results show that the variation of incidence angle affects not only the spectrum at the same detection angle, but also the spectrum around other detection angles. This is not exactly consistent with prevailing theory, which indicates that the angle of reflectance is as large as the angle of incidence, meaning that only the spectrum of a certain direction should be affected, or only a reflection should appear in the corresponding direction, while the other directions should have no protrusions; however, the reality is that the change in incident angle affects the specular reflection angle spectrum, as well as spectra at other detection angles, but the influence thereon varied. Figure 2.15 illustrates the problem with the spectrum of basalt without polarization at an incident angle of 40° in the B-band. According to the theory, only the spectral curve for a 40° viewing angle should change, and for the other curves they should be straight: however, only those at 0°, 10°, and 20° match this fact. They show an approximately linear form, while those at 30°, 50°, and 60° all contain a peak. This characteristic is also applicable to other graphs, therefore it can be concluded that changes in incident angle has a knock-on effect on the detection angle spectrum.

This phenomenon shows that, for rock surfaces under large angles of incidence, the mirror effect arises for all angles of elevation and not only when elevation angles match the angle of incidence, although the intensity of each spectrum is at its strongest in this case as well as when viewing angles are larger. The intensity of the reflection does not decrease uniformly with the dispersion of the incident angle, and the intensity of the reflection spectrum of a large angle is usually stronger than that at smaller angles. For example, for an angle of incidence of 40°, the degree of dispersion of viewing angles of 20° and 60° are the same, but the impact on the intensity of spectral detection at 60° is much stronger than that at 20°, which is almost negligible.

Secondly, at a large incident angle, although the surface of rock exhibits a mirror reflection in that direction, it does not exhibit discontinuities: this is not fully consistent with prevailing theory. Specular reflection occurs when the reflected light is on the same plane as the incident beam: this means that there is a discontinuity at an azimuth angle of 180°, but in fact, spectrum gradually rises within a plane azimuth angle of 360°. As for the spectrum of water, a discontinuity will occur.

2.4 Analysis of the Characteristics of the Multi-angle Nonpolarized Spectrum and Its Degree of Polarization

The bidirectional reflectance distribution function is used to define how the irradiance in a given incident direction affects the emissivity in a given outgoing direction. More generally, it describes how the incident light is distributed in all directions after passing through a certain surface reflection, which can be any kind reflection from the ideal specular one to the diffuse one, isotropic or anisotropic reflection. How the bidirectional reflectance distribution function changes with the scattering angle is mainly determined by the roughness of the sample surface. So, to assess the effects using a rough and a smooth rock of the same type, the black cloud obliquity of gneiss, was selected. We ground the rock surface with # 120 emery until its surface is largely smooth but still rough to the naked eye and number it H1 in this experiment; we number the other rock sample, which is smoothly polished to mirror effect, H5. The specific surface roughness of the two rocks is given in Chap. 4. Below, the two rocks are taken as examples to study the effects of multi-angle variations in the non-polarized spectrum and degree of polarization. The zenith angle of the incident light of both rocks is limited to 50°.

2.4.1 Multi-angle Characteristics of the Degree of Polarization of Rock Specimens

Figure 2.28 shows the degree of polarization DOP of rock H1 at an incident zenith angle of 50° and azimuth angle of 180°, which is maximised when the zenith angle is at 60° and 70° and gradually decreases on both sides.

As seen from Fig. 2.29, when the azimuth is 180°, specular reflection occurs on corresponding zenith angles of 50° and the maximum degree of polarization is obtained. The degree of polarization between 0° and 50° increases with increasing zenith angle and decreases rapidly between 50° and 80°. The degree of polarization at a zenith angle of 10° below tends to zero, therefore traditional remote sensing generally detects surfaces when facing vertically downwards. It is considered that the degree of polarization of a uniform surface is often less than 20% and the influence of polarization is usually ignored (Zhao et al. 2004; Wu et al. 2009). In terms of the spectral morphology of the degree of polarization, it is close to zero when the zenith angle decreases, and as the signal level to noise ratio is low, there is almost no information reflecting the feature itself. The change in degree of polarization spectra

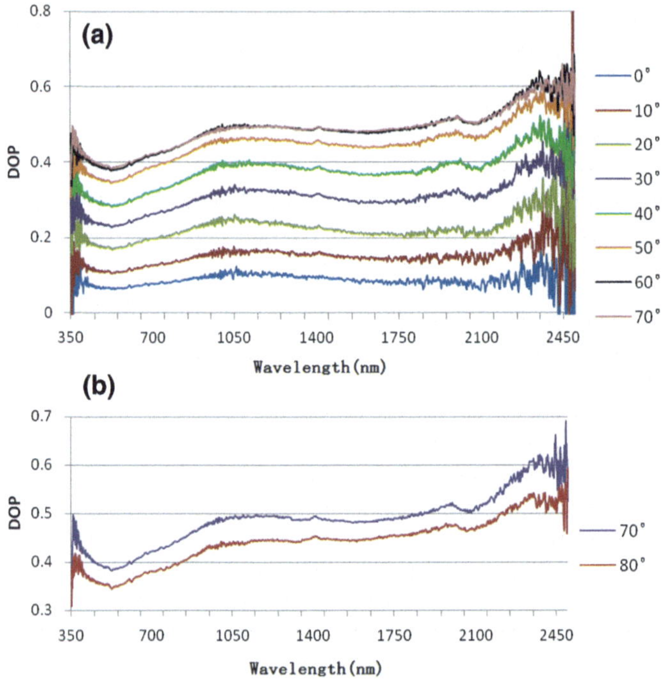

Fig. 2.28 Degree of polarization of rock H1 at different zenith angles when the incident zenith angle is 50° and the azimuth angle is 180° (DOP) (**a** 0°–70°; **b** 70°–80°)

is small at 50° having its average spectral value within the range 350 to 2500 nm to be 0.847, while its maximum is 0.873 and minimum is 0.827 (a difference of 0.046). It can be seen that, at 50° due to the strong specular reflection, polarization spectroscopic information of the ground feature is very limited and thus not suitable for remote sensing. Polarization remote sensing should also avoid specular reflection angles of smooth surfaces.

Figure 2.30 shows the variation of DOP at different azimuth angles of rock H1 in detection at a zenith angle of 50°. The degree of polarization reaches a maximum at 180° azimuth, gradually decreasing to a very small negative number on both sides. This is mainly caused by the background error in the system. At this point polarizations of 90° and 0° are very small, thus it should be considered that the degree of polarization is zero.

As seen from Fig. 2.31, the variation of the degree of polarization at an azimuth angle of 50° zenith angle is that DOP is largest at 180° and becomes smaller towards both sides until it is close to zero at around 20°; however, due to the fact that the surface roughness and the field of view cannot reach the ideal infinitesimal, the degree of polarization is very high in the range 180° ± 10°, which is relatively concentrated. The SNR is lower when the degree of polarization decreases and where the spectral morphology flattens and its value tends to zero.

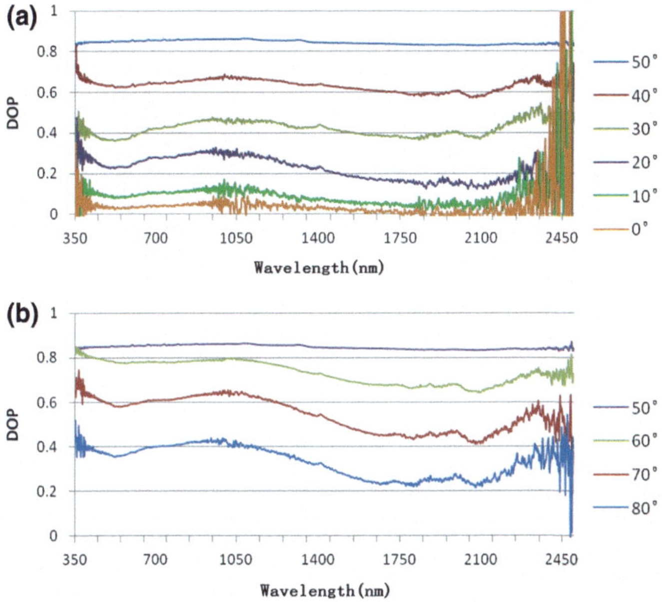

Fig. 2.29 Degree of polarization (DOP) of rock H5 at different zenith angles with an incident zenith angle of 50° and an azimuth angle of 180°

2.4.2 Multi-angle Characteristics of the Non-polarized Reflectance Spectra of Rocks

Figure 2.32 shows the non-polarized reflectance spectra of rock H1 at different detection angles. When the measured azimuth angle of rock H1 is constant, the measured reflectance ratio at a zenith angle of 0° is at a minimum, and the reflection ratio increases with increasing zenith angle. The larger the zenith angle, the greater the difference between the angles and vice versa.

Figure 2.33 shows the variation in detecting azimuth at a fixed zenith angle of 50°. It can be seen that the azimuth 180° represents the mirror reflection azimuth and the reflection ratio reaches its maximum: the further from the azimuth, the lower the reflectance ratio (beyond 270°, the reflectance ratio decreases slowly and is almost equal to that elsewhere).

Theoretically the intensity of reflected light is the same when the corresponding azimuth angle is centred on the specular reflection at 180° between the azimuth angle on both sides of the ranges 180°–270° and 180°–90°, while the same applies when the corresponding azimuth angle is concentric with the incident beam (0° or 360°) on both sides in the ranges of 0° to 90° and 360°–270°.

Since the observation platform is being blocked over a wide range at about 0°, this test only measures the approximate orientation of the forward side of the incident beam intensity, while the actual detection object is a rock that contains a mixture

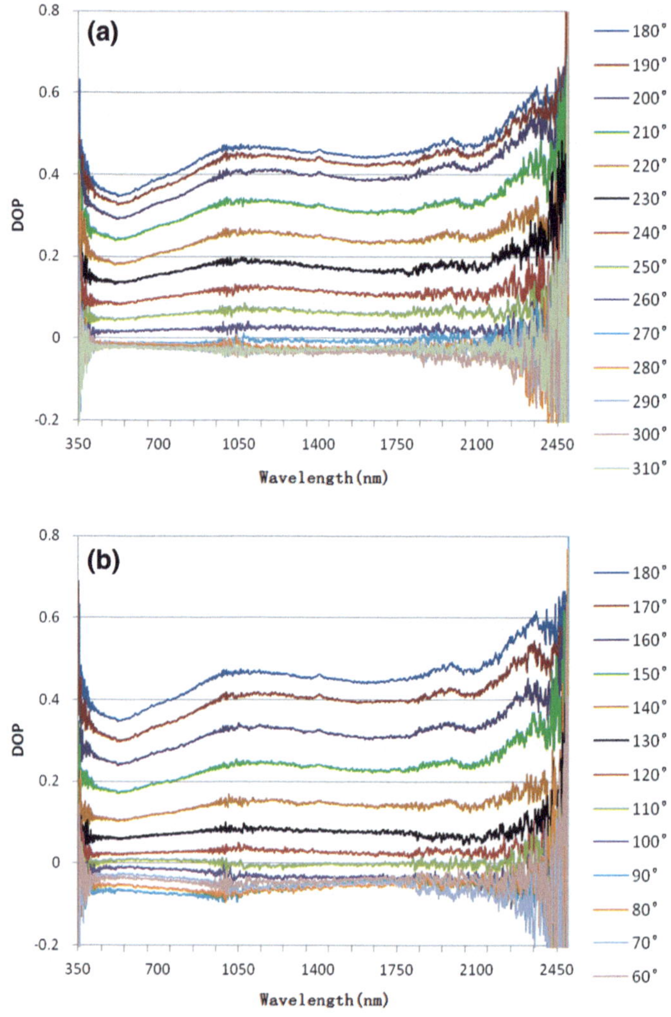

Fig. 2.30 Degree of polarization (DOP) of rock H1 at an incident zenith angle of 50° and detecting azimuth angle of 50°

of mineral components, there are some differences on both sides of the spectrum at a 180° azimuth angle due to the fact that the detection surface is not uniform on the scale involved in these experiments, yet, as shown in the figure, the azimuthal variation is the same.

Figure 2.34 shows the non-polarized spectrum of H5 rock measured at different zenith angles when the azimuth angle is fixed 180°. It can be seen, from the figure, that at the 180° azimuth, specular reflection occurs corresponding to the detection of the zenith angle of 50°, highlighting the maximum reflectance ratio. The reflectance

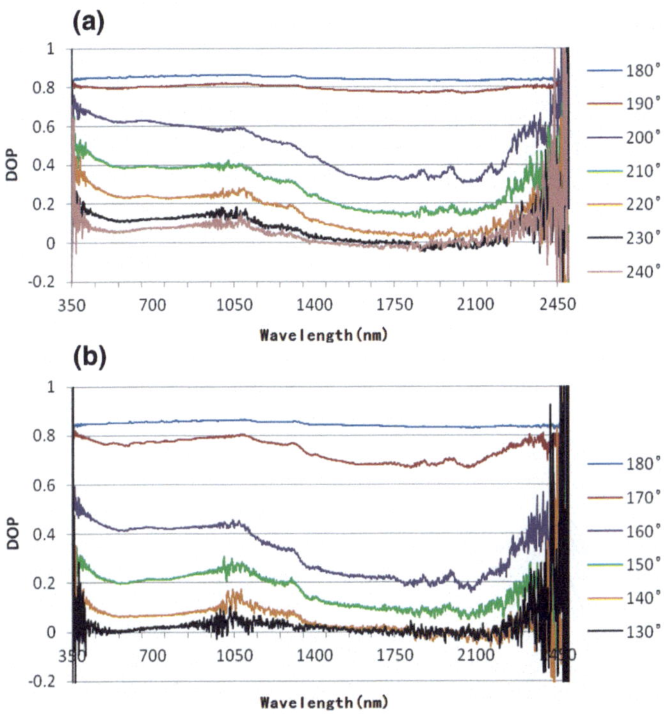

Fig. 2.31 DOP spectra of rock H5 at an incident zenith angle of 50° and azimuth angle of 50°

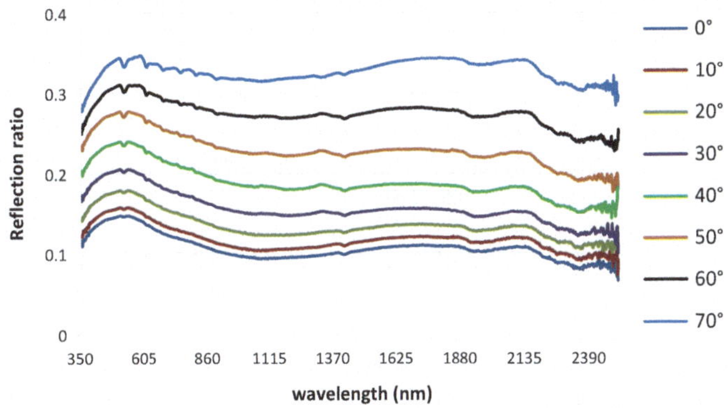

Fig. 2.32 Spectra at different zenith angles of rock H1 at an incident zenith angle of 50° and an azimuth angle of 180°

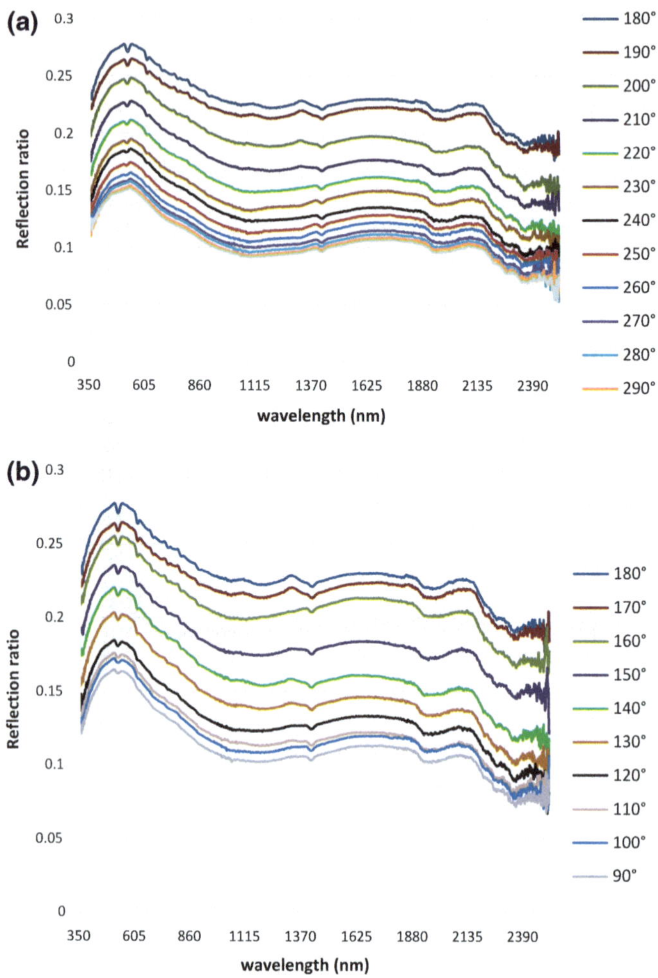

Fig. 2.33 Spectra at different zenith angles of rock H1 at an incident zenith angle of 50° and an azimuth angle of 50°

below 50° decreases as the detecting zenith angle decreases, while the reflectance above 50° increases as the zenith angle increases.

The difference in reflectance ratios is small between −20° and 20°, and the maximum difference with a stable spectrum is 0.030 within the range of 400–2400 nm. This explains the properties of the Lambert space being better demonstrated in the interval −20° to 20°.

From the spectra, spectral shape retention is better when the zenith angle is small. Different zenith angles only affect the level of spectral values but not the spectral shapes. The specular component begins to increase rapidly and surpasses the diffuse component largely for zenith angles of above 30°. The maximum radiation value that

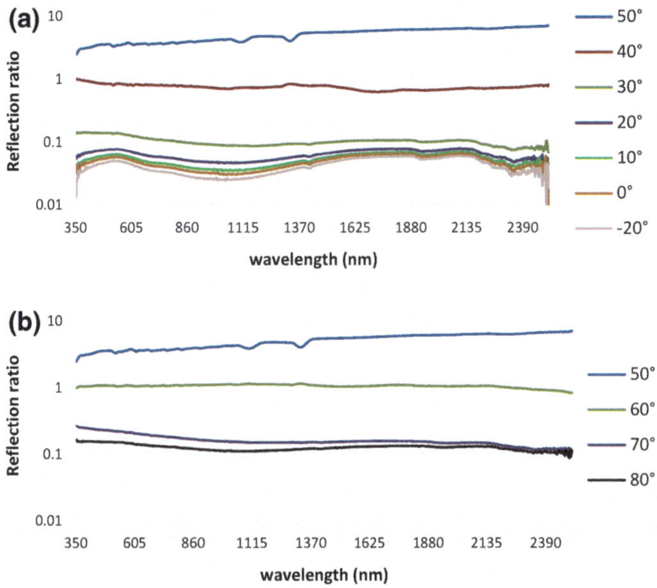

Fig. 2.34 Spectra of rock H5 at different zenith angles at an incident zenith angle of 50° and an azimuth angle of 180°

can be measured by ASD FS3 spectroscopy is more than twice the 100% reflective whiteboard radiation at a zenith angle of 0°, so the linearity of the instrumental response to strong specular reflection of smooth surfaces is relatively poor, however, as the reflectance value of viewing angles at 50°, and other nearby angles, is much higher than that at other angles, the influence of instrument measurement error on the analysis of this study is smaller. A high specular reflectance ratio also lowers the proportion of information of the characteristics of the ground object. Therefore, traditional remote sensing practitioners generally choose to avoid the orientation of the peak specular reflection.

As illustrated in Fig. 2.35, the reflectance ratio of rock H5 is maximised when the viewing azimuth angle is 180° and the zenith angles is 50°. The further from an azimuth of 180° towards the two sides, the lower the reflectance ratio. After passing 210° and 150° in each direction, the rate of change decreases, but the spectral morphology remains more consistent.

2.4.3 *Analysis of Other Angles*

In the experimental results cited above, only the measurements of a typical viewing zenith angle of 50° and azimuth angle of 180° are listed. These angles are representative because they correspond to the angular orientation of the mirror reflection and the

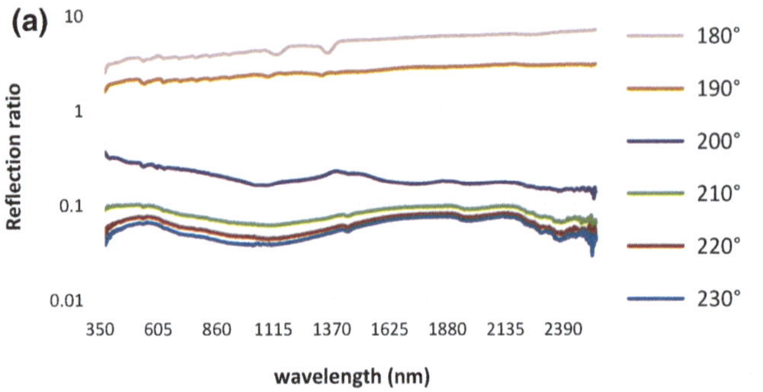

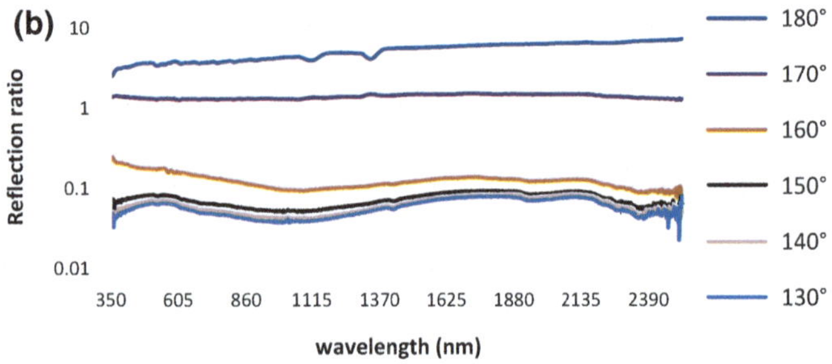

Fig. 2.35 shows the spectrum at different zenith angles of rock H5 at an incident zenith angle of 50° and an azimuth angle at 50°

intensity of light is relatively strong. In the experiment, other detected azimuth angles and other detected zenith angles (30°, 40°, and 60°) in the range of forward scattering are also measured, showing that the reflectance difference at different azimuth angles is relatively small when the zenith angle is less than 10°. The signal to noise ratio of the degree of polarization is as low as approaching zero, while the pattern of performance of zenith angle corresponding to the variation of different azimuth angle can be clearly distinguished and is quite consistent. As for the circumstances of other zenith angles that change with different azimuth angles, the reflectance does not change to any significant extent when the azimuth angle exceeds 210°, and the degree of polarization tends to zero. The pattern of the variation of the other azimuth angle that can be significantly differentiated corresponding to the variation of the zenith angle is the same as that when the azimuth angle is 180°.

2.4.4 *Multi-angle Variation of the Degree of Polarized and Non-polarized Spectra*

According to the analysis above, the pattern of the nonpolarized spectral variation with the zenith angle is such that, for rock H5 with its smooth surface, strong specular reflection peaks occur at the observed zenith angle with the same size as those formed with the incident beam: these decline, in sequence, on both sides of the peak. For rock H1, with its rough surface, specular reflection peaks are not observed and the amplitude gradually increases between 0° and 70°.

As for variation in the degree of polarization with changing zenith angle, it first rises between 0° and 80°, reaching its highest value and then decreasing thereafter (the peak value for rock surfaces of different roughnesses also changes therewith, and the smoother the surface, the higher the peak).

A common feature of the variation of the degree of polarized and non-polarized spectra changing with zenith angle is that their values are lowest at 0° zenith angle, and gradually increase with increasing zenith angle. Affected by surface roughness, the peak can be observed when the surface is smooth, and then declines with increasing zenith angle. When the surface is rough, the peak may not be observed. The degree of polarization angle and the angle with the peak value of a non-polarized spectrum is not completely consistent.

The nature of a non-polarized spectrum, and the change in degree of polarization, with the detecting azimuth angle is consistent: both reach their maxima at a 180° azimuth angle and decrease on both sides away from 180°. The degree of polarization tends to zero at a faster rate than in a non-polarized spectrum.

2.4.5 Application of the Degree of Polarization and Multi-angle Variation Non-polarized Spectrum

The characteristics of multi-angle reflection have a close relationship with surface structure, texture, roughness, and other surface features. The bidirectional reflectance ratio changes with wavelength and is mainly determined by the reflectivity at different wavelengths, which means that the morphological composition affects the expression of spectral features. For features with smooth, flat surfaces, the variation of the bidirectional reflectance distribution function corresponding to the scattering angle is mainly determined by the surface roughness. The example of the multi-angle non-polarized spectral measurements provided here illustrates this point. With increased viewing zenith angle and consequently the rise in amplitude of the non-polarized spectrum, the shape of the spectrum remained unchanged. The variation in multi-angle degree of polarization can be applied to other suitable types of ground objects; of course, these trends are only applicable to opaque mineral rocks. For transparent mineral rocks, refraction and transmission components predominate, while reflection and scattering components only play a minor role, and the physical processes governing the behaviour of light at the interface is different.

References

Li MZ (2006) Spectral analysis technology and its application. Science Press, Beijing (in Chinese)

Li X, Zheng XB, Xun LN et al (2008) Realisation of field BRDF acquisition by multiangular measurement system. Opto-Electron Eng 35(1):66–70 (in Chinese)

Tong QX (1990) Spectra and its characteristics analysis of typical ground objects in China. Science Press, Beijing (in Chinese)

Wu TX, Yan L, Xiang Y et al (2009) Polarization reflection effect of plane rough surface under vertical observation. J Infrared Millimeter Waves 28(02):151–155 (in Chinese)

Ye YT, Rao JZ, Xiao J et al (2005) Course in optics. Tsinghua University Press, Beijing (in Chinese)

Yuan Y, Sun CM, Zhang XB (2010) Measuring and modelling the spectral bidirectional reflection distribution function of space target's surface material. Acta Physica Sinica 59(3):2097–2103 (in Chinese)

Zhang D, Wang X, Coburn C et al (2013) Design and experiment of ground-based agriculture-oriented multi angle observation device. Trans Chin Soc Agric Mach 44(1):174–177 (in Chinese)

Zhao HY, Zhao H, Yan L et al (2004) Model of reflection spectra of rock surface in 2π-space. Acta Geol Sin 78(3):843–847 (in Chinese)

Zhao ZY, Dai JM, Li Y (2007) Error analysis and calibration of BRDF measure system. Infrared Technol 29(10):579–583 (in Chinese)

Chapter 3
Remote Sensing of Ground Objects 2: Multispectral Chemical Characteristics of Surface Polarization Reflection

In this chapter, multi-spectral chemical characteristics, which are the second most important feature of polarization remote sensing, are introduced in terms of surface polarization reflection. In the formulation the various new concepts and equations we shall not aim at the maximum generality possible but limit ourselves, rather, by the situations which the problems recounted in this chapter actually need. In view of this, we shall devote this chapter to the study of multi-spectral chemical characteristics.

The chapter is outlined as follows:

In Sect. 3.1 we establish and analyse the relationship between the degree of polarization (DOP) spectra and non-polarized spectra to explore the nature of the chemical characteristics of surface multi-spectral polarization reflection. To accomplish effective probing of various types of typical ground objects, in Sect. 3.2 we develop, explicitly, the reflectance spectrum of a rock surface in 2π-spatial orientation and further obtain a theoretical representation of the chemical characteristics of its multi-spectral polarization. In Sect. 3.3 we obtain, in particular, the spectral parameter inversion of rock, which can achieve recognition of surface features using the chemical characteristics of multi-spectral polarization. In Sect. 3.4, the relationship between rock compositions and DOP is discussed, thus, we use the characteristics of DOP on a rough structure to explore the geological composition and structure of rocks.

3.1 Degree of Polarization (DOP) Spectra and Non-polarized Spectra

Polarization is another type of useful information source used in remote sensing and it is also a property applying to transverse waves that specifies the geometrical orientation of the oscillations. In a certain transverse wave, the direction of the oscillation is perpendicular to the direction of motion of the wave. It is noted that polarization does not exist independently but has a certain relationship with the intensity in traditional measurements. This section mainly focuses on the analysis and

L. Yan et al., *Polarization Remote Sensing Physics*, Springer Remote Sensing/Photogrammetry, https://doi.org/10.1007/978-981-15-2886-6_3

comparison of the relationship between the polarization spectrum and the traditional measurement of reflectance spectrum (non-polarized spectrum), and further explores the nature of the chemical characteristics in the region of multi-spectral polarized reflection of ground objects.

3.1.1 Spectral Characteristics of the DOP

Figures 3.1 and 3.2 show the spectra of the DOP and the non-polarized reflectance spectral measurement of two random rocks in outdoor conditions. The DOP in the 350–2500 nm wave range represents an abundance of crests and troughs, and is closely related to the non-polarized reflectance spectrum in conventional measurement.

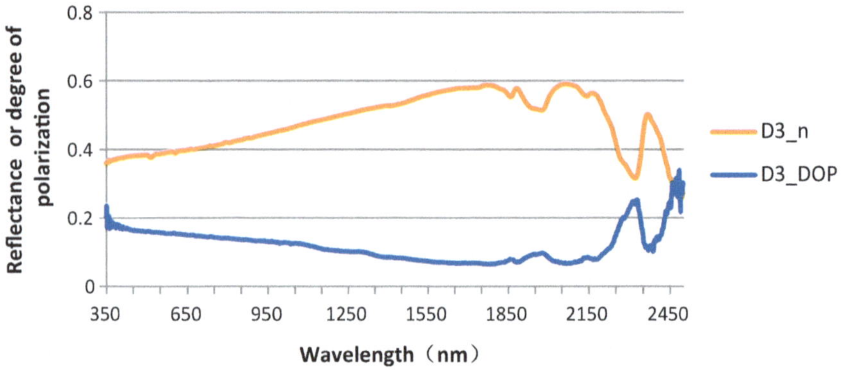

Fig. 3.1 DOP and non-polarized (n) reflectance spectra of calcite dolomite (calcite dolomite, D3)

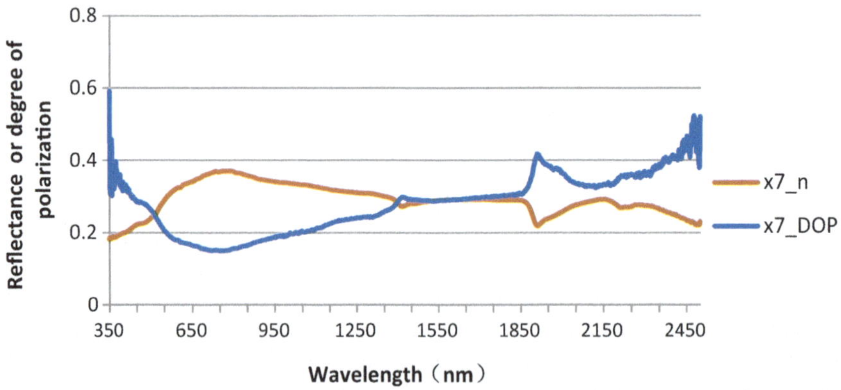

Fig. 3.2 DOP and non-polarized (n) reflectance spectrum of tuff (tuff, x7)

3.1.2 Qualitative Spectral Analysis of DOP and Non-polarized Data

Figures 3.1 and 3.2 represent the reflectance spectrum and the DOP: all crests on the reflectance ratio curve correspond to the troughs on the DOP curve. When the reflectance is at a trough, the DOP will be at a crest. This can be explained by the basic definition of the DOP. Natural light shines on a dielectric surface and is then refracted and reflected thereby, undergoing partial polarization relative to the incident light. The basic definition of the DOP is the proportion of completely polarized light over the total intensity of partially polarized light:

$$DOP = \frac{I_P}{I_{sum}} \tag{3.1}$$

I_{sum} is the light intensity defined in traditional measurements, therefore, the DOP has an inverse relationship with the traditional intensity reflectance, however, molecular intensity I_P changes with the varying wavelength and viewing angles, especially, the loss of light intensity is usually affected by absorption, reflection, refraction, and other activities of the polariser, therefore, the inverse relationship is still not satisfied, which means the constant inverse coefficient does not exist in any cases and it would only show a qualitative relationship with both crest and trough on a certain spectrum.

3.1.3 Quantitative Spectral Analysis of DOP and Non-polarized Data

Figures 3.3 and 3.4 illustrate a calcite dolomite (D3, sample number) polarization

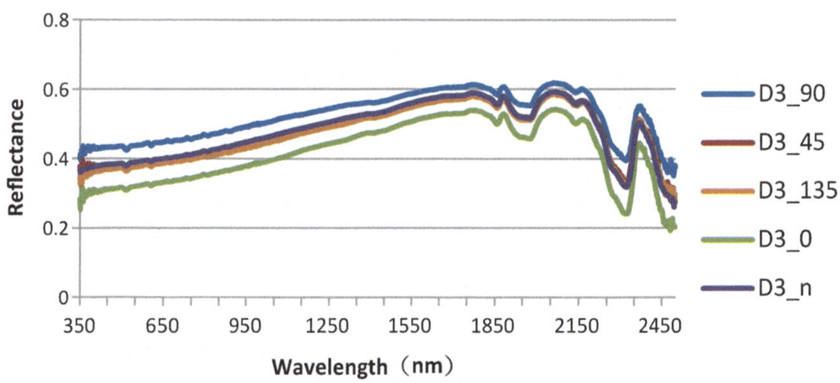

Fig. 3.3 Reflectance spectrum of calcite dolomite (D3) with no polarization (n), 90° polarization (90), 135° polarization (135), 45° polarization (45), 0° polarization (0)

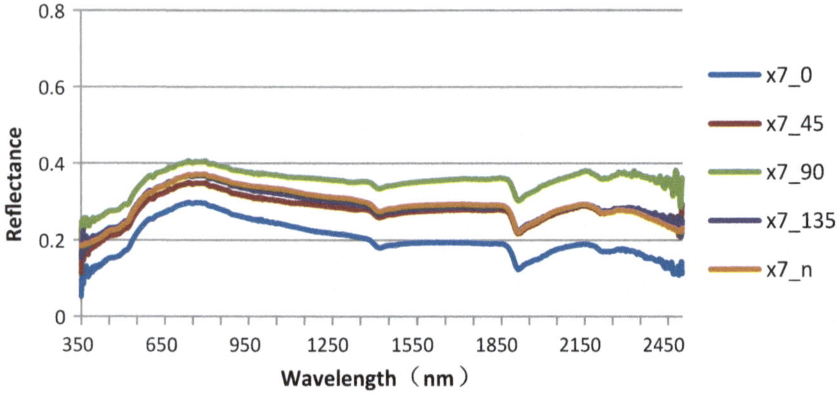

Fig. 3.4 Reflectance spectra of tuff (x7) with no polarization (n), 90° polarization (90), 135° polarization (135), 45° polarization (45), 0° polarization (0)

curve and a tuff (x7, sample number) polarization curve, respectively, it is seen that the polarization curves at 135° and 45° are almost equal to the non-polarized hyperspectral curve, which are between 0° and 90° polarization curves. Zhao Yunsheng (Zhao et al. 2006, 2005a, b), Song Kaishan (Song et al. 2005, 2007) and others proposed the following theoretical relationship through observation of poplar leaves, corn blades, peridotite and an oil spill on water:

1. When the angle between the polarization direction of the incidence beam of rock and the transmission axis is at 45°, the intensity of the post-incident beam is half of the intensity of original beam, thus, the arithmetic mean of the intensity of light in the transmission axis and the transmissive extinction direction is equal to the intensity when the angle between the polarization direction of the incidence beam and the transmission axis is at 45°. That is,

$$I_{45} = \frac{1}{2} I_{sum} = \frac{1}{2}(I_0 + I_{90}) \tag{3.2}$$

2. In the 45° direction polarization reflection, the bidirectional reflectance and the arithmetic mean of the light intensity in the transmittance direction and transmissive extinction direction of the rock are equivalent at corresponding zenith angle, azimuth angle and viewing angle on the corresponding band.

$$R_{45} = R_{reflectance} = \frac{1}{2}(R_0 + R_{90}) \tag{3.3}$$

The above conclusion summarises the quantitative relationship between multi-angle polarization reflection and bidirectional reflectance, which can improve the accuracy of a probe and retrievals from geo-object information by combining existing quantitative remote sensing methods. Additionally, this method can also be helpful when exploring new ideas about parameter inversion for rock media.

The quantitative relationship between the reflectance and DOP, as well as the spectral relationship can be used as a standard to test spectral quality. In practice in the future, if there are some types of spectra that do not meet the above description, the factors need to be carefully studied to verify the uncertainties.

3.1.4 The Choice of Characteristic Bands for Multi-angle Polarization Observations

For the qualitative and quantitative analysis of the relationship between the non-polarized polarization spectra, polarization reflection spectra, and the DOP, Figs. 3.5 and 3.6 list some other spectra of common ground objects for test and verification.

1. Leaf (cloves) reflectance spectra and polarization spectra
2. Water (clear water) spectra and polarization spectra.

The quantitative and qualitative relationships summarised above are commonly seen in nature.

The reflection and absorption of light of the ground objects is the basis of measuring their characteristics of inversion by reflectance spectroscopy. From the above analysis, there is an inverse relationship between DOP spectroscopy and the spectra of non-polarized reflectance spectroscopy. When non-polarized reflectance is at its crests, the corresponding DOP is at its troughs; non-polarized reflectance is the characteristic band of absorption troughs, and the corresponding DOP is at its crests.

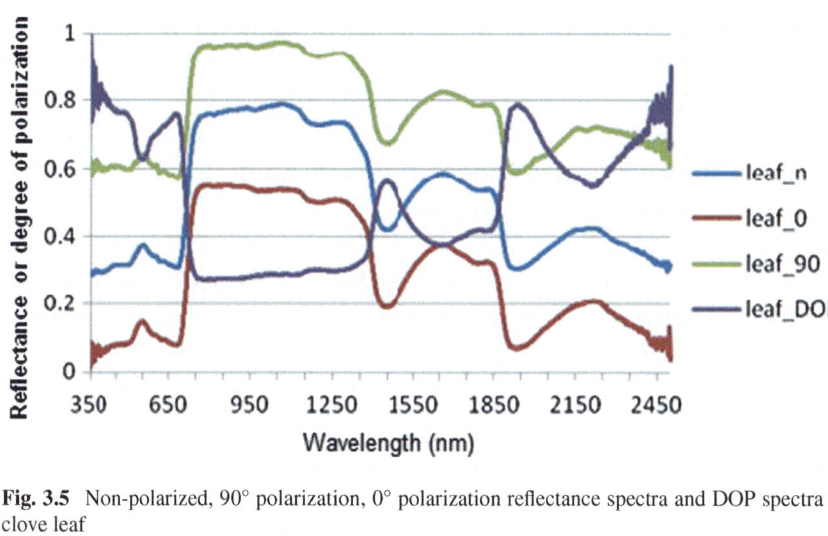

Fig. 3.5 Non-polarized, 90° polarization, 0° polarization reflectance spectra and DOP spectra of clove leaf

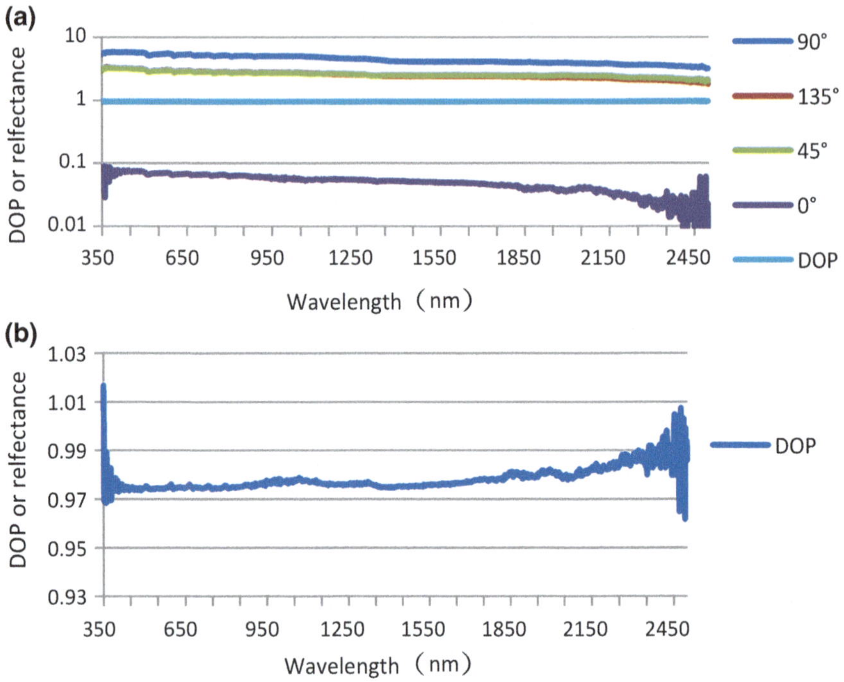

Fig. 3.6 90°, 45°, 135°, and 0° polarized reflectance and DOP spectra of clear water. *Note* **b** shows details pertaining to the DOP in (**a**)

With this relationship, the utilisation of DOP parameters in the analysis of polarization remote sensing of ground objects can draw on the selection of characteristic absorption/reflection bands used in traditional remote sensing to provide a reference for the analysis of new data.

3.2 Reflectance Spectral Model in the 2π-Spatial Orientation of a Rock Surface

This section discusses the functional relationship of the model (including non-polarized incident beam, 0° and 90° polarization) of multi-angle reflectance spectrum in the 2π-spatial orientation when the beam is incident at a large angle. In other words, this section explains what happens when a beam of light (including polarized and non-polarized beams) is incident at an angle of greater than 30° onto the surface of the rock, and its reflectance spectrum in 2π-spatial orientation.

To be able to explain the model of the reflectance spectrum of a rock surface in 2π-spatial orientation, several physical quantities must be mentioned: in physics, the energy transmitted per unit solid angle perpendicularly from each square centimetre of the surface per second is defined as the radiative power (or radiation flow density W) of the surface, usually limited to a given wavelength between λ and $d + \lambda$ at a spacing ensuring the spectral brightness in units of W cm^{-2} nm^{-1} sr^{-1}.

According to the above definition, the author defines the reflection intensity Z (reflected energy received in the unit solid angle) as the energy reflected into a unit solid angle from each square centimetre of the surface per second. This has the same units as radiation intensity, only with a weaker strength of reflection of the surface for the object in the 2π-spatial orientation. The importance of establishing this variable is such that the value determined in the experiment is Z instead of B, although they have a difference of $\cos\beta$(β-zenith angle) between them. In the probing part of the experiment, a photoelectric converter is equipped to calculate the Z-value through current measurement because it has a constant relationship with the intensity of current I. Based on these variables and definitions, the authors established the multi-angle reflectance spectrum model of the rock surface in a 2π-spatial orientation.

3.2.1 Model Description and Experimental Results

Figures 3.7, 3.8, 3.9, 3.10, 3.11 and 3.12 show reflectance spectra of a coarse-grained granite in the B (760–1100 nm) band with incident angles of 10°–60° (zenith angle calculated as 0°) and viewing angles of 0°–60° without polarization. The horizontal axis is the variation of azimuth angle from 0° to 360° while the vertical axis is the intensity of spectral reflectance of ground object Z.

Fig. 3.7 Spectra of coarse-grained granite on the B-band without a polariser, at an incident angle of 10°

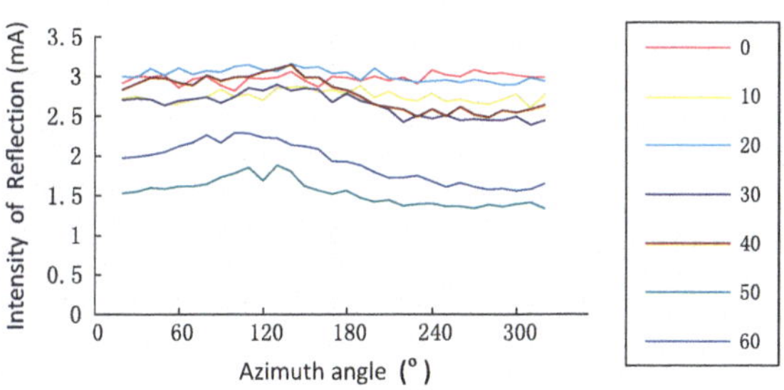

Fig. 3.8 Spectra of coarse-grained granite on the B-band without a polariser, at an incident angle of 20°

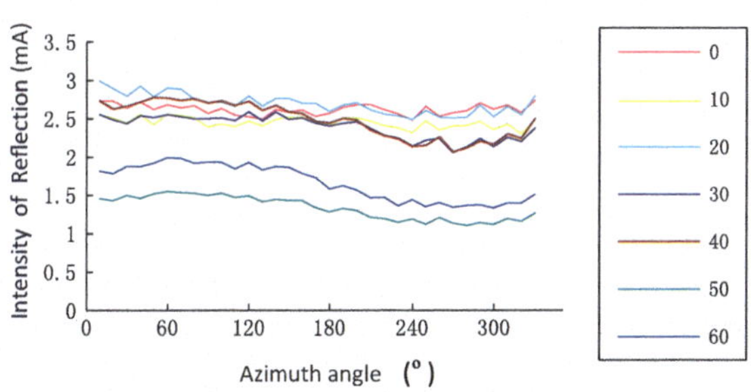

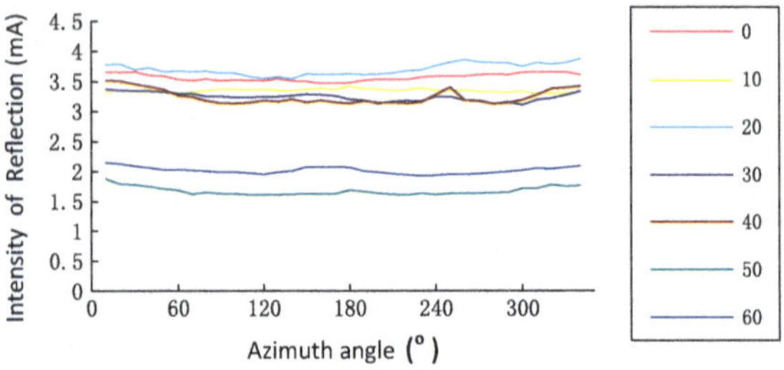

Fig. 3.9 Spectra of coarse-grained granite on the B-band without a polariser, at an incident angle of 30°

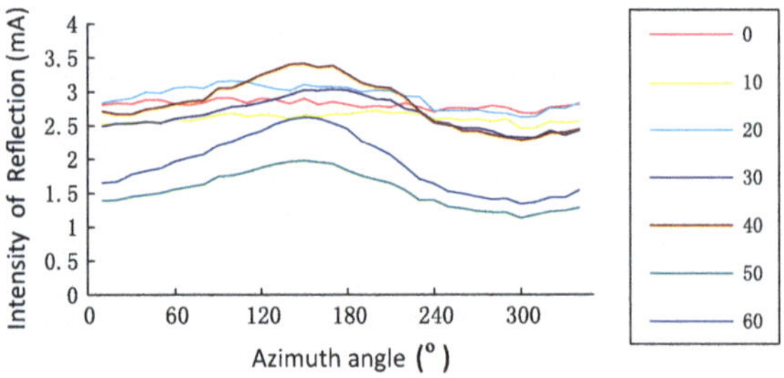

Fig. 3.10 Spectra of coarse-grained granite on the B-band without a polariser, at an incident angle of 40°

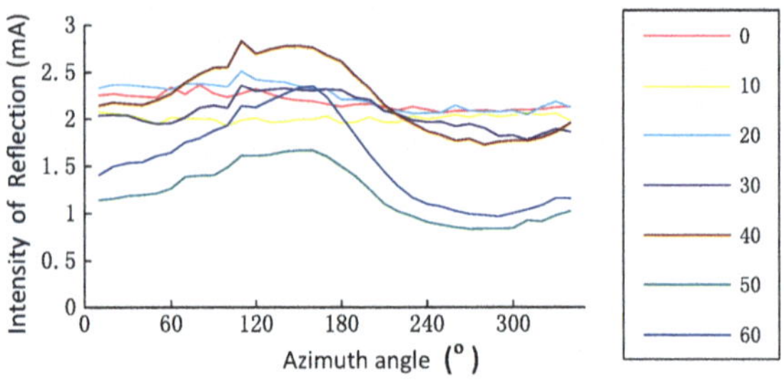

Fig. 3.11 Spectra of coarse-grained granite on the B-band without a polariser, at an incident angle of 50°

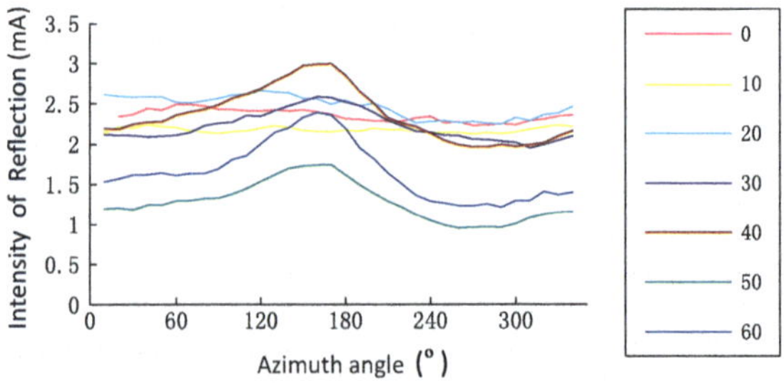

Fig. 3.12 Spectra of coarse-grained granite on the B-band without a polariser, at an incident angle of 60°

Figures 3.13, 3.14, 3.15, 3.16, 3.17 and 3.18 show 0° polarization (polarization adjustment on incident light) reflection spectral curves of a coarse-grained granite under the same conditions, while the horizontal axis is the variation of azimuth angle from 0° to 360° and the vertical axis is the intensity of spectral reflectance of ground object *Z*. Each colour represents a different height of the viewing angle.

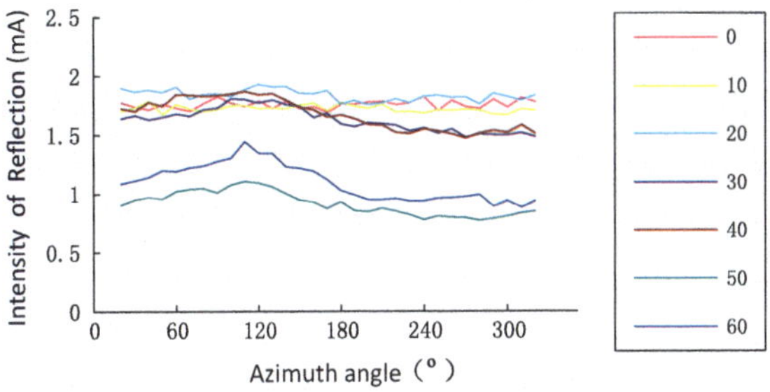

Fig. 3.13 Spectra of coarse-grained granite on the B-band under 0° polarization, at an incident angle of 10°

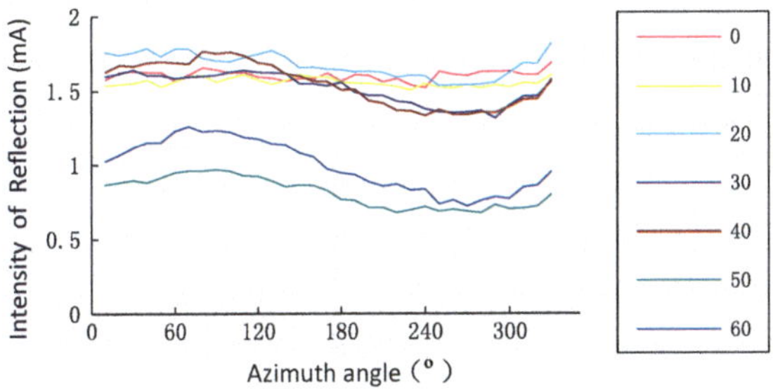

Fig. 3.14 Spectra of coarse-grained granite on the B-band under 0° polarization, at an incident angle of 20°

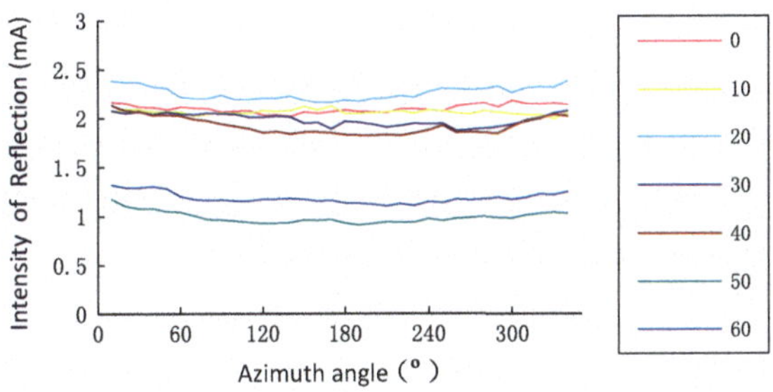

Fig. 3.15 Spectra of coarse-grained granite on the B-band under 0° polarization, at an incident angle of 30°

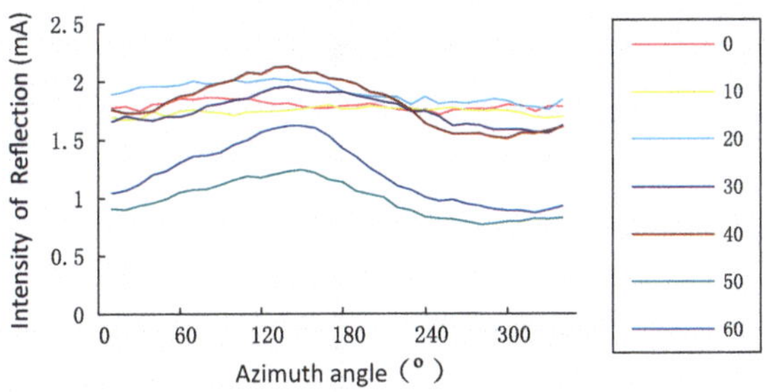

Fig. 3.16 Spectra of coarse-grained granite on the B-band under 0° polarization, at an incident angle of 40°

Fig. 3.17 Spectra of coarse-grained granite on the B-band under 0° polarization, at an incident angle of 50°

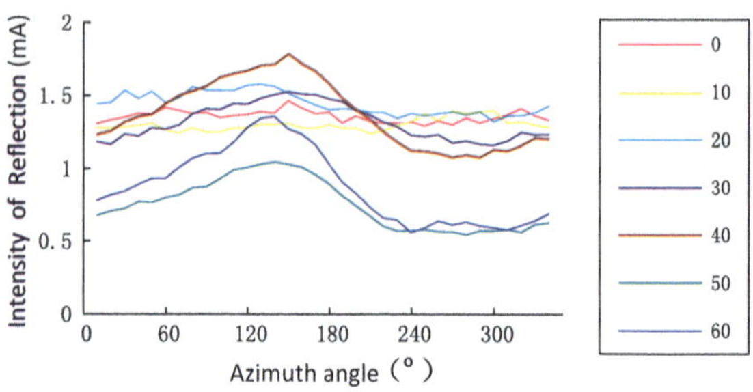

Fig. 3.18 Spectra of coarse-grained granite on the B-band under 0° polarization, at an incident angle of 60°

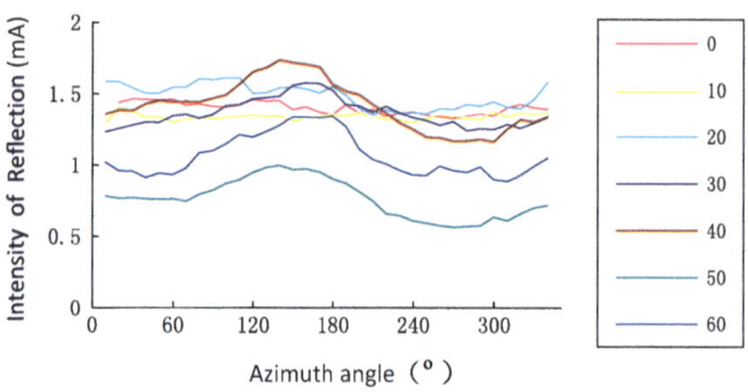

Figures 3.19, 3.20, 3.21, 3.22, 3.23 and 3.24 show the 90° polarization(polarization adjustment on incident light) reflection spectral curves of coarse-grained granite

Fig. 3.19 Spectra of coarse-grained granite on the B-band under 90° polarization, at an incident angle of 10°

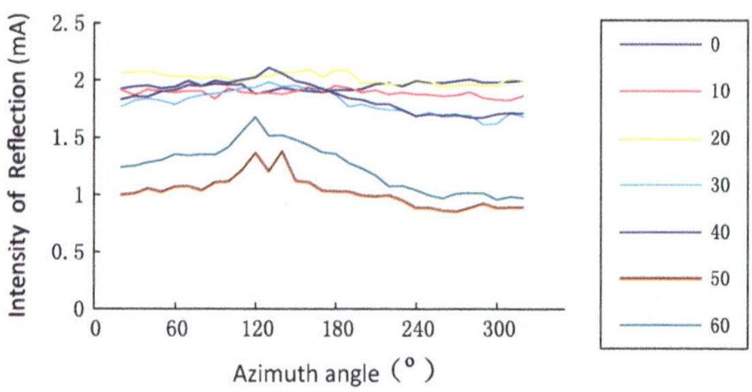

Fig. 3.20 Spectra of coarse-grained granite on the B-band under 90° polarization, at an incident angle of 20°

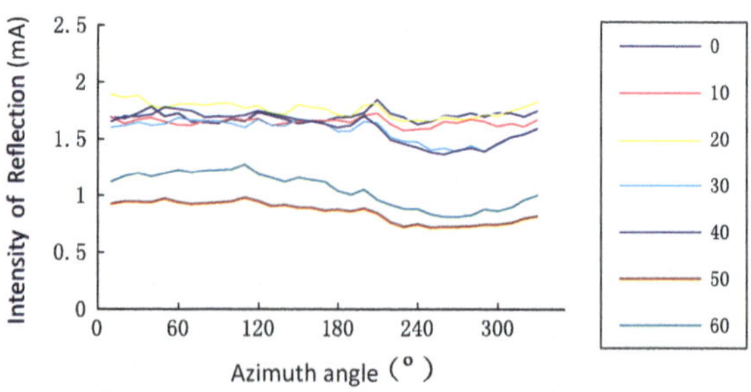

Fig. 3.21 Spectra of coarse-grained granite on the B-band under 90° polarization, at an incident angle of 30°

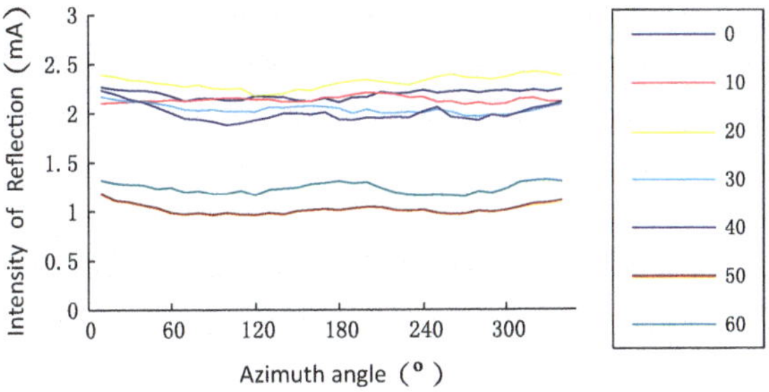

Fig. 3.22 Spectra of coarse-grained granite on the B-band under 90° polarization, at an incident angle of 40°

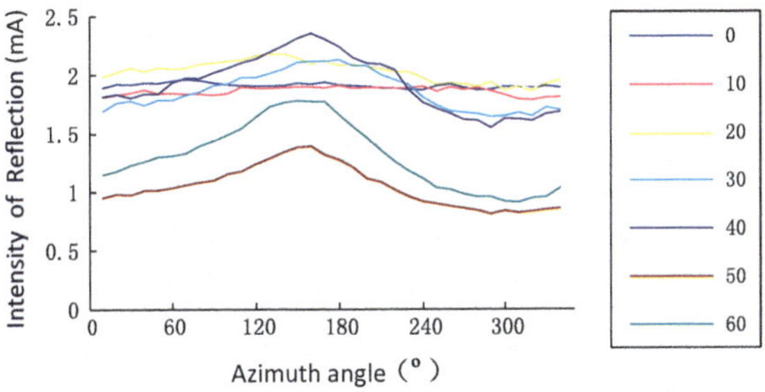

Fig. 3.23 Spectra of coarse-grained granite on the B-band under 90° polarization, at an incident angle of 50°

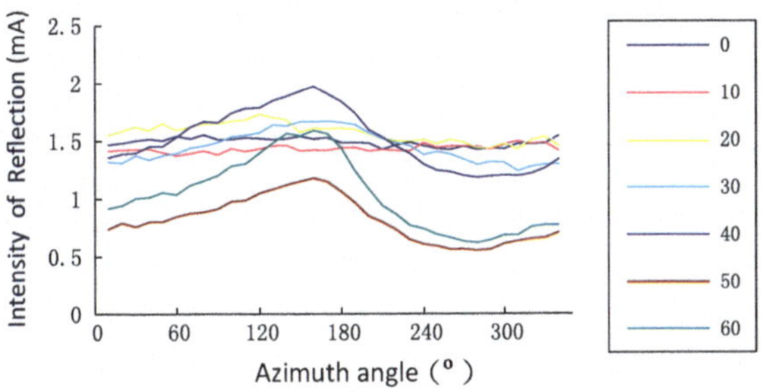

Fig. 3.24 Spectra of coarse-grained granite on the B-band under 90° polarization, at an incident angle of 60°

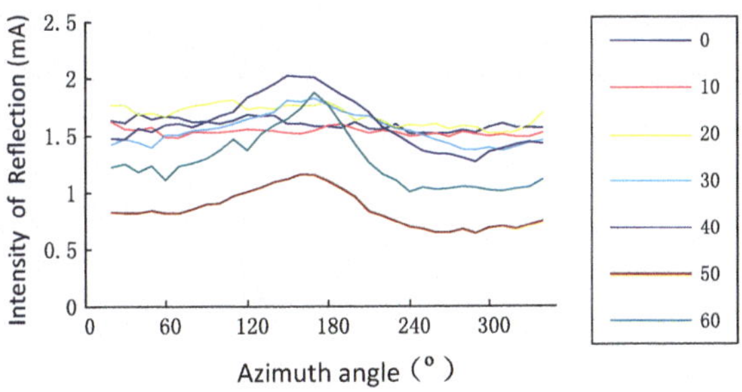

under the same conditions, with azimuth angles varying from 0° to 360° in the horizontal axis and reflection intensity of spectrum of ground object in the vertical axis respectively.

3.2.2 The Relationship Between Reflection Intensity Spectra Z and Azimuth Angle θ

This section will analyse the relationship between the reflection intensity spectra Z and azimuth angle θ, focusing on the mechanism regarding qualitative and quantitative aspects thereof.

1. The intensity of spectral reflectance Z in 2π-spatial orientation

As can be seen from Figs. 3.7, 3.8, 3.9, 3.10, 3.11, 3.12, 3.13, 3.14, 3.15, 3.16, 3.17, 3.18, 3.19, 3.20, 3.21, 3.22, 3.23 and 3.24, there are some obvious differences in the reflectance spectra of coarse-grained granite under non-polarized, 0° polarization, and 90° polarization in the 2π-spatial orientation, especially at large incident angles. Similarly, the differences of reflection spectra are also similar to other rocks (20 rock samples measured); however, no significant difference is shown in the spectra in three conditions of non-polarized, 0° polarization, and 90° polarization, which means that the impact of polarization on the spectral curve in 2π-spatial orientation is insignificant.

At the same time, all the curves exhibit the following characteristics regarding all the type of rocks: some relatively flat curves such as the spectral curves at viewing angles of 0°, 10°, and 20° are mainly straight lines, indicating that their reflectance does not vary with azimuth angle. Theoretically, the 0° spectral curve is a straight line without fluctuation because the spatial position of the 0° viewing angle does not change with the azimuth angle.

The spectral curve for a viewing angle of 10° is shown in Fig. 3.17 and its relationship with the azimuthal plane is illustrated in a plane diagram (Fig. 3.25). The

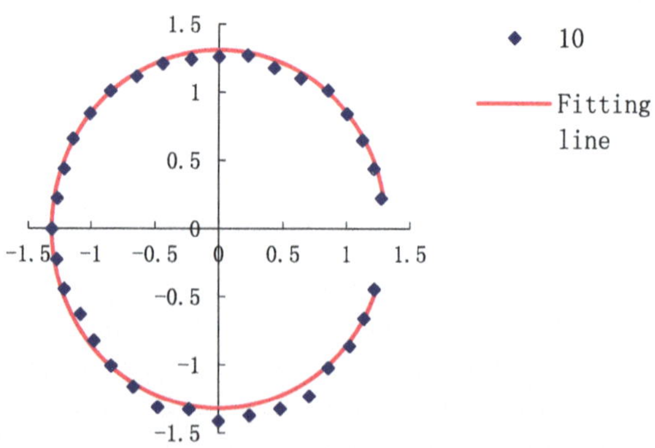

Fig. 3.25 Spectra of coarse-grained granite on the B-band at 50° incident angle, 0° polarization, and 10° viewing angle

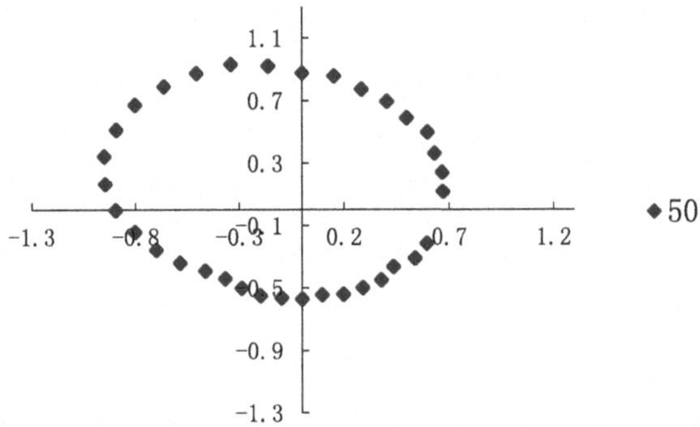

Fig. 3.26 Spectra of coarse-grained granite on the B-band at 50° incident angle, 0° polarization, and 50° viewing angle

dots are observed values while the red circle is drawn using the mean value 1.2145. For Fig. 3.17, a peak value is obtained from the spectral curve in the vicinity of the interval from 160° to 200° when the viewing angle is varied from 30° to 60°, and the range of fluctuation varies with viewing angles, where the smallest peak value is seen in the 30° curve and greater peak values emerge at 40°, 50°, and 60° curves accordingly; additionally, the distribution of all characteristic values are asymmetrical. Figure 3.26 shows a plane diagram of the relationship between spectral curves at a viewing angle of 50° (Fig. 3.7) and the azimuth angle accordingly. Fortunately, the shape of this spectral variation with azimuth angle approximates to an ellipse. The author tried to estimate the major and minor axes of the ellipse and major axis $a = 0.825$, minor axis $b = 0.672$, and the eccentricity $e = 0.58$. Similarly, the 40° and 60° curves also have characteristic peaks and an elliptical shape albeit with different curvatures. This indicated that rocks irradiated with an incident beam at a large angle show an elliptical distribution in their reflectance spectra in the 2π-spatial orientation.

Throughout the comparison of spectral curves of all rocks, the following may be deduced from their common features: the reflection spectra under the same viewing zenith angle will in the shape of ellipse (a circle being a special case), and discontinuities will not appear because of specular reflection.

2. Under the same detection zenith angle, why do reflectance spectra of rocks follow elliptical distributions?

For viewing angle at the same height, why both the reflection spectra and polarization reflection spectral curves appear to be elliptical in terms of their reflections? When the object is a standard Lambertian surface, due to the reflection intensity of the spectrum being constant, the spectral reflectance in 2π-spatial orientation will be a standard sphere and thus they show a linear curve; however, at large incident angles, the surface of the object undergoes specular reflection simultaneously so that the intensity reflected from the object is enhanced in the opposite incident direction (direction of reflection), therefore, the aforementioned straight line is stretched in this range and shows peak value for the intensity of reflection in the region 160°–200°.

Next, the corresponding physical model is reconstructed based on the theories above: it is known that the Fresnel formula describing a beam in the direction of reflection can be derived from Maxwell's electromagnetic theory of light, however, before the establishment of Maxwell's theory (1823), Fresnel had derived his formula based on classical elastic wave theory (Zhang 1985). This event is a milestone and it is reasonable to think light waves act as an elastic substance.

Thus, one can assume that an elastic and circular cross-section of a ring (with radius a), at both ends of its diameter AB, inserting a pair of opposite forces F along the direction of the diameter, as a result specular reflection is synthesised into a full-diffuse reflector (Lambertian reflection). In this case it can be proved that the ring is deformed to an ellipse. This is because we have assumed that the ring was cut along the horizontal diameter due to the symmetrical nature of the load, shear forces on cross-sections C and D are equal to zero (vertical diameter: AB and CD) except for axial force F_{N0} and moment M_0. A balanced condition $F_{N0} = \frac{F}{2}$ can be easily obtained, which means that only M_0 is superfluous. The ring is symmetrical to both the vertical diameter AB and the horizontal diameter CD, and the first quadrant of the ring, arc DA, of the ring can be selected. Since the angle of rotation of asymmetrical sections, A and D are equal to zero, therefore cross-section A can be set as a fixed end, and we set the zero degree rotation angle of cross-section D as a deformation compatibility condition to establish the equation:

$$\delta_{11} M_0 + \Delta_{1F} = 0 \tag{3.4}$$

In the Eq. 3.4, Δ_{1F} is the angle of rotation of cross-section D in the basic statically indeterminate system only when $F_{\mathrm{N0}} = \frac{F}{2}$. δ_{11} is the rotation angle of cross-section D when $M_0 = 1$ and acting alone.

The detailed process of how to calculate Δ_{1F} and δ_{11} is as follows: given the centre of the circle and an angle $\varnothing$ with the horizontal direction OC, then:

$$M = \frac{Fa}{2}(1 - \cos\varnothing) \tag{3.5}$$

$$M_1 = -1 \tag{3.6}$$

Thus,

$$\Delta_{1F} = \int_0^{\frac{\pi}{2}} \frac{MM_1}{EI} a d\varnothing = \frac{Fa^2}{2EI} \int\limits_0^{\frac{\pi}{2}} (1 - \cos\varnothing)(-1) d\varnothing = -\frac{Fa^2}{2EI}\left(\frac{\pi}{2} - 1\right) \tag{3.7}$$

$$\delta_{11} = \int_0^{\frac{\pi}{2}} \frac{M_1 M_1}{EI} a d\varnothing = \frac{a}{EI} \int\limits_0^{\frac{\pi}{2}} (-1)^2 d\varnothing = \frac{\pi a}{2EI} \tag{3.8}$$

Taking Δ_{1F} and δ_{11} into Function (3.4) we obtain:

$$M_0 = Fa\left(\frac{1}{2} - \frac{1}{\pi}\right) \tag{3.9}$$

When get M_0, then to calculate the moment of the collective effect of $\frac{F}{2}$ and M_0.

$$M(\varnothing) = \frac{Fa}{2}(1 - \cos\varnothing) - Fa\left(\frac{1}{2} - \frac{1}{\pi}\right) = Fa\left(\frac{1}{\pi} - \frac{\cos\varnothing}{2}\right) \tag{3.10}$$

This is one of the time spots (or moments) within a quarter circle, which causes the circle to deform and end up finally in the shape of an ellipse under a balance of forces. The change in the length of diameter AB can now be determined from force F, which is the relative displacement Δ_{AB} of points A and B caused by force F. To find the displacement, a unit force is imposed on points A and B, then assuming that $F = 1$, the moment within the ring under unit force can be obtained by:

$$M_0(\varnothing) = a\left(\frac{1}{\pi} - \frac{cos\varnothing}{2}\right)\left(0 \le \varnothing \le \frac{\pi}{2}\right) \tag{3.11}$$

Using Mohr's integral to obtain the relative displacement Δ_{AB} of points A and B, the integration should be applied throughout the ring. Therefore,

$$\Delta_{AB} = 4\int_0^{\frac{\pi}{2}} \frac{M(\varnothing)M_0(\varnothing)}{EI} ad\varnothing = \frac{4Fa^3}{EI}\int_0^{\frac{\pi}{2}} \left(\frac{1}{\pi} - \frac{cos\varnothing}{2}\right)^2 d\varnothing = \frac{Fa^3}{EI}\left(\frac{\pi}{4} - \frac{2}{\pi}\right) \tag{3.12}$$

where, EI is the bending stiffness of the material. Important factors that the make the spectrum of the rock elliptical with a different curvature are: the nature of the rock, viewing angle, angle of incidence, and so on.

3. Mathematical expression of the reflection intensity spectra and the azimuth angle

For rocks in the aforementioned three states, there is an elliptical geometric relationship between the intensity of the reflectance spectra and the azimuth angle. What then are the connections between the ellipses? For this, the author draws a series of standard elliptical curves with different curvature but the same distance P (where P is the distance between the left focus and the right alignment, which is different from the standard functions of an ellipse.) (Fig. 3.27). Equation 3.13 is the function used to generate the seven curves seen in Fig. 3.27:

$$y = \frac{ep}{1 + ecos\theta} \tag{3.13}$$

When $P = 8$, e (eccentricity) is 0.2, 0.4, 0.5, 0.6, 0.7, 0.8, and 0.9, respectively. As can be seen, these curves contrast with the experimentally measured curve. The crest of the curve at around 180° increases as the eccentricity increases, and for ellipses of eccentricity $e < 0.5$, the amplitude thereof is smaller (dropping curves

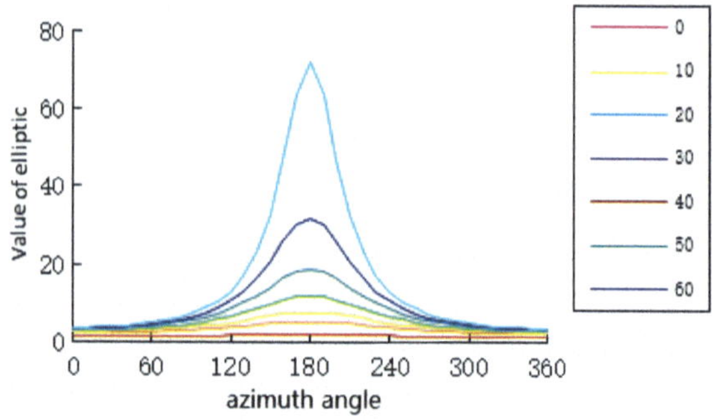

Fig. 3.27 Spectra of elliptical function $y = \frac{ep}{1+ecos\theta}$ when $P = 8$, eccentricity e of 0.2, and 0.4–0.9

with eccentricity $e = 0.1$, and 0.3 for clarity). For eccentricity $e < 0.5$, the amplitude of the crest is bigger.

It should be noted that Eq. 3.13 is not a standard function denoting an ellipse in the polar coordinate system, and the elliptic function equation:

$$y = \frac{ep}{1 - ecos\theta} \tag{3.14}$$

Equation 3.14 is the standard function for an ellipse (established by left focus and left alignment when $p = \frac{b^2}{c}$); however, if the right focus and right alignment are used to construct the elliptic equation, according to the definition given:

$$\frac{r}{p - r\cos\theta} = e \tag{3.15}$$

This is simplified to

$$r = \frac{ep}{1 + e\cos\theta} \tag{3.16}$$

While $p = \frac{b^2}{c}$ (as in Eq. 3.13).

Figures 3.28 and 3.29 are the two smoothest spectral curves among all those measured values.

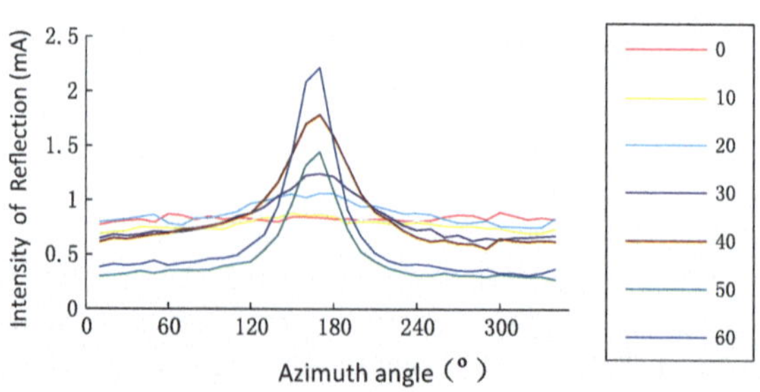

Fig. 3.28 Spectra of granite porphyry on the B-band without a polariser, at an incident angle of 50°

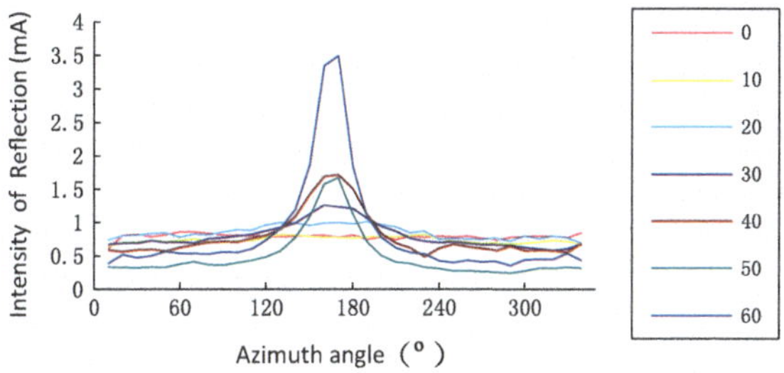

Fig. 3.29 Spectra of granite porphyry on the B-band without a polariser, at an incident angle of 60°

Comparing the experimental data shown in Figs. 3.27, 3.28 and 3.29, it can be concluded that the spectral curve with the smallest viewing angle could be fitted with the curve of the smallest eccentricity, while the curves with the largest viewing angle could be fitted with an elliptical curve of the smallest eccentricity; however, the problem has not been completely resolved. One significant difference is that, in Fig. 3.27, curves of smaller curvature are always beneath the curves with larger curvature while the measurement shows that the curves with smaller curvature are always above those of larger curvature, therefore, the author reconstructed the equation of the ellipse (Eq. 3.17) when the eccentricity e varied from 0.1 to 0.9 (Fig. 3.30) such that P varies with eccentricity, but their product remains 1.

$$y = \frac{1}{1 + e\cos\theta} \tag{3.17}$$

As the coordinate axis is relatively large, the small curvature plot with greater curvature is not obvious because the peak value is too large at eccentricity $e = 0.9$ and curve appears compressed. Figure 3.31 only represents corresponding curves of

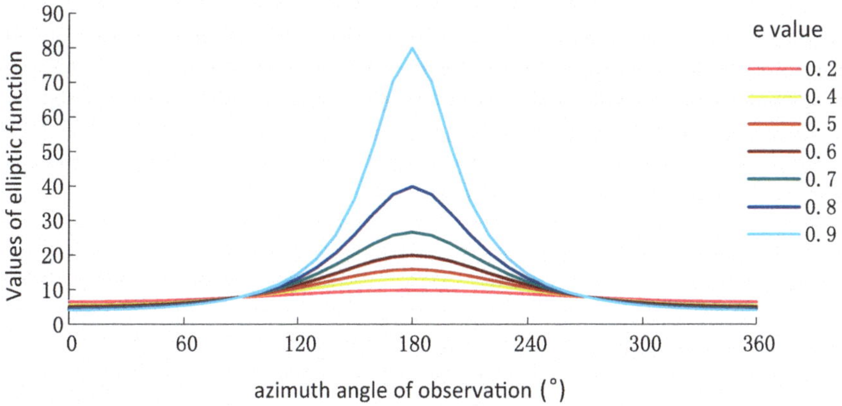

Fig. 3.30 Corresponding curves of elliptic function (Eq. 3.17) when eccentricity e is respectively 0.2, and 0.4–0.9

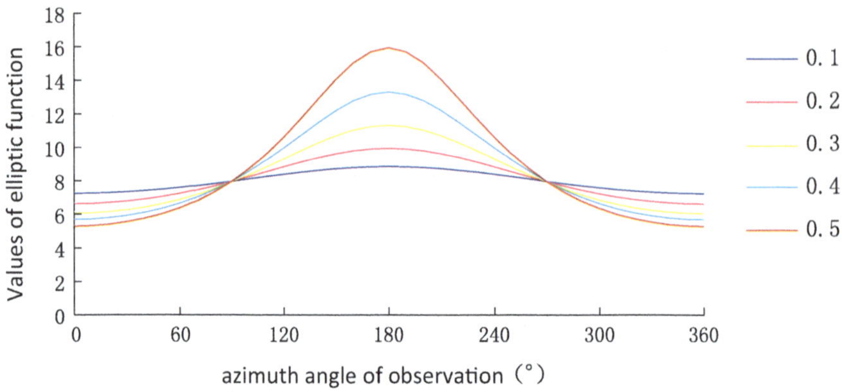

Fig. 3.31 Corresponding curves of elliptic function (Eq. 3.17) for eccentricity *e* from 0.1 to 0.5

e from 0.1 to 0.5 instead of relatively larger value of *e*, which is helpful when trying to recognise the detail therein.

3.2.3 Relationship Between the Reflection Intensity Spectra and Viewing Angle

In this section, we focus on the relationship between reflection intensity of spectrum and viewing angle, and analyse the mechanism governing qualitative and quantitative aspects thereof.

1. Mathematical expression of reflection intensity spectra and viewing angle

According to the definition of the radiation intensity, the radiation energy of unit area through the unit solid angle per unit time from the outer surface of the unit area, the radiation from the outer surface of the unit area refers to the unit area along the direction of radiation. It is the surface area of the object in the unit area of the orthographic projection onto a plane orthogonal to the direction of radiation (Zhang et al. 1997), and, the radiation intensity is:

$$L = \frac{d\varnothing}{dwdScos\beta} \tag{3.18}$$

where, L is the radiation intensity; $d\varnothing$ is radiation flux; dw is radiation solid angle; dS is the surface area of radiation; and β is the angle between the direction of measurement and the surface normal direction (or normal line).

For Lambertian reflectance, L is a constant, meaning that the same intensity is obtained in all directions. According to the above equation, Eq. 3.19 can be obtained:

$$d\varnothing = LdwdScos\beta \tag{3.19}$$

The radiant flux per unit solid angle (the radiant energy per unit time) is directly proportionate to $\cos\beta$, where β is the viewing angle measured in the experiment. The intensity Z measured in the experiment is usually an accurate reflection of radiation flux per unit solid angle, so:

$$Z = \frac{ep}{1 + e\cos\theta}\cos\beta \tag{3.20}$$

According to the spectral data observed, the spectral curves at 0°, 10°, and 20° tend to overlap, regardless of the type of rock.

Figures 3.9, 3.15, and 3.21 are good demonstrations of Lambertian characteristics since the curves do not vary with the change in azimuth angle. Changing the data collected in Figs. 3.9, 3.15, and 3.21 into those in Tables 3.1, 3.2 and 3.3, respectively can explain another phenomenon. Thus, taking the mean values as observation values and taking the 0° curve as a theoretical value, as well as deriving the values along other viewing angles based on the theoretical model, gives their absolute error and relative error.

From Tables 3.1, 3.2 and 3.3 the relative errors are remarkably similar, especially at viewing angles of 40°, 50°, and 60°, respectively. In addition, for Tables 3.2 and 3.3, the relative errors are within 1%, which is within the entire range of the relative error values of viewing angles. Analysis of relative errors demonstrates that the model is correct although it may suffer other shortcomings in practice.

Secondly, the problem whereby the spectral curve at 50° has a significant difference between its observed and theoretical values suggests that the curve which

Table 3.1 Data of the coarse-grained granite spectra in the B-band without a polariser, at 30° incidence angle

	cos 0°	cos 10°	cos 20°	cos 30°	cos 40°	cos 50°	cos 60°
Theoretical value	3.575	3.521	3.361	3.096	2.739	2.299	1.788
Observed value	3.575	3.374	3.707	3.248	3.247	1.680	2.027
Absolute error	0	−0.147	0.346	0.152	0.508	−0.619	0.240
Relative error (%)	0	−4.37	9.34	4.69	15.67	−36.81	11.83

Table 3.2 Data of the coarse-grained granite spectra in the B-band under 0° polarization, at 30° incidence angle

	cos 0°	cos 10°	cos 20°	cos 30°	cos 40°	cos 50°	cos 60°
Theoretical value	2.103	2.071	1.977	1.821	1.611	1.352	1.052
Observation value	2.103	2.068	2.259	1.987	1.923	0.989	1.188
Absolute error	0	−0.003	0.282	0.165	0.313	−0.364	0.136
Relative error (%)	0	−0.16	12.5	8.33	16.25	−36.77	11.48

Table 3.3 Data of the coarse-grained granite spectra in the B-band under 90° polarization, at 30° incidence angle

	cos 0°	cos 10°	cos 20°	cos 30°	cos 40°	cos 50°	cos 60°
Theoretical value	2.194	2.161	2.062	1.900	1.681	1.411	1.098
Observation value	2.194	2.141	2.324	2.045	2.004	1.023	1.240
Absolute error	0	−0.020	0.262	0.145	0.323	−0.387	0.143
Relative error (%)	0	−0.94	11.27	7.11	16.12	−37.85	11.52

lies above the spectral curve for 60° is beneath it in actual conditions. Moreover, all rocks showed the same property, which was attributed to two key reasons: the detector of the measuring instrument at 50° is inaccurate and non-compliant with normal multi-angle reflectance behaviours. Finally, as each band is composed of data from seven different detectors, the authors prefer the former reason.

2. Physical mechanism underpinning the relationship between reflection intensity spectra and viewing angle

This section explains the physical mechanism underpinning the problems proposed above; in other words, we explain the relationship between reflection intensity Z and the radiation intensity B. The energy transmitted into the unit solid angle perpendicularly from each square centimetre of the surface per second is defined as the surface radiation intensity B_0; however, this definition is not restricted to the results relating to radiation from the surface, and it can also be applied to reflection, scattering, and even energy transmitted. In fact, regarding radiation as being emitted or absorbed by the surface is idealistic due to the following mechanism: assuming that points P and P' receive radiation emitted from a small surface S simultaneously, and $PO = P^{'}O^{'}$ (Fig. 3.32), then for points P and P′, each of energy of the light cone is filled with radiation. Since point P is on a perpendicular to surface S, so the radiation intensity of bin S to point P is B_0. As for point P′, it can be thought that the radiation comes from a small surface $S^{'}$, which is perpendicular to the optical axis of the cone $P^{'}S^{'}$ with a normal $P^{'}S$. For point $P^{'}$, the solid angle of this light cone $d\Omega$ does not change. Let point $\mathrm{P}^{'}$ form an angle θ with P, so the energy transferred $[B(\theta)Sd\Omega]$ from bin S to $P^{'}$ is equal to the energy transferred $[B(\theta)Sd\Omega]$ from bin $S^{'}$ to $P^{'}$ per unit time (as bin

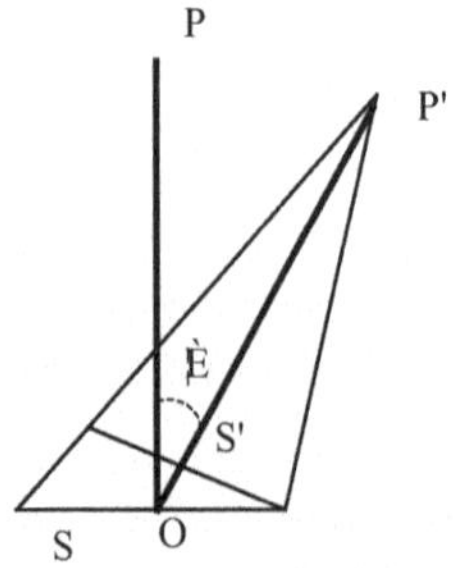

Fig. 3.2 Physical mechanism analysis between the reflection intensity spectrum and the viewing angle

S is very small, the distance from bin S' to P' is $P'O$ thus the intensity of radiation is also B_0). The energy transferred $[B(\theta)Sd\Omega]$ from bin S to P' is substantially equal to $[ZSd\Omega]$.

Thus,

$$B(\theta)Sd\Omega = B_0S'd\Omega \tag{3.21}$$

$$B(\theta)Sd\Omega = B_0S'd\Omega \tag{3.22}$$

$$B(\theta)Sd\Omega = ZS'd\Omega \tag{3.23}$$

And $S' = Scos\theta$, therefore

$$\mathrm{Z} = B_0\cos\theta \tag{3.24}$$

For the same object, generally, it is clear that the radiation intensity B_0 is fixed, and therefore it can be deduced that the reflection intensity of spectrum is proportional to the cosine of the viewing angle.

3.3 Spectral Parameter Inversion

In the previous section, the multi-angle reflection spectrum model of rock surface at large incident angles in 2π-spatial orientation was discussed. In this section the physical implications of the reflection spectrum model will be discussed as a whole, and the fitted results will be compared with those measured.

3.3.1 Overall physical implications of the reflectance spectrum model of rocks

The functional relationship between the reflection intensity, viewing angle, and azimuth angle of the reflection spectrum model is discussed above, and an explanation given based on the aforementioned physical mechanisms. Overall physical implication of the model, or those of the *e* and *p* values, will be discussed next.

At a large incident angle, the multi-angle reflection spectrum model of rock surfaces in 2π-spatial orientation can be illustrated by Eq. 3.20.

Geometrically, the *e* value represents elliptical eccentricity, hence if the reflection spectrum of the rock surface shows obvious specular features, it forms a very flat oval which also means that its *e* value is large and even tends to 1; if the specular reflection spectrum does not demonstrate specular effects, or is minimal, then the value of *e* tends to 0. In graphs, the *e* value depicts the degree of fluctuation of each curve. Similarly, the *p* value in this model represents the starting height or the reflectivity of the reflection spectrum. Different rocks have different reflectivities in the same band, hence if the reflectivity of rock is high, its p value will be large and vice versa.

The implications of changes in θ and β, which is the azimuth and elevation angle in 2π-spatial orientation respectively, are very clear in physical terms and all positions within the 2π-spatial orientation are determined by the two factors above.

Finally, there are two footnotes to this discussion of the model:

1. The model is still effective at small incident angles. When the incident beam enters at a small angle, there is almost no specular effect on the rock surface but it mostly shows Lambertian characteristics. In fact, the model is fully adapted to the situation with a small e value, where the curve is near-flat and the spectral intensity will decrease with increasing viewing angle.
2. The polarization features will be erased due to the same measurements being recorded along all directions no matter whether the incident beam is polarized or natural. Although the model is applicable to both polarized and natural light, the e and p values under the same viewing angle are still different. The difference of the characteristics of polarization is not significant in this 2π-spatial orientation but the vertical plane of the reflected light is. The reason why e and p are different under the same viewing angle is that the intensity of the incident beam has changed when a polariser is employed, thus affecting the e and p values.

3.3.2 Fitted Results and Actual Observations

Some typical curves (Fig. 3.33) are selected to show the spectra of granite porphyry at an incident angle of 50° in the B-band without a polariser, and further illustrate the following fitted results (Figs. 3.34, 3.35, 3.36, 3.37, 3.38, 3.39 and 3.40).

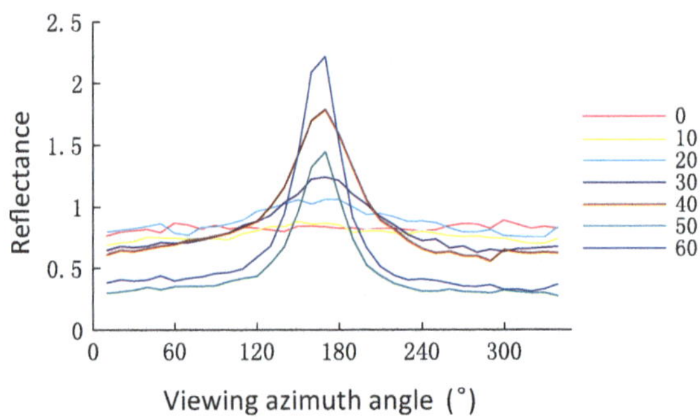

Fig. 3.33 Spectra of granite porphyry at an incident angle of 50°, in the B-band, without a polariser

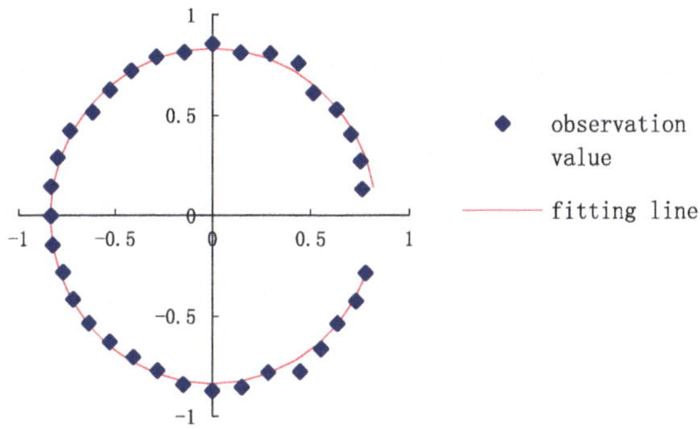

Fig. 3.34 Fitting diagram of the spectra when the viewing angle is at 0°, with a radius of 0.835

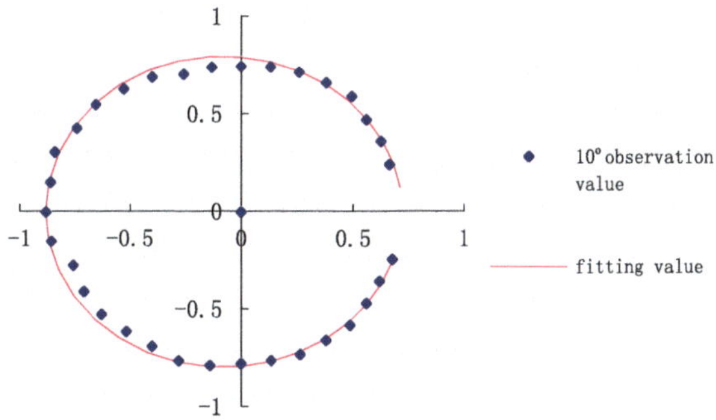

Fig. 3.35 Fitting diagram of the spectra when the viewing angle is 10°, the fitted ellipse has $e = 0.1$ and $p = 8.05$

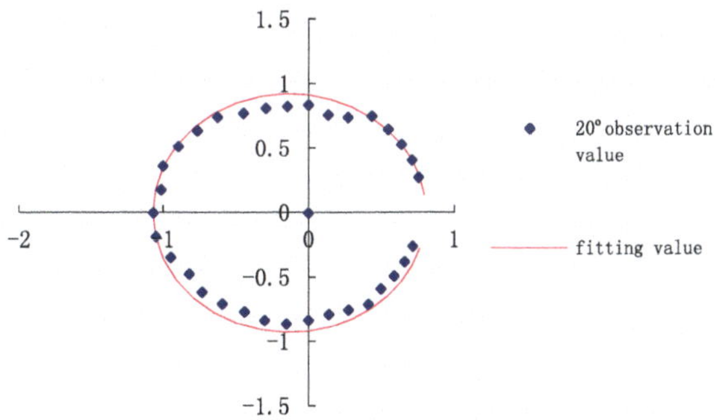

Fig. 3.36 Fitting diagram of the spectra when the viewing angle is 20°, the fitted ellipse has $e = 0.14$ and $p = 6.969$

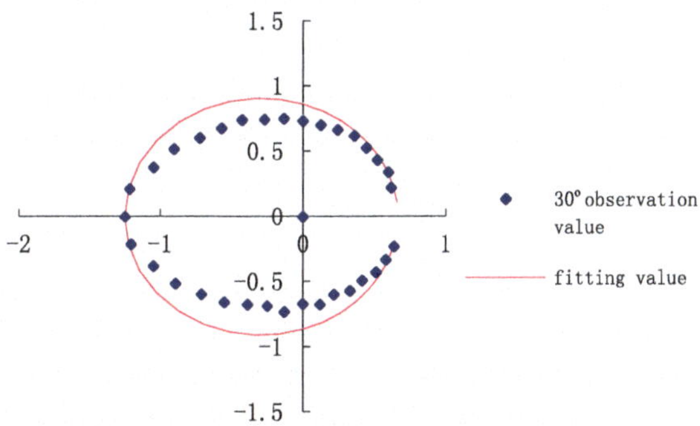

Fig. 3.37 Fitting diagram of the spectra when the viewing angle is 30°, the fitted ellipse has $e = 0.307$ and $p = 3.25$

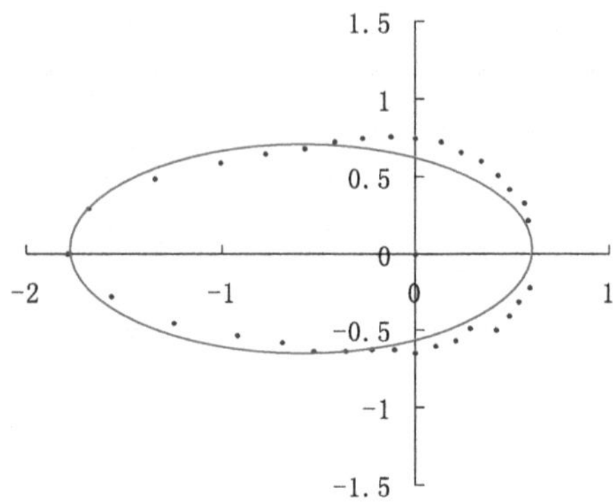

Fig. 3.38 Fitting diagram of the spectra when the viewing angle is 40°, the fitted ellipse has $e = 0.794$ and $p = 0.76$

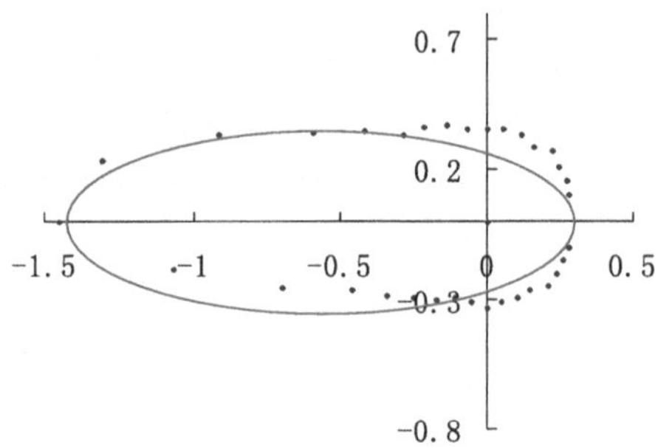

Fig. 3.39 Fitting diagram of the spectra when the viewing angle is 50°, the fitted ellipse has $e = 0.90$ and $p = 0.28$

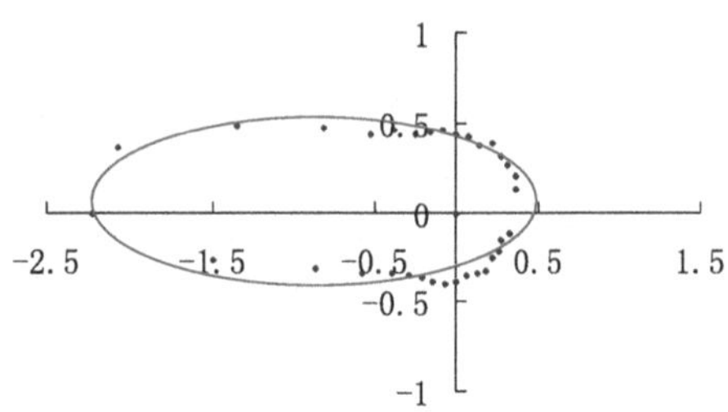

Fig. 3.40 Fitting diagram of the spectra when the viewing angle is 60°, the fitted ellipse has $e = 0.924$ and $p = 0.414$

3.3.3 The General Meaning of e *and* p

At a large incident angle, multi-angle reflection spectrum model of rock surfaces in the 2π-spatial orientation can be in fact applied to both polarized and natural light, but the e and p values at the same viewing angle are different (for a theoretical explanation, refer to Eq. 3.20).

Here, we want to emphasise the importance of both e and p: firstly, e and p can be used to describe the characteristics of the reflection spectrum of ground objects in 2π-spatial orientation: secondly, e and p could be different due to different rock compositions, even under the same conditions. These differences can help us to distinguish between different types of rock. In fact, different ground objects such as soil, plants, water, and rocks are significantly different in their physico-chemical properties, therefore, the reflectance spectra of object surfaces in 2π-space are significantly different, as are their e and p values.

3.4 Relationship Between the Composition of a Rock and the DOP

Polarization information includes the properties of target objects, which is another information source of remote sensing that is well known. These properties are determined by the components, composition, density, and surface veins of ground objects. This section aims to examine the relationship between the composition of the rocks and the DOP in the visible to near-infrared band and to investigate whether, or not, polarization spectroscopy could be used to explore the compositions of rock.

As described above, many factors affect the polarization spectrum of rocks. To highlight the effects of chemical composition, all other factors should be kept constant (Xing and Liu 1999). Selecting rock samples and considering one or two main component gradients while the other minor components scarcely change, we then grind the rock samples once under the same conditions and finally measure them under the same geometrical measurement conditions.

Accordingly, many characteristic absorption spectra of rock samples appear more often in the mid-infrared band and less so in the visible and near-infrared bands, based on the sampling conditions. Finally, the limestone-dolomite series (carbonate rocks) are selected: the main components changed are Ca^{2+} and Mg^{2+} contents, and the corresponding pulse centre of significant absorption characteristics in the vicinity of 2300 nm.

3.4.1 The Composition of Rock Samples and Spectral Characteristics

The limestone-dolomite series consist of four rock samples, namely oolitic limestone, dolomitic limestone, dolomite lime, and dolomite, named D1, D2, D3, and D4 (see Fig. 3.1), respectively. Here, oolitic limestone and dolomite lime samples were obtained from Western Hills of Liuzhuang in Dahe Township, Kuquan City of Hebei Province, belonging to the Cambrian Zhangxia; dolomite and dolomitic limestone samples were taken from South of the Lingkou village of the Luquan City, belonging to Gaoyuzhuang group of the Great Wall system. All samples were carbonate rocks. Limestone is mainly composed of calcite $CaCO_3$, and others are composed of dolomite $CaMg(CO_3)_2$. We cut a flat surface out of the four samples then polished it with #120 silicon carbide paper, making sure the time and process are consistent among all different samples, then undertook subsequent measurement of their hyperspectral and polarization spectral data. Additionally, a small piece of sample that is consistent with the surface area of the sample spectra is intercepted then, its specific components measured with sequential X-ray fluorescence spectrometer (analytical errors at <1% of the main elements) in the laboratory. Table 3.4 summarises the results of these measurements.

As shown in Table 3.4, the main differences of the components of the sample series are the contents of Ca^{2+} and Mg^{2+}. Oolitic limestone and dolomitic limestone have almost the same main contents (Ca^{2+} and Mg^{2+}). Next, further investigation is made based on the results above. The actual measurement conditions are as follows: the incident zenith angle of the light is 50°, the viewing zenith angle is 50°, and the azimuth angle relative to the direction of incident light is 180°. Spectral measurements results are shown in Fig. 3.41.

As can be seen from Fig. 3.41, the spectral curve of the dolomite series is higher than that of the limestone series, in which the dolomite curve is significantly higher than the dolomite lime curve. Two curves within the limestone series almost match each other. Note that there is an obvious difference in the shape of the curve for dolomitic limestone in the 350–1050 nm band. The overall spectra-sampled characteristic is closely correlated to the total difference in the Ca^{2+} and Mg^{2+} contents, as summarised above, i.e., the Ca^{2+} and Mg^{2+} contents of oolitic limestone and dolomitic limestone are similar and the Ca^{2+} content is lower in dolomite; the higher Mg^{2+} content in calcite dolomite produced higher spectral values.

To illustrate the relationship in detail, next we take the spectral-sampled values of non-characteristic absorption bands (at 950, 1550, 2050, and 2350 nm) as examples (Fig. 3.42).

So what is the relationship between the compositions and the morphological differences of spectra and absorption bands of oolitic limestone and dolomitic limestone before the 850 nm wavelength? This will be analysed in the next section.

3.4.2 *Analysis of the Characteristic Bands of Non-polarized Spectra*

A red area appears on the surface of the dolomitic limestone, while the other three rock surfaces have a uniform grey or white appearance (refer to Fig. 3.1). The Fe_2O_3 content (0.33%) in dolomitic limestone is significantly higher than in the other three rocks (0.05%). Dolomitic limestone exhibits a significant blue-shift compared to that in the other three samples, and it covered a broad absorption band in the vicinity of 870 nm due to the Fe^{3+} content (Yan et al. 2003).

Figure 3.41 summarises the characteristic absorption spectra of the four kinds of rocks as follows: 520, 1400, 1860, 1900–2050, 2130, and 2300–2350 nm. According

Table 3.4 Component analysis of limestone-dolomite sample series (%)

Rock sample	SiO_2	Al_2O_3	Fe_2O_3	CaO	MgO	K_2O	Na_2O	MnO_2	TiO_2	P_2O_5	LOI
Oolitic limestone D1	0.68	0.24	0.05	54.06	0.95	0.04	<0.01	<0.001	0.008	0.152	43.72
Dolomitic limestone D2	1.91	0.58	0.33	54.22	0.82	0.18	0.05	0.007	0.024	0.199	41.72
Dolomite lime D3	0.95	0.14	0.05	35.04	13.08	0.01	<0.01	0.006	<0.001	0.193	46.24
Dolomite D4	0.56	0.11	0.05	38.29	14.29	0.01	<0.01	0.006	0.002	0.217	46.55

Note LOI indicates loss on ignition. After measuring the loss on ignition of samples using X-ray fluorescence spectrometry, the mass percentage of the main metal oxides in carbonate rocks is detected

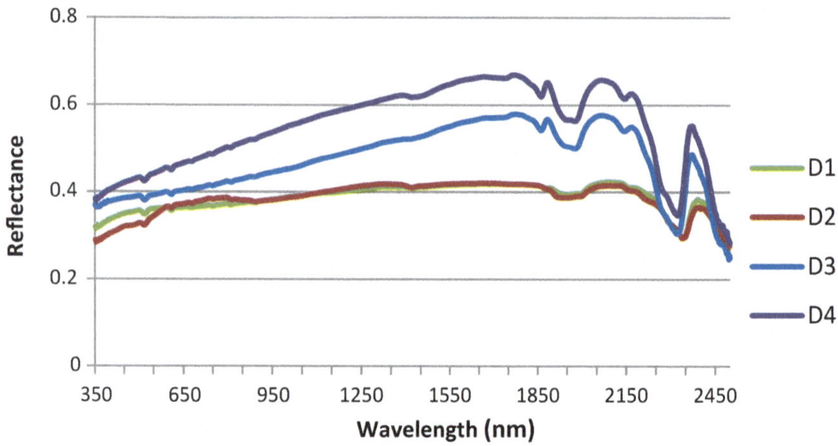

Fig. 3.41 Spectra of the limestone-dolomite sample series (350–2500 nm)

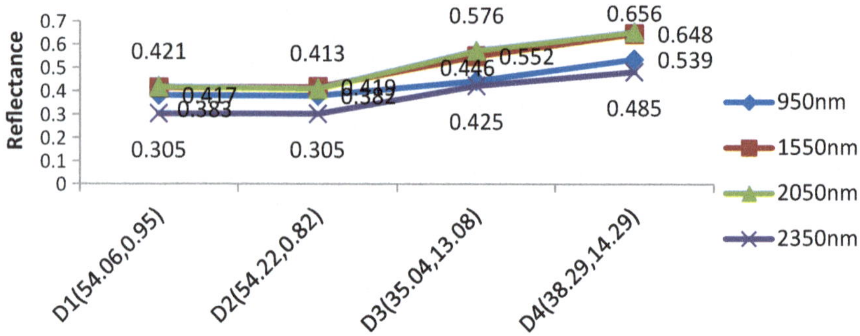

Fig. 3.42 A detailed analysis of the results of non-characteristic absorption bands

to the experimental process, the 520 nm band often appears in many sample observations and seems to be related to the light source system, but this is not discussed here.

The multiple-frequency or total frequency of water or carbonate caused by vibration appeared after the 1.3 um band and the water is rare in carbonates. Therefore, in this experiment, the water absorption band at 1400 nm is very low. Characteristic bands mainly appeared between 1.6 and 2.5 μm, and Hunt and Salisbury (1970) argued that they were generated by the multiple-frequency and total-frequency of the internal vibration of carbonate groups, or the interaction between the vibration of carbonate groups and the lattice (see Table 3.5).

In this study, the characteristics of four visible bands of dolomite are obvious, while the two bands (1860 and 2130 nm) of limestone are unclear.

Some other recent studies by Yan (2003) and Wang et al. (2009) summarised the spectral characteristics of the samples as follows: oolitic limestone at water bands of

Table 3.5 Four notable feature bands [after Hunt and Salisbury (1970)]

Band (μm)	I	II	III	IV
Calcite	1.88	2.0	2.16	2.35
Dolomite	1.86 or 1.87	1.99	2.14 or 2.16	2.33 or 2.34

1.45 and 1.9 μm and carbonate at 2.33 μm; dolomitic limestone at the water bands of 1.4 and 1.9 μm, and carbonate at 2.0 and 2.33 μm. A general conclusion was drawn that the band around 1.9 μm was likely to be arise from the presence of water.

Previous studies (Chen et al. 1998) in the Xinjiang Keping area indicated that the absorption peaks of dolomite and limestone in the vicinity of 2.35 μm differ: the absorption peak of dolomite shifted 10 nm to the direction of shorter wavelengths with respect to limestone. This conclusion can be roughly seen from Fig. 3.41. This overall band is the strongest characteristic absorption band in the spectrum, as discussed below.

3.4.3 Analysis of the Ca^{2+}, Mg^{2+} Characteristic Bands and the Composition of Non-polarized Reflectance Spectra

The trough values and corresponding wavelengths of the 2300–2350 nm band are summarised in Table 3.6.

Correlation analysis of CaO and MgO contents, and minimum trough values and the corresponding wavelengths, is conducted. The correlation coefficients (Deng 1993) are given in Table 3.7.

As shown in Table 3.7, the correlation coefficient between the minimum trough value corresponding to a given wavelength position and the CaO, and MgO, contents is high, at $\gamma_{0.01,2} = 0.990$, and the correlation coefficient between wavelength position and CaO, and MgO, contents $r > \gamma_{0.01,2}$ illustrates that the CaO, and MgO, contents are the main factors affecting the spectral wavelength. Wavelength position has a positive correlation with the CaO content, such that the position of the

Table 3.6 CaO and MgO content, minimum trough values of the non-polarized reflectance spectra and corresponding wavelengths

	CaO content (%)	MgO content (%)	Min. trough value	Corresponding wavelength (nm)
D1	54.06	0.95	0.2955	2339
D2	54.22	0.82	0.2952	2338
D3	35.04	13.08	0.3049	2321
D4	38.29	14.29	0.3535	2322

Table 3.7 Correlation coefficients between CaO, and MgO, contents and minimum trough values of the non-polarized reflectance spectra and corresponding wavelengths

Correlation coefficient r	Min. trough value	Wavelength
CaO	−0.60254	0.994879451
MgO	0.747424	−0.992959256

wavelength shifts to a longer wavelength when the CaO content is high, and vice versa; however, the MgO content is negatively correlated, such that the position of the wavelength shifts to shorter value when the MgO content is low, and vice versa. The correlation with the reflectance trough values is not significant.

Spectral continuum removal (Salisbury et al. 1991; Xu et al. 2005; Clark 1983) can effectively eliminate the influence of the baseline to give a standard reflection curve with absorption peak features. Here we use this method to extract effective characteristic bands, to determine the spectral envelope R_{up_c} (Yan 2003), and divide the envelope by the original spectrum R value to extract the characteristic absorption peak, where R/R_{up_c} is generally normalised between 0 and 1. Finally, the continuum-removed diagram for four types of rocks is shown in Fig. 3.43.

Reading the depth of trough in 2300–2350 nm band from Fig. 3.43, some results can be summarised (Table 3.8).

Further correlation analysis has been undertaken for the trough depth and contents of CaO and MgO: the results are summarised in Table 3.9.

The correlation coefficient r of the wavelength position corresponding to the trough and the CaO and MgO content is relatively high ($r > \gamma_{0.01,2} = 0.980$) indicating that CaO and MgO contents are the main factors affecting the depth of the trough. The depth of trough is negatively correlated with the CaO content and positively correlated with the MgO content, thus the lower the CaO content, the higher

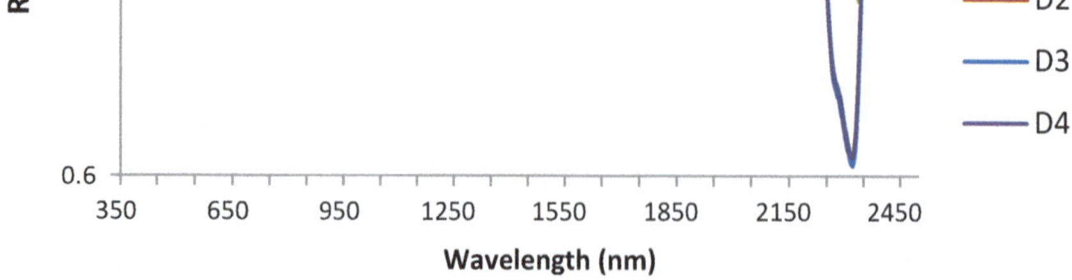

Fig. 3.43 Hyperspectral data of the dolomite-limestone samples after continuum removal

Table 3.8 CaO and MgO contents and the position and trough depth of continuum-removed non-polarized spectral reflectance of characteristic bands

	CaO	MgO	Wavelength	Trough depth
D1	54.06	0.95	2339	0.231488
D2	54.22	0.82	2338	0.199474
D3	35.04	13.08	2321	0.389827
D4	38.29	14.29	2322	0.381371

Table 3.9 Correlation coefficients between contents of CaO and MgO and trough depth of continuum-removed non-polarized reflectance spectra

Correlation coefficient *r*	Trough depth
CaO	−0.98757
MgO	0.987053

the MgO content will be, and the deeper the absorption trough in the 2300–2350 nm band.

This theoretically explains why the depth and wavelength position of the characteristic absorption band between 2300 and 2350 nm have a strong correlation with the CaO and MgO contents. Next, we analyse the spectral characteristics of the DOP.

3.4.4 Correlation Analysis Between Characteristic Bands of DOP and Content of Different Compositions

Figure 3.44 shows the DOP of samples in the limestone-dolomite series. Due to the

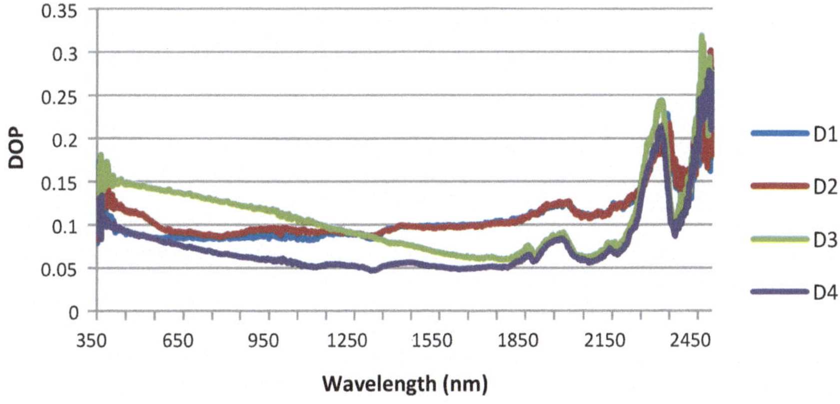

Fig. 3.44 DOP of the limestone-dolomite sample series

polarising prism being used here, the weak light intensity suggested the use of 90° and 0° polarization spectra in calculation of the DOP. The signal to noise ratio of the DOP data was also increased. As can be seen from the spectra, there are consistent waves between the spectra of the four samples. Additionally, the moving-average method is adopted here into weaken the effects of noise.

The down envelope DOPdown_c of the spectra shown in Fig. 3.44 is calculated and then divided by the DOP spectra: the results are illustrated in Fig. 3.45.

The depth of the troughs of characteristic absorption band is deduced and the corresponding wavelength positions between 2300 and 2350 nm are as listed in Table 3.10.

Likewise, the correlation between the depth of troughs and the CaO, and MgO, contents is deduced and the results are as listed in Table 3.11.

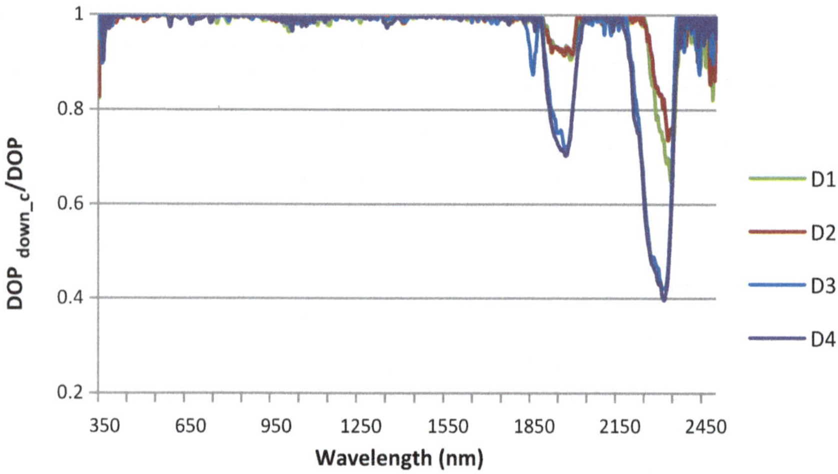

Fig. 3.45 Polarization spectra of the dolomite-limestone sample series after continuum removal

Table 3.10 Trough depths of the characteristic absorption bands after continuum removal

	D1	D2	D3	D4
Wavelength (nm)	2344	2334	2323	2321
Trough depth	0.350183	0.264137	0.578864	0.603101

Table 3.11 Correlation coefficients between depths and wavelengths of the DOP characteristics after continuum removal

Correlation coefficient *r*	Trough depth	Wavelength
CaO content	−0.96139	0.900281
MgO content	0.979317	−0.92089

$\gamma_{0.05,2} = 0.950$ and $\gamma_{0.01,2} = 0.900$ are found in the critical value table for correlation coefficients: the correlation coefficient between the troughs depth and the CaO, and MgO, contents ($\gamma > \gamma_{0.05,2}$) indicated that it is statistically significant (at the 0.05 significance level) to establish such regression models. Thereafter, the correlation coefficient between the wavelength and CaO, and MgO, contents $\gamma > \gamma_{0.01,2}$, showed that it is statistically significant to establish such regression models at the 0.10 significance level; that is, the CaO, and MgO, contents are strongly correlated to the crest and trough depth of DOP and its corresponding wavelength. The lower the content of CaO and/or the higher the content of MgO present, the higher the peak value of the DOP spectra in the 2300–2350 nm band, and the longer the corresponding wavelength of the peak tends to be. This relationship is consistent with that presented in the hyperspectral analysis, the only difference being that the characteristic band usually appears as an absorption trough on hyperspectral while appearing as a peak on the DOP spectra.

References

Chen SP, Tong QX, Guo HD (1998) Research on remote sensing information mechanism. Science Press, Beijing (in Chinese)

Clark RN (1983) Spectral properties of mixtures of montmorillonite and dark carbon grains: implications for remote sensing minerals containing chemically and physically adsorbed water. J Geophys Res 88(12):10633–10635

Deng B (1993) Statistical processing method for analyzing test data. Tsinghua Press, Beijing (in Chinese)

Hunt GR, Salisbury JW (1970) Visible and near-infrared spectra of minerals and rocks: I silicate minerals. Mod Geol 1:283–300

Salisbury JW, Walter LS, Vergo N et al (1991) Infrared (2.1–2.5 μm) spectra of minerals. The Johns Hopkins University Press, Baltimore

Song KS, Zhao YS, Zhang B (2005) Relationship between polarized and bi-directional reflectance—case study of poplar tree leave reflectance data collected in the lab. J Grad Sch Chin Acad Sci 22(2):164–169 (in Chinese)

Song KS, Zhang B, Zhao YS et al (2007) Study of polarized reflectance of corn leaf and its relationship with laboratory measurements of bi-directional reflectance. J Remote Sens 11(5):632–640 (in Chinese)

Wang JD, Zhang LX, Liu QH et al (2009) Spectral knowledge base of typical ground objects in China. Science Press, Beijing (in Chinese)

Xing LX, Liu JY (1999) The studies of the relation of the rocks spectral reflectance and its chemical compositions. Remote Sens Technol Appl 14(03):24–29 (in Chinese)

Xu YJ, Hu GD, Zhang ZF (2005) Continuum removal and its application to the spectrum classification of field object. Geogr Geo-Inf Sci 21(6):11–14 (in Chinese)

Yan DW (2003) Introduction and improvement of IDL visualization tools. Machinery Industry Press, Beijing (in Chinese)

Yan SX, Zhang B, Zhao YC et al (2003) A review of visible and near-infrared spectroscopy of minerals and rocks. Remote Sens Technol Appl 18(4):191–201 (in Chinese)

Zhang ZX (1985) Polarization of light. Higher Education Press, Beijing (in Chinese)

Zhang XC, Huang ZC, Zhao YH (1997) Remote sensing digital image processing. Zhejing University Press, Hangzhou (in Chinese)

Zhao YS, Wu TX, Song KS et al (2005a) Research on quantitative relation between multi-angle polarized reflectance and bi-directional reflectance of peridotite. Min RD 25(3):63–66 (in Chinese)

Zhao YS, Wu TX, Hu XL et al (2005b) Study on quantitative relation between multi-angle polarized reflectance and bidirectional reflectance. J Infrared Millimeter Waves 24(6):441–444 (in Chinese)

Zhao YS, Wu TX, Luo YJ et al (2006) Research on quantitative relation between polarized bidirectional reflectance and bidirectional reflectance of water-surface oil spill. J Remote Sens 10(3):294–298 (in Chinese)

Chapter 4
Remote Sensing of Ground Objects 3: Influence of Roughness and Density Structure Characteristics on the Surface Polarization Reflection

Rock is treated as an observation target to introduce the third important feature of polarization remote sensing of ground objects: the roughness and density structure characteristics thereof. These include: examining the rock surface roughness and its relationship with the multi-angle degree of polarization spectrum to find how the roughness of ground objects affects the characteristics of the polarization structure; analysing how the surface roughness affects the degree of polarization (DOP) of different scattering angles to deduce the fundamental mechanism underpinning how the surface roughness influences the polarization scattering angle, and eventually, the DOP; exploring the reflectance ratio and its relationship with the density of a ground object to provide a methodology for distal end and non-contact polarization measurements of the density of ground objects, so as to establish new theoretical bases for a distal extraterrestrial probe.

4.1 Rock Surface Roughness and Its Spectral Relationship with the Multi-angle Degree of Polarization

The characteristics of multi-angle reflectance are largely dependent on the surface structure, texture, roughness, and other surface conditions of the ground objects. The surface roughness directly affects the spectral values and the multi-angle scattering process. We carried out a study of the relationship between surface roughness and the degree of polarization.

4.1.1 Selection and Processing of Samples

Natural surfaces lie between the ideal specular surface mirror and the ideal Lambertian surface. The reflectance spectrum is a superposition of the specular component

L. Yan et al., *Polarization Remote Sensing Physics*, Springer Remote Sensing/Photogrammetry, https://doi.org/10.1007/978-981-15-2886-6_4

and the diffuse reflection component. The smoothness to which a rock can be polished is related to the hardness of its phenocrysts. The surface roughness is achieved by grinding and polishing by different abrasive substances of different gradations with the help of machining: the abrasive substances are mostly powdery grains with many edges and sufficient hardness or toughness so that they can be used to grind and polish more effectively. In gem processing, the appropriate utilisation of abrasive grains of different sizes is summarised in the following table, where the granularity (radius of a grain in μm) is numbered in decreasing order.

Abrasive substances with the same granularity used on rock surfaces of different size of crystalline inclusions and hardness can result in different surface roughnesses after grinding. To ensure the surface roughness is uniform, highly-polished, and finished after grinding, samples are required to be hard and of uniform texture. Finally, a common rock available on the decorative market is selected.

The samples are collected from "Zhongshan" in Paifang village, Shijiazhuang City, Hebei Province, and are known locally as "evergreen stone" within the biotite plagioclase gneiss geology classification. The rocks are green and hard, have a uniform structure and are highly-polished as finished. These rocks are treated as granites in the decorative market, and are referred to as gneiss herein. We cut a large piece of the rock sample into fine sheets, and then divided it into five pieces. We ground four pieces with silicon carbide of sizes #120 (diameter 125–100 μm), 240# (diameter 63–50 μm), W40 (diameter 40–28 μm), and W28 (diameter 28–20 μm), respectively, and leave one piece for polishing, then number them h1, h2, h3, h4, and h5 to encompass the several levels of abrasion quoted in Table 4.1. Since rock surfaces are non-uniform, the gradient of the selection of abrasive substances should be relatively large. If the gradient is too small, errors caused by factors such as uneven surface roughness, lack of cohesiveness during grinding, and the placement of the samples will exceed the level difference in the surface roughness, resulting in difficulties in distinguishing between measurement results and thus leading to an inability to achieve the purpose of the experiment.

There are special instruments available to measure the surface roughness of samples, which are divided into two types—mechanical stylus and light interferometry—according to the detection method used. The detection range of a mechanical

Table 4.1 Appropriate scope of granularity in gem processing

Granularity number	Appropriate scope
46#-80#	Aniseed cutting, rough chamfering
100#-120#	Small cutting, rough chamfering, pre-shaping
150#-180#	Forming coarse grinding, pre-shaping, perforating, stone cutting, grinding and finishing
240#-W40	Forming fine grinding, perforating
W28-W14	Forming grinding, rough polishing
W10-W0.5	Fine polishing

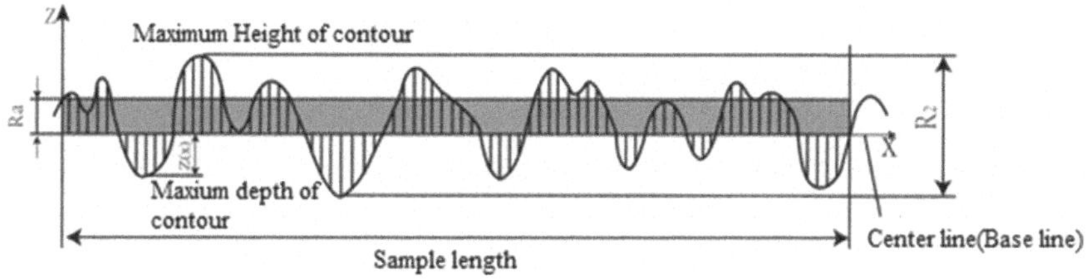

Fig. 4.1 Common profile parameters used in the assessment of surface structure

stylus is large, and thus it is more commonly used. This research uses a needle-detection type after spectral measurements, and then gives a contour arithmetic mean difference Ra which refers to the arithmetic mean of the contour offset absolute value with the length of 1 (Yu 1997), as shown in Fig. 4.1. The formula for Ra is:

$$\mathrm{Ra} = \frac{1}{l}\int_{0}^{l} |y(x)|dx \approx \frac{1}{n}\sum_{i=1}^{n}|y_i| \tag{4.1}$$

We select seven measurement points on each rock surface, and measure each point ten times before calculating the mean to finally obtain the surface roughness Ra. The results are: h1: 4.08 μm, h2: 2.91 μm, h3: 1.47 μm, h4: 0.86 μm, and h5: 0.33 μm.

The surface roughness mentioned above is based on geometric parameters. The evaluation of surface smoothness is related to the wavelength of the incident light. It is mainly based on the scattering characteristics of the surface to determine differences between a smooth, or rough, surface. For example, a surface is considered smooth when specular reflection occurs and rough if diffuse reflection occurs. Such apparent roughness is difficult to be unified and given a quantitative description, thus the criteria are diversified. A set of criteria for the classification of surface roughness may be given as follows: smooth surfaces are ($h < \frac{\lambda}{25\cos\theta}$), medium-rough surfaces are ($\frac{\lambda}{25\cos\theta} < h < \frac{\lambda}{8\cos\theta}$), while rough surfaces are ($h > \frac{\lambda}{8\cos\theta}$) where *h* represents the relative height of two points on the surface, λ is the wavelength of the incident light, and θ is the angle of incidence, (Xie et al. 2005). Measurement of Ra can be used to replace *h* according to these criteria, while criterion calculation is shown in Table 4.2 at a zenith angle of 50°.

After comparison, h5 can be considered a medium rough surface according to the criterion proposed here, while the other four rocks have rough surfaces.

Table 4.2 Criteria for the classification of surface roughness

Wavelength (μm)	Medium rough surfaces $\frac{\lambda}{8\cos\theta}$ (μm)	Rough surfaces $\frac{\lambda}{25\cos\theta}$ (μm)
0.35	0.068	0.022
2.5	0.486	0.156

4.1.2 Multi-angle Characteristics of the Degree of Polarization Spectra

Section 2.4.1, Chap. 2 has already summarised the characteristics of changes of DOP varying with the viewing zenith angle and the viewing azimuth angle. In this section the variation of multi-angle polarization reflectance of objects with different surface roughnesses is supplemented to verify the aforementioned behaviour.

1. How does DOP vary with different viewing zenith angles

Figures 2.30 and 2.31 (Sect. 2.4.1) show the variation in DOP of the roughest sample h1 and the smoothest sample h5 against the change in zenith angles at an azimuth angle of 180° (specular reflection position) and summarises the behaviour as increasing between 0° and 80° and declining thereafter. Peak values and corresponding angles are different with different surface roughnesses of rocks, while Figs. 4.2, 4.3 and 4.4 show additional information about the other three rock samples.

As depicted, the surface roughness influences the occurrence of specular reflection and its peak values. The smoother the surface, the greater the peak amplitude of specular reflection, and the closer the angle is to the angle of specular reflection; the

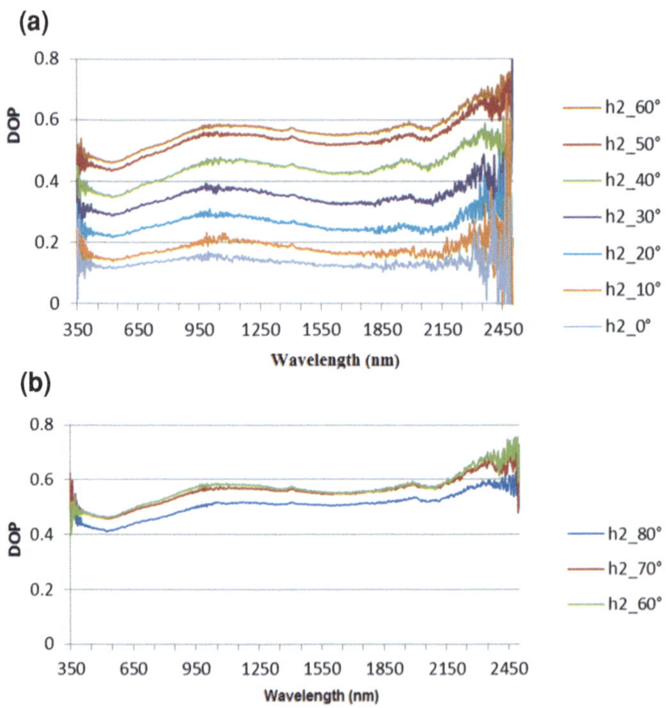

Fig. 4.2 DOP of Rock h2 at an incident zenith angle of 50°, viewing azimuth angle of 180°, and different viewing zenith angles (**a** 0°–60°; **b** 60°–80°)

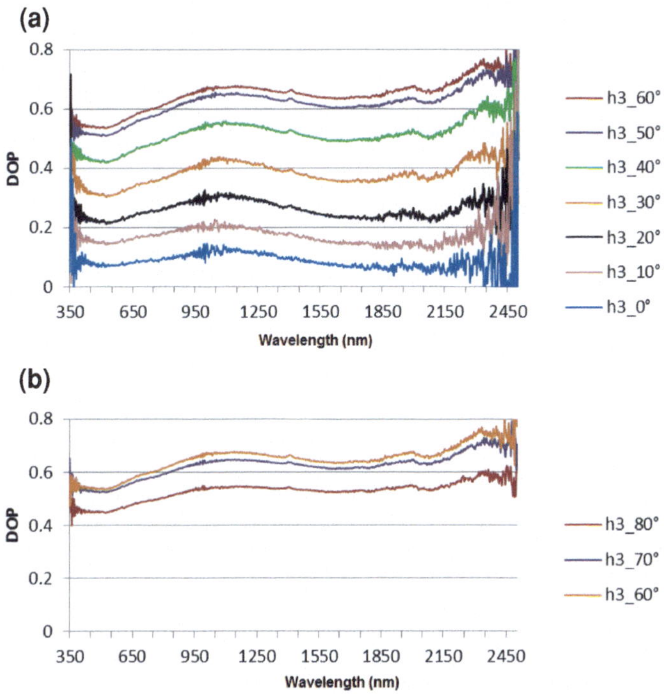

Fig. 4.3 DOP of Rock h3 at an incident zenith angle of 50°, viewing azimuth angle of 180°, and different viewing zenith angles (**a** 0°–60°; **b** 60°–80°)

rougher the surface: the lower the peak amplitude of specular reflection, the further the angle from the surface normal deviating towards the zenith angle of 90°.

2. Changes in DOP upon azimuthal variation

Figures 2.32 and 2.33 (Sect. 2.4.1) show how the DOP of rocks h1 and h5 change with viewing azimuth variation at specular reflection and zenith angle of 50°. The DOP is highest at an azimuth angle of 180°, and then gradually decreases towards each side thereof, tending to zero. Figures 4.5 to show the conditions for the other three rocks (Figs. 4.6 and 4.7).

4.1.3 Analysis of the Typical Azimuth Angle of Specular Reflection

Figure 4.8 shows the comparison of the DOP of the five rocks at a typical angular orientation of specular reflection when the viewing zenith angle is at 50° and the viewing azimuth angle is 180°. Apparently, when the surface roughness of rocks h1 to h5 decreases, their DOP increases, thus, the influence of the surface roughness is evident.

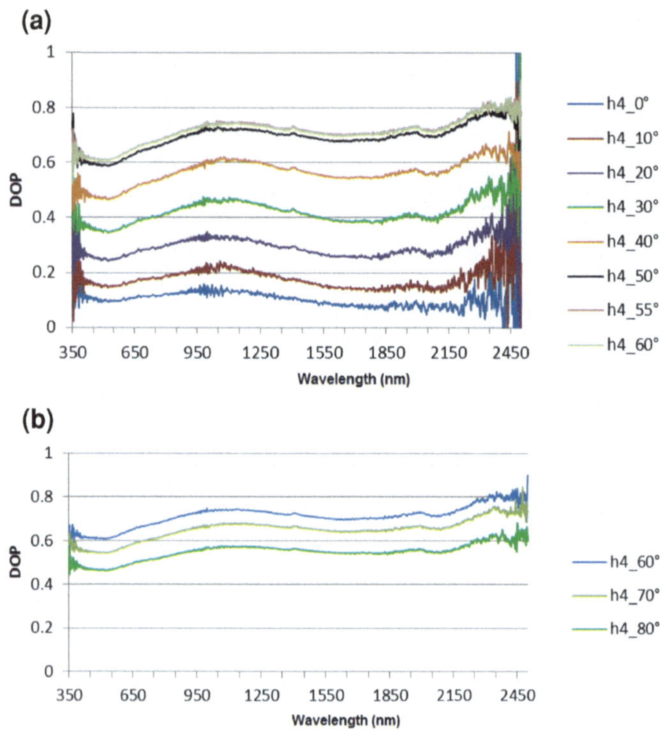

Fig. 4.4 DOP of Rock h4 at an incident zenith angle of 50°, viewing azimuth angle of 180°, and different viewing zenith angles (**a** 0°–60°; **b** 60°–80°)

4.2 Surface Roughness Effects on the DOP of Different Scattering Angles

The influence of surface roughness on the degree of polarization of different zenith angles and viewing azimuth angles is first analysed before analysing the physical mechanism of light scattering of rough surfaces.

4.2.1 Surface Roughness Effects on the DOP of Different Zenith Angles

This section will analyse the quantitative relationship of surface roughness, peak values of the DOP, and the corresponding angles.

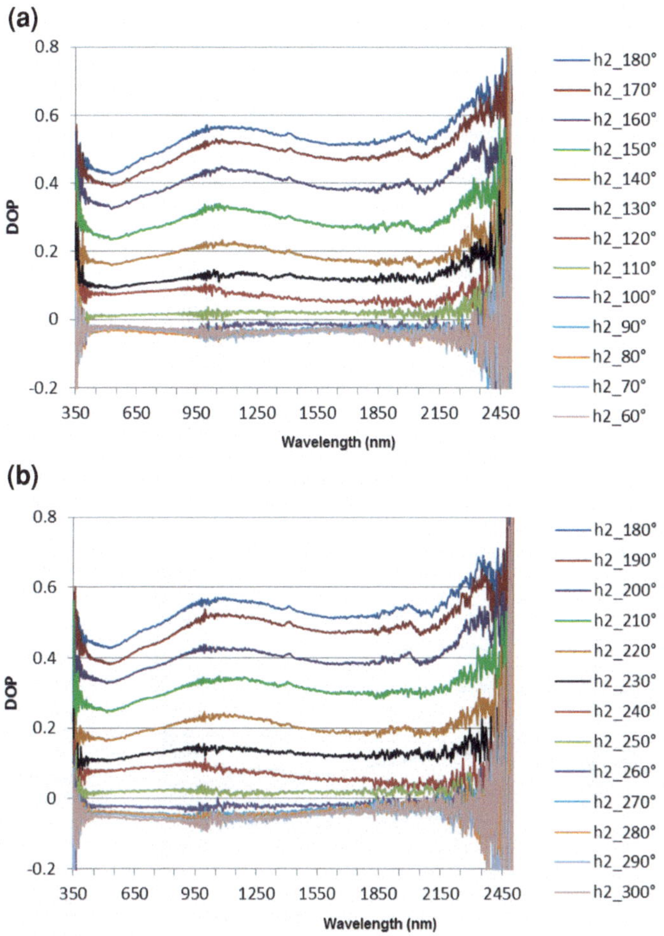

Fig. 4.5 (DOP of Rock h2 at an incident zenith angle of 50°, viewing zenith angle of 50°, and different viewing azimuth angles (**a** 180°–60°; **b** 180°–300°)

1. Quantitative analysis of the relationship between the surface roughness and the peak value of the DOP

To analyse the relationship between the surface roughness and the DOP quantitatively, the vicinity of a characteristic band (520 nm) of the DOP spectrum is selected, wherein lies a trough. The rock is green containing light green and dark green phenocrysts. We calculate the average values of the DOP at 520 nm $\pm$ 10 nm to remove the effects of measurement noise. Figure 4.9 depicts the DOP values of the five rocks

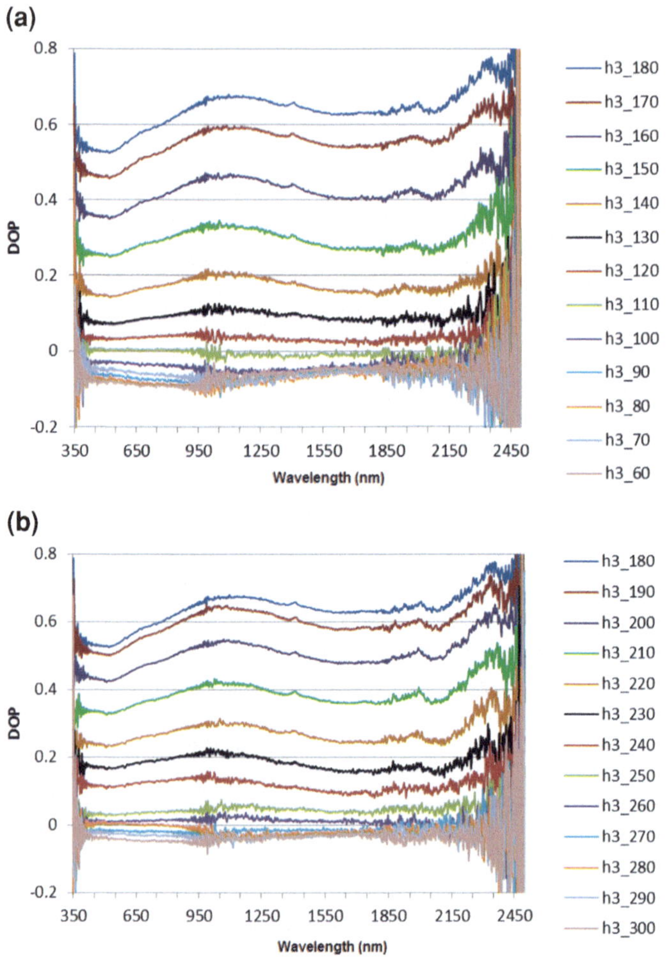

Fig. 4.6 DOP of Rock h3 at an incident zenith angle of 50°, viewing zenith angle of 50°, and different viewing azimuth angles (**a** 180°–60°; **b** 180°–300°)

within these wavelengths, which also reflects the characteristics of the behaviour of the DOP of the rocks affected by surface roughness at different zenith angles.

Smooth rocks such as h5 have a higher peak value of the DOP, the width of the crest of DOP at azimuths of between 0° and 80° is narrow; rougher rocks such as h1 have lower peak DOP values, and the widths of these peak values are larger.

After analysing the peak value of the DOP and the surface roughness of the five rocks, a power regression equation is discovered to link them:

$$y = 0.604x^{-0.297} \tag{4.2}$$

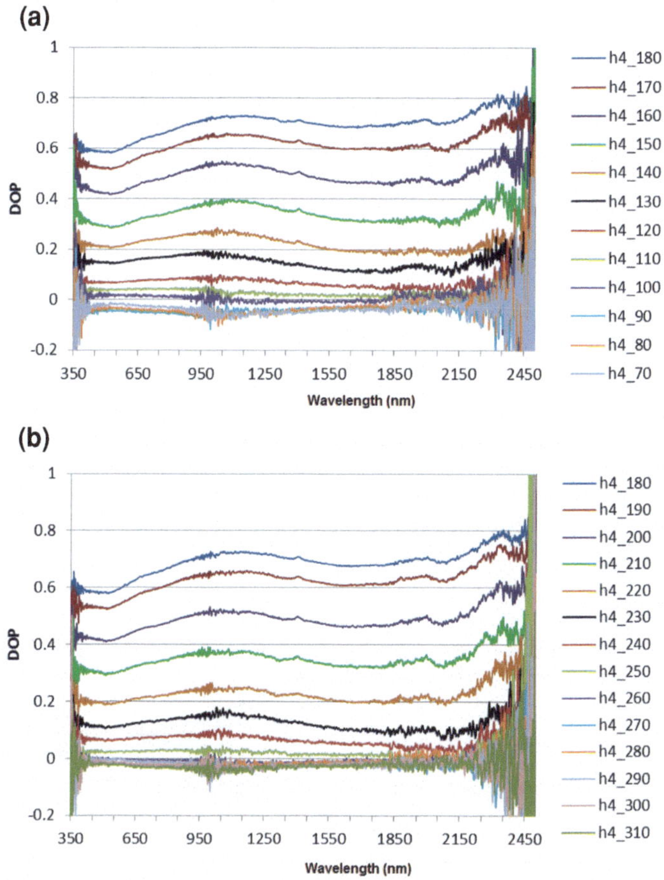

Fig. 4.7 DOP of Rock h4 at an incident zenith angle of 50°, viewing zenith angle of 50°, and different viewing azimuth angles (**a** 180°–70°; **b** 180°–310°)

Here variable X is surface roughness, Y is peak DOP. The coefficient of determination R^2 therewith is 0.9854. An *F*-test is conducted on the regression model and significant correlation (Fig. 4.10) between the peak value of the DOP and surface roughness is found.

2. Quantitative analysis of the relationship between surface roughness and the angles with the peak values of DOP

The angle with peak values of DOP can be seen in Fig. 4.9: fort h1 to h5 they are 60° to 70°, 60°, 60°, 60°, and 50°, respectively. The peak value of h1 at 60° is equal to that at 70° which can be linearly interpolated to 65°, and its DOP values are as shown in Fig. 4.11. Similarly, the curve for rock h4, measured at a zenith angle of 60°, is consistent with that of 55° and is thus interpolated to 57.5°. The angle accuracy of comprehensive evaluation of the peak is within 5°.

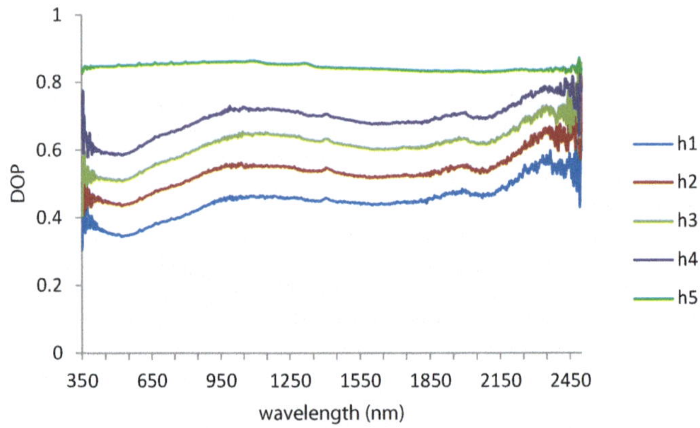

Fig. 4.8 DOP of rock samples at an incident zenith angle of 50°, viewing zenith angle of 50°, and viewing azimuth angle of 180°

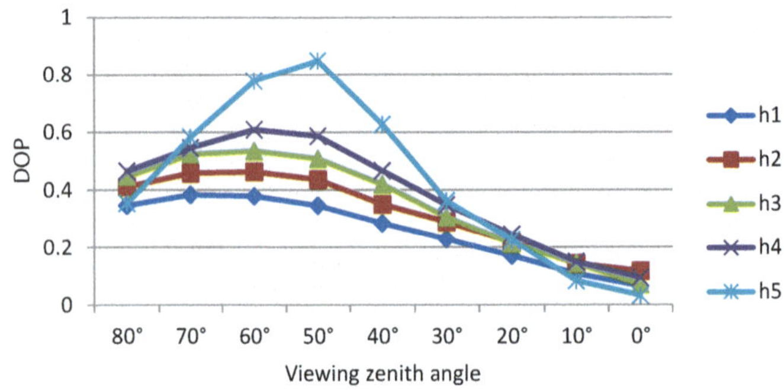

Fig. 4.9 DOPs of the five rocks at 520 nm ± 10 nm varying with viewing zenith angle

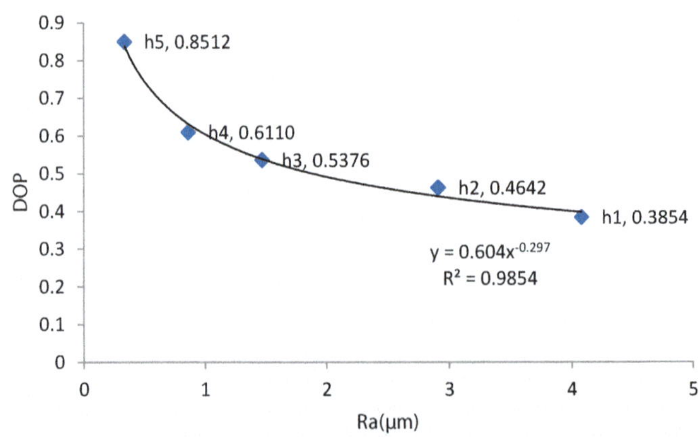

Fig. 4.10 Correlation analysis of the peak values of the DOP of the five rocks in the incident main plane ($\varphi r = 180°$) and the surface roughness (Ra)

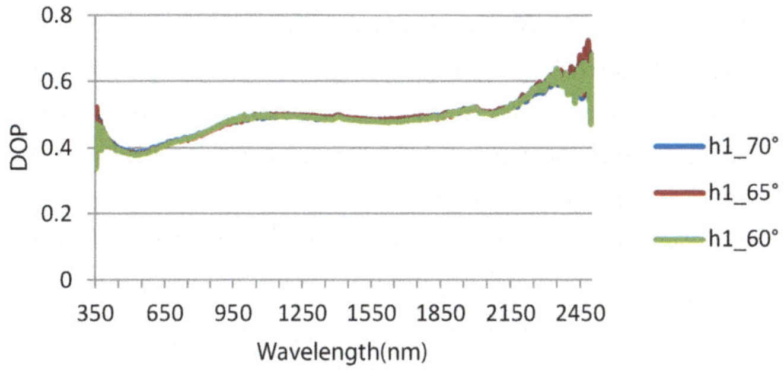

Fig. 4.11 DOP measurements of rock h1 at viewing zenith angles of 60°, 65°, and 70°

As for the analysis of the relationship between the angle corresponding to the peak value and the surface roughness, when surface roughness tends to zero, the angle corresponding to the peak value approaches 50° with nearly 100% specular reflection. When the surface roughness tends to infinity, the DOP shows no peak with a nearly Lambertian surface. Natural surfaces are ranged in between Lambertian and perfect specular reflection, usually following the law that when surface roughness increases, the angle corresponding to the peak value increases which is of course less than 90°. The correlation model is given by:

$$y = 3.4194x + 51.584\ (y < 90^\circ) \tag{4.3}$$

The model coefficient of determination R^2 is 0.8177 (Fig. 4.12). We conducted an F-test on the regression model and significant correlation was found between the angle corresponding to the peak value of DOP and the surface roughness.

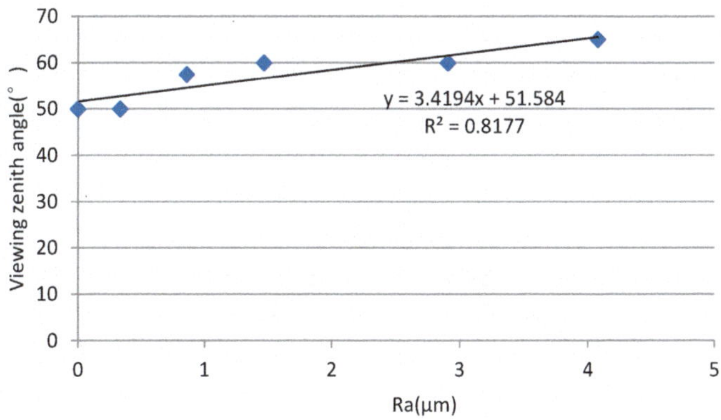

Fig. 4.12 Analysis of the correlation between the angle corresponding to the peak value of DOP and the surface roughness of the five rocks

4.2.2 *Effect of Surface Roughness on the DOP at Different Viewing Azimuth Angles*

To analyse the relationship between the surface roughness and the DOP quantitatively, again the same characteristic bands near 520 nm in the DOP spectra were selected. The DOP values within 520 nm ± 10 nm were averaged to remove the effects of measurement noise. Figure 4.13 shows the DOP values of five rocks at this wavelength, and the general characteristics of the influence of surface roughness on the DOP values at different viewing azimuth angles. Smoother rocks, such as h5, have higher peak values of the DOP, and the peak width of the DOP at the range of azimuth angles examined is narrow; as for the rougher rocks such as h1, they have lower peak values of DOP, the peak width of the DOP at the range of azimuth angles examined is wider, however, the change in peak width is not clear for the five rocks tested here, and there may be two reasons for this: first is that the gradient of sample preparation changes is not steep enough to cause measurable change; the other reason is that the DOP is measured by analysing polarization spectra at 90° and 0° (both of which lie far from the azimuthal area of 180°), where there is little light and the measured value has fallen within the range of measurement error with a low signal to noise ratio, thus resulting in large spectral errors therein. Therefore, in this case the quantitative analysis of the peak width is not discussed further.

Correlation analysis is conducted between the peak value of DOP of the surface roughness at a 180° azimuth angle and a power function is deduced:

$$y = 0.5822x^{-0.333} \tag{4.4}$$

The coefficient of determination R^2 is 0.9843 (Fig. 4.14). An F-test is conducted for the regression model and a significant correlation between the peak value of DOP and surface roughness is discovered.

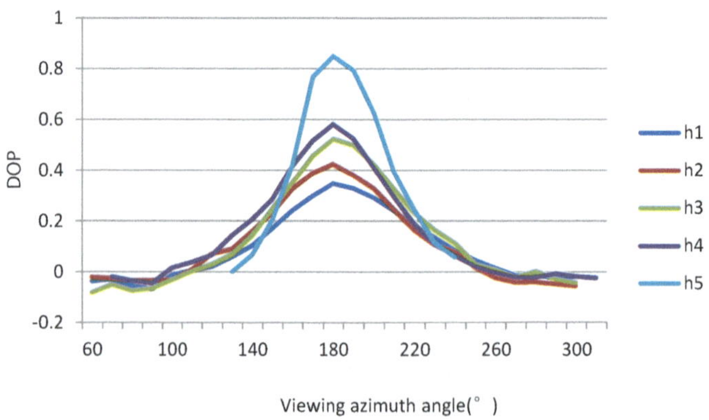

Fig. 4.13 changes in the DOP of the five rocks with changes in viewing azimuth angle at 520 ± 10 nm

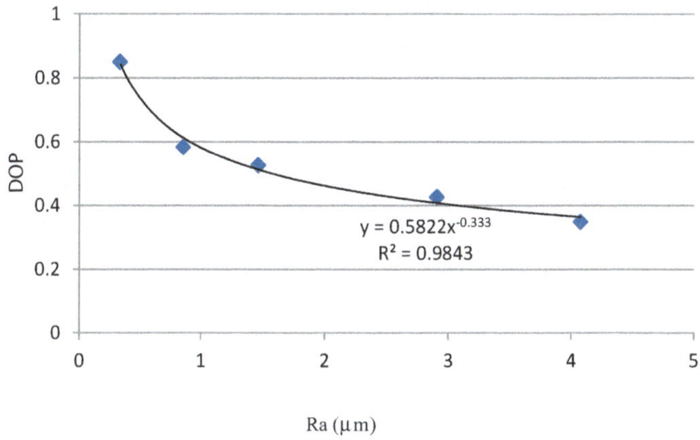

Fig. 4.14 Correlation analysis of the peak value of DOP and the surface roughness (Ra) of five rock samples at the specular reflection zenith angle

4.2.3 Mechanism Analysis of Light Scattering from Rough Surfaces

Natural surfaces are in between the ideal specular surface and the ideal Lambertian surface. The reflectance spectrum is a superposition of the specular component and the diffuse reflection component (Zheng 1999), while the smoother the surface the greater the proportion of specular reflection. The phenomenon of peak shift from the corresponding angle of specular reflection caused by surface roughness is due to shadowing and masking effects of rough surfaces (Yang and Wu 2009). Shadowing refers to the incident beam on the microscopic plane being intercepted, while masking refers to the case where the reflected light in the viewing direction is stopped. When light is incident on a roughened surface, generally there will be shaded areas on the surface, and relative to the direction of viewing angle (scattering angle) some scattered waves are blocked by rough surfaces from reaching the observation point, therefore the shadowing and masking effects of rough surfaces must be considered. With increased surface roughness, shadowing effects will be more pronounced.

The necessary condition for off-specular peak phenomenon is that the surface roughness is greater than, or equal to, the wavelength ($\delta/\lambda \geq 1$) (Torrance and Sparrow 1967; Ourlier et al. 2001). Here, the arithmetic mean deviation *Ra* of the outline of the rocks measured are h1: 4.08 μm, h2: 2.91 μm, h3: 1.47 μm, h4: 0.86 μm, and h5: 0.33 μm. With respect to the previous section concerning the 520 nm band, off-specular peaks should appear in h4 to h1, which has been consistent with observed phenomena. In this section, specular reflection takes place when the zenith angle is 50° and the azimuth angle is at 180°, where the reflectance ratio is at its highest and the DOP at its maximum. This means that the surface of rock h5 is smoother, while unusually high reflection peaks are not observed in unpolished rocks. Meanwhile, the rock reflectance in the range around the angle corresponding with peak values is also high, meaning that the microscopic fluctuations on polished rock surfaces remain rough relative to the wavelength with which they are measured.

Rock surfaces also lie between the ideal specular surface and the ideal Lambertian surface with the reflectance spectrum characteristics being a superposition of the specular component and the diffuse reflection component. While some rock surfaces are close to the ideal specular surface, some are close to the ideal Lambertian surface, they vary over a wide range: however, due to physical characteristics and processing condition constraints, the same type of rock has a limited range of surface roughness.

4.3 Relationship Between Reflectance Rate and Density of Ground Objects

Rock density can be calculated by polarization reflectance spectra. The refractive index of rock is first determined by the polarized reflectance spectroscopy, and then it is used to obtain its density.

4.3.1 Relationship Between Polarization Reflection Spectra and Density of Rocks

The relationship between the DOP and density is illustrated as follows: firstly, the DOP *P* of the reflectance spectra of rock surfaces is decided by the following function:

$$\mathrm{P} = \frac{2\cos\alpha\cos\beta\sin\alpha\sin\beta}{\cos^2\alpha\cos^2\beta + \sin^2\alpha\sin^2\beta} = \frac{2}{\frac{1}{tg\alpha tg\beta} + tg\alpha tg\beta} \tag{4.5}$$

Using the law of refraction, the refraction angle in the above function can be eliminated with the refractive index, whereby,

$$\mathrm{P} = \frac{2\cos\alpha\sqrt{1-\frac{\sin^2\alpha}{N^2}}\sin\alpha\frac{\sin\alpha}{N}}{\cos^2\alpha\frac{N^2-\sin^2\alpha}{N^2} + \sin^2\alpha\frac{\sin^2\alpha}{N^2}} = \frac{2\sin\alpha tg\alpha\sqrt{N^2-\sin^2\alpha}}{N^2-\sin^2\alpha+\sin^2\alpha tg^2\alpha} \tag{4.6}$$

Function (4.6) demonstrates that, when the incident angle and DOP are known, the refractive index of a rock can be calculated, which can then be used to deduce the density of the rock through use of the Lorentz-Lorenz refraction formula:

$$\frac{\mathrm{n}^2-1}{\mathrm{n}^2+2} \times \frac{1}{\rho} = \text{constant} = \gamma_{\mathrm{L.L}} \tag{4.7}$$

where n is the refractive index, ρ is the density, and the constant is usually 0.12. For example, the refractive index of augite is 1.713, its density can then be calculated as 3.330 g cm^{-3}. This density falls into the range of 3.23–3.52 g cm^{-3}, accordingly, the refractive index can be used to estimate the density of minerals.

In 1880 Lorenz and Lorentz proposed Eq. (4.7) at the same time. Lorentz based his studies on electromagnetic theory, while Lorenz based his research on the propagation of light. This formula has a rigorous derivation and according to this theory, refraction is measured at a wavelength of infinity, but in practice it is generally measured with yellow light, while the constant is approximately 0.12.

With the above theoretical foundation, the refractive index and the density of rock can be calculated from the DOP: however, due to the DOP being measured using rough rock surfaces instead of smooth surfaces, the results of the two calculations differ. In addition, the DOP in the specular direction is often greater, while being smaller in other directions: if the value is substituted by their mean then the difference will be smaller. These reasons require further investigation, meanwhile, Table 4.3 lists the main mineral densities of seven types of rock.

In Table 4.3, the data presented are mineral densities: for example, the main mineral in peridotite is olivine, and therefore the density of the main mineral in Table 4.3 is the density of olivine. Indeed, the mineral composition of many rocks is complex and unlikely to contain a single mineral, so the results may not be uniform and more detailed work is needed, such as upgrading equipment to improve the performance of the instrument, while at the same time researching the functional relationship between the DOP and the density of rocks. In short, these are the following steps required: the study of the polarization characteristics of the reflectance spectrum of rock surfaces has value. In theory, it shows the functional relationship between the DOP and the density of rocks. In practical applications, such functional relationships need to be verified.

Table 4.3 Several rock densities calculated by polarization spectroscopy

Density	Name						
	Peridotite	Pyroxenite	Gabbro	Diorite	Quartz porphyry	Syenite	Serpentinite
Calculation of the mineral densities with DOP P	3.3–3.5	3.23–3.52	2.60–2.76	3.1–3.4	2.5–2.8	2.56–2.58	2.5–2.65
Actual density	3.27–3.48	3.02–3.45	3.02–3.45	2.5–3.3	–	2.5–3.3	to 2.57
Mean error (%)	0.7	4.1	20.7	10.7		12.8	2

4.3.2 *Discussion of the Utilisation of Polarization Spectroscopy to Explore the Lunar Surface Density*

First of all, a method for the measurement of surface density of stars is given in theory, followed by a feasible research programme. The theory and methods are to be tested and verified.

1. Theoretical basis

Through the measurements and calculations of most rocks on Earth, the average ratio of refraction value *K* of the surface rocks and minerals is found, by calculation, to be approximately 0.21. On Earth, the *K* values of most rocks and minerals are around 0.21 despite them all having different refractive indices and densities. Important components of the earth's crust, such as SiO_2, Al_2O_3, MgO, Na_2O, K_2O, CaO, and P_2O_5 have a ratio of refraction (*K* values) between 0.20 to 0.23, while that of FeO is smaller (at 0.188), yet Fe_2O_3 is at 0.290, and H_2O is at 0.340 (liquid), and after the synthesis of these rock and mineral components, the average *K* value is found to be 0.21.

The average refraction ratio of 0.21 represents the overall *K* value of rocks and minerals on Earth and according to the Gladstone-Dyer refraction formulae, and the Lorentz-Lorenz refraction equation, the following two equations can be obtained:

$$\frac{\mathrm{n}-1}{\rho} = \mathrm{K}_{\mathrm{Earth}} = 0.21 \tag{4.8}$$

$$\frac{\mathrm{n}^2-1}{\mathrm{n}^2+2} \times \frac{1}{\rho} = \text{constant} = \gamma_{\mathrm{L.L}} = 0.12 \tag{4.9}$$

From the above simultaneous Eqs. (4.8) and (4.9),

$$4n^2 - 7n + 1 = 0 \tag{4.10}$$

Solving this gives:

$$n_1 = \frac{7+\sqrt{33}}{8} \tag{4.11}$$

$$n_2 = \frac{7-\sqrt{33}}{8} < 1 \quad \text{(Eliminate forthe refractive index of air is equal to 1)} \tag{4.12}$$

Substitute this into the original function (4.8) to get:

$$\rho = \frac{n-1}{0.21} = \frac{\frac{7+\sqrt{33}}{8}-1}{0.21} = \frac{\sqrt{33}-1}{1.68} \approx 2.824\,\mathrm{g/cm^3} \tag{4.13}$$

This density of the Earth's surface is coherent with Bullen's data. Bullen proposed a distribution model of the Earth's density in depth in 1963, which (1) satisfies the average density of the Earth (5.517 g/cm^3); (2) Calculated the Earth's rotation around its axis of inertia; (3) calculated the density of ultrabasic rocks under the Moho; (4) established the P-wave and S-wave velocity-depth curve; (5) Surface wave data; and (6) used free-oscillation data in a recent model. At a depth of less than 1000 m, the density of this layer of the ellipsoidal shell can be considered as an average of 2.9 g/cm^3. The estimation of the density of Earth is calculated through the fitting of boundary conditions, and all existing observations particularly those pertaining to seismic discontinuity surfaces.

Results of the two methods of calculations give 2.824 g/cm^3 and 2.9 g/cm^3 respectively, generally consistent with each other, proving that it is feasible to use the K value to estimate the surface density of the Earth. In fact, this K value also reflects the overall behaviour of rocks comprising the Earth's surface. Therefore, it is also feasible to estimate the surface density of celestial bodies using their K values.

The above reasoning is correct in theory, but is it feasible in practice? Merely obtaining the density of a substance from its refractive index has been explored in many previous studies, but calculating it from the DOP of the substance is seldom tried. The DOP of a substance can be measured from afar, while physically measuring the density of a substance requires direct contact. If the above theory is correct, then the average surface densities of celestial bodies can be estimated by first obtaining their general DOP.

The following explains why the surface density of celestial bodies such as the Moon calculated through its polarization spectra is more accurate: first of all, the Moon itself does not emit light and reflects light from the Sun. According to the Fresnel formula, this ensures that the light uploaded to the sensor from the Moon has a different polarization in the vertical plane of propagation.

Secondly, there is no water and atmosphere (the lunar atmosphere is very thin) on the surface of the Moon, these two conditions ensure that the DOP of the observed spectra reflected by the Moon is entirely determined by the property of surface rock. As such, the detector can conduct polarization spectra probing at different points (for example 100 points) when circulating around the Moon, to determine the refractive indices and the densities of different (100 types of) lunar rock at different points using the polarized reflection spectra.

When the DOP distribution is large, the average K value of lunar surface rocks can be calculated, and thus it is possible estimate the density of the lunar surface using the method described above. The advantage of this approach is that it does not require direct contact with the surface of the Moon to estimate the densities of the lunar surface and its constituent rocks.

Currently, scientists know that the average surface density of the Moon is 3.33 g/cm^3, and that of the Earth is 5.5 g/cm^3. The density of the moon rock samples that astronauts on Apollo 11 and 12 brought back is 3.2–3.4 g/cm^3, and is higher than that of most terrestrial rocks (the average Earth crustal rock density is 2.7–2.8 g/cm^3).

The moon is an exceptional case: it is known that the surface density of the Earth is 2.9 g/cm^3, and with increasing depth, the pressure increase as does the density, especially in the inner core. For the Earth, its average overall density is 5.5 g/cm^3. In summation, generally for a celestial body, its surface density should be vastly lower than its average density, however for the Moon, its surface density (3.2–3.4 g/cm^3) differs little from its average density, in other words its surface density is not much smaller than its average density.

There are two explanations for this anomaly: Dr. Harold Urey and other scientists believe that this is due to the centre of gravity hollow of the Moon. Dr. Wilkins of the Royal Astronomical Society said in the book "Our mysterious spaceship moon", that the moon is estimated to have a hollow volume of about 14 million cubic miles. Dr. Sean C. Solomon, a scientist at Massachusetts Institute of Technology, studied the moon's gravity and suggested that the moon's interior may be hollow.

Another theory is that the moon's interior is not composed of heavier elements such as iron and nickel, and instead may be in a partially molten state or somehow plastic.

Moon rock samples brought back by Apollo 11 and 12 astronauts may have their own peculiarities, and are not representative of the entire lunar surface. Rocks and minerals on Earth, for example, have larger K values and densities when their Fe_2O_3 content is high, and according to the Gladstone-Dale refraction formula, such rocks have a density of 5.34 g/cm^{3}which is almost equal to the average density of the Earth. Yet we are unable to ignore the presence of other types of rocks with smaller densities and conclude that the Earth's surface density is just 5.34 g/cm^3. Therefore, rocks brought back from Apollo 11 and 12 do not represent the general density of moon rocks due to the limited number of samples and limited area of sampling.

Using the characteristics of polarized reflection light to deduce the surface density of the Moon, if the deduced results of 100 or more probe positions are the same, then it can be concluded that the surface density of the Moon is 3.2–3.4 g/cm^3, otherwise further discussion and verification are required.

Thus, a method for determining the density of the stellar surface is proposed in theory, and its correctness and feasibility can be explored through experimentation, measurements and analysis of the DOP, and measurement of the refractive index and density of various rocks and minerals on Earth in the laboratory. If the use of polarized reflectance spectra can accurately calculate densities of rocks, it will provide a powerful method and means to calculate the density of rocks on any non-luminous celestial bodies. Instead of direct contact with the celestial body, remote sensing can be employed for the calculations of their physical properties such as surface density and refractive index. At present, the surface density of the moon has no viable means with which it may be calculated. Using polarization spectra to calculate the surface density of moon would be perfect if the results are accurate, as remote sensing is suitable for bodies with which we are unable to achieve direct contact.

2. Feasible plans

We select all common mineral rocks on the ground as experimental subjects and measure their degree of polarization, refractive index, and density before analysing the functional relationships among the three properties to assess compliance with the theory.

In conclusion, this study is based on theory and all previous results of calculations and analysis of experiments on existing rock and mineral samples to answer the following questions: is it possible to deduce the refractive index of samples through the measurement of their DOP? What of their accuracy and errors?

Is it possible to deduce the density of samples from their refractive index? For minerals and rocks, are the Gladstone-Dale formula and Lorenz-Lorentz equations feasible? What constant can be used to demonstrate the properties of samples to better effect? Is there a more suitable formula of expression thereof? Is it possible to deduce the density of rocks and minerals directly from their DOP? What of the functional relationship and its accuracy?

If answers to the above questions are yes, then in theory and practice it can be proven that the method for determining the surface density of celestial bodies is correct and can be applied in practice.

A viable research programme should aim to conduct qualitative and quantitative analyses of data obtained from the measurements of the polarization reflectance spectra of rocks and minerals, based on the measured results of previous studies to identify formulae and methods able to calculate the surface density of celestial bodies. The basic technical route is as follows:

1. Collation and analysis of existing data

For a significant number of mineral rocks, their refractive index and density data are tractable, thus only their DOP need be determined. Currently such results are broad and complex, requiring systematic collation and analysis. These experimental data bring huge benefit to validation trials.

2. Sample selection and analysis

Then we should select some common ground rocks such as: peridotite, pyroxenite, basalt, gabbro, diorite, quartz porphyry, granite porphyry, coarse-grained granite, syenite, conglomerate, breccia, purple shale, diatomaceous earth, bauxite, oil shale, serpentine, marble, slate, garnet sericite schist, and biotite plagioclase gneiss. These rocks are common on Earth, thus the calculation and analysis of their corresponding refractive index, DOP and density help validate our theories.

3. Experimental data measurement and analysis

Use existing polarization spectroscopes to measure samples and then analyse the data, in past experiments, more attention was dedicated to the characteristics of polarization spectra of rocks and minerals in 2π-spatial orientation: data were often qualitative and vague. Using the physical properties of rocks and minerals to examine the relationship between their refractive indices and densities would be more convincing. If such theories are proven to be correct, they would expand the basic applied research domain of remote sensing.

4.4 Estimation of Substance Composition of Lunar Surface Rocks

How to use the characteristics of polarization in the reflectance spectrum of substances to inversely deduce their composition? The third chapter summarises the statistical relationship between the DOP and chemical composition from the experimental point of view: based on the physical mechanism underpinning such behaviour, we have to start with the refractive index of the rocks. Since only the refractive index data can be retrieved from polarization reflection spectra of rocks, other information cannot be obtained.

To date, the relationship between the refractive index and the chemical composition and structure of substances has rarely been investigated. From the late-1950s to the early-1960s, Russian petrographer E. A. Kuznetsov proposed a method of measurement of the double refraction dispersion ratio, believing that the double refractive index ratio of the rocks and minerals at different wavelengths is related to the chemical composition and crystal structure of minerals and rocks. Kuznetsov's proposal raised scepticism, both at home and abroad, and did not receive promotion and recognition.

Why is the reflectivity dispersion of substances helpful in understanding their chemical composition and physical structure? This is because the phenomenon of light dispersion demonstrates that the refractive index is closely associated with the frequency of light waves. This is not reflected in Maxwell's electromagnetic theory. Light dispersion has the following characteristics:

1. The shorter the wavelength, the greater the refractive index.
2. When wavelength is constant, the refractive index of different substances tends to be greater, and $\frac{dn}{d\lambda}$ is larger;
3. There is no simple similarity seen between dispersion curves of different substances;
4. Anomalous dispersion and absorption of light are closely related.

Figure 4.15 shows a graphical representation of how the refractive index changes with wavelength. In the non-absorption spectra range, refractive index decreases when wavelength increases; when wavelength increases to *R*, the curve starts to decline more rapidly until it reaches the absorption spectra, light can no longer

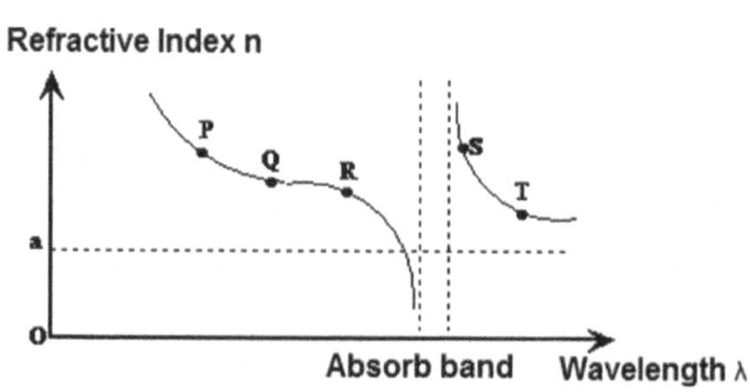

Fig. 4.15 Changes in refractive index with wavelength λ

pass through. Beyond the absorption region, the refractive index suddenly increases significantly. As the wavelength continues to increase, the refractive index decreases again, and when wavelength exceeds 0.1 cm, the refractive index remains constant and is no longer dependent on the wavelength.

Different elements correspond to different absorption bands, therefore, anomalous dispersion usually appears in the absorption region of the refractive index. Thus, the changes in the refractive index of a substance changes with wavelength and is related to chemical composition, while different substances have different dispersion rates.

Of course, the above discussion is only possible in idealised states involving only one mineral. In fact the mineral composition of rocks is complex, therefore it is necessary to examine the relationships between the refractive index and the chemical composition of substances to provide theoretical and technical support for lunar resource exploration.

From the above discussion, polarization measurements can obtain more information that is helpful in distinguishing target objects, while multi-angle remote sensing improves single direction observations (near vertical) of traditional remote sensing to allow acquisition of three-dimensional information about target objects. Therefore, their organic integration provides a new research direction for space probe design, and helps to expand and improve the means of remote sensing (Zhao and Zhao 2006).

Of course, the technology of multi-angle polarization probes is still relatively new: the design of its optical system is still in the exploratory stages of research, with a lot of practical issues needing trialled.

1. Polarization remote sensing technology in collaboration with hyper spectral analysis

To examine the relationship between the wavelength and the refractive index of rocks, the wavelength has to be divided into segments of a few nanometres each to complement any hyperspectral remote sensing. Currently, hyperspectral researchers focus on the spectral reflectance characteristics of different objects that vary with wavelength, while the refractive index of different objects is not studied in any depth. Therefore, it is possible to assess how the refractive index changes with wavelength through the polarized reflectance hyperspectra to reveal their chemical composition, and provide more comprehensive, specific experimental data for further exploration of lunar surface material.

2. The complexity of the lunar spectrum

The sun always changes its elevation and azimuth angle, while remote sensing detectors are also constantly changing their positions in the sky. To a certain extent, these changes complicate the temporal and spatial variation of the spectra in 2π-spatial orientation.

While there is always unevenness on the lunar surface, these changes in elevation and gradient are in fact mixed in remote sensing images.

3. Strengthening research linking spectral, and image, information

With further advances in remote sensing technologies and applications, using only spectral information (one-dimensional information) cannot satisfy modern scientific and engineering needs. People not only require detailed spectral information, they also hope to get a direct image of the desired wavelength range. This resulted in a new generation of imaging spectroscope that associates spectral information with image information. For the moon, how to utilise such three-dimensional data (in the spatial and spectral dimensions) to achieve better information analysis in remote sensing, will a task for future research.

4. Physical and mathematical models of the polarization spectral characteristics of lunar substances

The correspondence between the polarization spectral characteristics and lunar substances establishes physical and mathematical models required to understand the mechanism underpinning the spectral characteristics of lunar substances. With the inversion of lunar environment parameters, quantitative analysis is helpful in improving the application of such effects.

5. Establishment of a multi-angle polarization spectral database based on lunar hyper-spectroscopy

Through indoor measurements and outdoor corrections of means, typical polarization spectroscopy of ground objects will be established. After comparing the measured data and sample data in a database, a short-cut for the large-scale exploration and identification of various rock distributions can be created. We also aim to enrich and improve the existing information systems pertaining to the lunar spectrum and gradually develop this into part of a geographic information system, so that duplications in measurement can be reduced and the sharing of spectral data can be improved.

References

Ourlier C, Berginc G, Saillard J (2001) The theoretical study of the Kirchhoff integral from a two-dimensional randomly rough surface with shadowing effect: application to the backscattering coefficient for a perfectly-conducting surface. Waves Random Media 11(1):91–118

Torrance KE, Sparrow EM (1967) Theory for off-specular reflection from roughened surface. J Opt Soc Am 57(9):1105–1114

Xie M, Xu H, Zou Y et al (2005) Experimental methodology of moorstone's surface BRDF. J Eng Thermophys 26(04):683–685 (in Chinese)

Yang J, Wu F (2009) Experimental research on visible light scattering characteristics of rough surfaces. China Meas Test 35(2):125–128 (in Chinese)

Yu HQ (1997) Standard and application of surface roughness. China Metrology Publishing House, Beijing (in Chinese)

Zhao LL, Zhao YS (2006) The study of exploring major minerals on the lunar surface with multi-angle polarization technology. Prog Geophys 21(3):1003–1007 (in Chinese)

Zheng XB (1999) Review on research and applications of light scattering from randomly rough surfaces. Chin J Quantum Electron 16(2):97–103 (in Chinese)

Chapter 5
Remote Sensing of Land Surfaces 4: Signal-to-Background High Contrast Ratio Filtering of Polarimetric Reflections

This chapter focuses on water and soil, two of the four primary observational subjects of the Earth's surface (rocks, water, soil, and vegetation) to introduce the fourth important characteristic of polarization reflection of land surface: the high signal-to-background ratio filtering feature. This includes: polarization reflection radio filtering of water bodies in dark backgrounds (weak light intensification) to explain the nature of data filtering in polarimetric remote sensing when it intensifies weak light; strong water bodies' sun-glint polarization separation (bright light attenuation) and the measurement of the density of water to explain the nature of data filtering in polarimetric remote sensing when it attenuates bright light; soil polarimetric reflection filtering characteristics, to show the polarimetric remote sensing high signal-to-background ratio filtering feature; the high signal-to-background ratio filtering relationship between soil water content and polarization to show polarization remote sensing's ability in high signal-to-background ratio filtering.

5.1 Polarimetric Reflection Radio Filtering of Water Bodies with Dark Backgrounds (Weak Light Intensification)

Remote sensing has long been used in radiology and photometry, the birth of polarization remote sensing is an inevitable outcome following the development of space remote sensing technology. Explanation of traditional remote sensing relies on a subject's individual spectral characteristics and the outcome it receives is a two-dimensional signal while polarimetric remote sensing can provide a three-dimensional signal, which helps to lay a foundation for the retrieval of a subject's spatial structure.

Water is the source of life and exerts critical influences on the ecosystem. Studying the characteristics of water's polarized reflection has important theoretical significance as well as wide practical applications that can contribute to environmental

L. Yan et al., *Polarization Remote Sensing Physics*, Springer Remote Sensing/Photogrammetry, https://doi.org/10.1007/978-981-15-2886-6_5

conservation and the monitoring of water resources. Water's polarized reflection characteristics are primarily related to wavelength and source-view geometry but polarized reflection is strongest in the specularly reflected direction, especially on water surfaces so smooth that they are close to specular reflections. Polarized reflection information in other directions is negligible; however, polarized reflections caused by waves on water surfaces need more detailed and complex explanation.

At the visible and near-infrared region of ocean colour remote sensing, water-leaving radiance refers to the upward radiation emitted from a water surface, also known as ocean colour, that occurs after light enters water and undergoes backscattering within: it carries useful information about the water and is an optical expression of the collected data from the water. Water-leaving radiance in ocean colour products includes two physical quantities: water-leaving radiance L_w (upward irradiance from the water surface (0+), units: W cm^{-2} μm^{-1} sr^{-1}) and water-leaving reflectance ρ_w (the ratio of upward and downward irradiance from the water surface), as shown in Fig. 5.1. Water-leaving radiance is the basic quantity of ocean colour remote sensing and can be used to calculate parameters such as the inherent optical properties of bodies of water, the inversion of ocean colour component concentrations, photosynthetically active radiation, primary productivity, and red tide index: however, the top of atmosphere (TOA) signal measured by satellite sensors includes atmospheric scattering and absorption, surface specular reflection, reflection of whitecaps, and so on. The presence of these factors will affect the accuracy of remote sensing when used to monitor water.

Fig. 5.1 Data composition arriving at satellite sensors

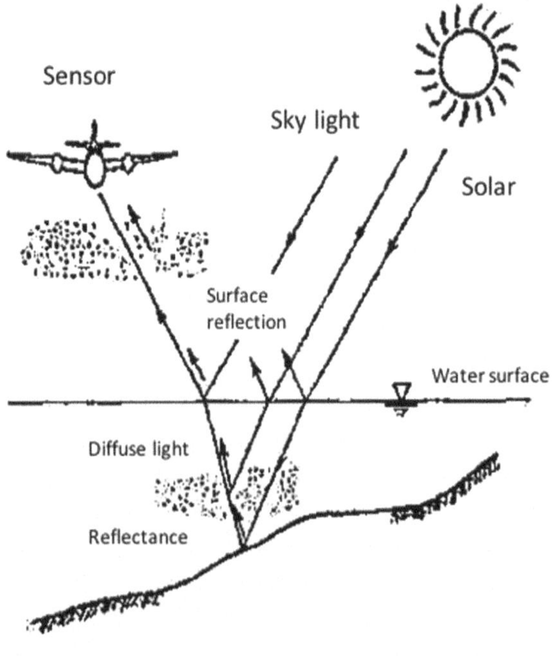

The reflectivity of clean water in the visible and near-infrared region is relatively small and its spectral characteristics are not obvious. In optical remote sensing images, water generally is presented as dark, which has caused difficulties when using spectroscopy to identify water remote sensing features and in parameter inversion of water quality data. When studying the polarized spectrum of water, we found that, during multi-angle observations of water, at visible and near-infrared bands, water's polarimetric reflectance is much larger than its reflectance in an unbiased state and its polarization images are much brighter than its intensity images. This shows that using multi-angle polarization remote sensing to probe water has its advantages since it overcomes the issue of low reflection values when using optical remote sensing to probe water, improving the ability of remote sensing to identify and differentiate water-related data while also improving the accuracy of parameter inversion.

When light is obliquely incident to water surface, part of it will experience reflection while the other part will be refracted into the water body. Let α be the angle of incidence and β be the angle of refraction such that the incident beam, reflected beam, and refracted beam surface forms the incident plane. Regardless of the direction in which the incident beam is vibrating; its electrical field vector can always be expressed by $E_\perp$ which is perpendicular to the incident plane and $E_\parallel$ which is parallel to the incident plane, provided that the corresponding reflection beam's electrical field vectors are $E'_\perp$ and $E'_\parallel$.

When a beam of natural light reflects and refracts at the border of two different mediums, the direction of propagation of the reflection beam and refraction beam is decided by the laws of reflection and refraction, however, its orientation of vibration, also known as polarization, follows the electromagnetic theory of light and is determined by the boundary conditions of electromagnetic fields.

Without considering directionality, we get:

$$\frac{E'_\parallel}{E_\parallel} = \frac{tg(\alpha - \beta)}{tg(\alpha + \beta)} = \frac{E'_\perp}{E_\perp} \cdot \frac{\cos(\alpha + \beta)}{\cos(\alpha - \beta)} \tag{5.1}$$

When $\alpha = 0°$, we get:

$$\frac{E'_\parallel}{E_\parallel} = \frac{E'_\perp}{E_\perp} \tag{5.2}$$

Given that $E_\perp = E_\parallel$, the above equation shows that the parallel component $E'_\perp$ and the perpendicular component $E'_\parallel$ of the reflection beam's electric vector are equivalent, however, these two components are unrelated as reflected light after synthesis remains natural light. As such, when the incident beam is perpendicular to the surface of the water, there is no polarization in the reflected beam.

In the case of $0° < \alpha < 90°$, $|\cos(\alpha + \beta)| < \cos(\alpha - \beta)$,

When $0° < \alpha < 90°$, $|\cos(\alpha + \beta)| < \cos(\alpha - \beta)$, then

$$\frac{E'_{\parallel}}{E_{\parallel}} < \frac{E'_{\perp}}{E_{\perp}} \tag{5.3}$$

The physical effects of the interface on the two components of the incident beam ($E_{\perp}$ and $E_{\parallel}$) are different. Regardless of the state of polarization of the incident beam, Eq. (5.3) shows that, within the electrical field vector of the reflection beam, the value of the parallel component is always smaller than the value of its perpendicular counterpart. Structurally, the two components are aligned in different directions, the electrical field vectors of polarized light, with different amplitudes, project onto these two directions vectors, so these two components remain unrelated, cannot be integrated into one vector, and are only partially polarized. As such, its state of polarization is different from that of the incident beam, which means that when the incident beam is natural light (non-polarized light), the incident beam experiences a reflection from the water surface and thus partial polarization.

We conducted multi-angle polarization measurement on natural water using ASD's Vis/NIR spectrometer (spectral range 350 nm to 2500 m). The experiment was conducted on Peking University's Weiming Lake as the water is clear and without ripples and the weather is clear and cloudless. Data collection was conducted at noon (local time). The experiment adopted the Thompson prism, which is less volatile with regard to its effect on the polarization spectrum, as a polariser. Data collection pertaining to multi-angle polarization in the chosen water body was conducted thus: spectrometer probes were placed at different probe zenith angles, spectrometers were used to measure reflectivity in three different situations, namely, horizontal polarization, vertical polarization, and no polarization. The viewing zenith angle was varied from 0° to 45° in increments of 15°.

Figure 5.2 shows the water reflectance spectrum under horizontal polarization, vertical polarization, and no polarization at a viewing zenith angle of 0° (nadir), 30°, and 45°, respectively. The red line represents degree of polarization (DOP) spectrum and the DOP itself is a result of calculation. The horizontal axis represents wavelength and the vertical axis represents reflectivity.

Image 5.2a shows the reflectivity of the lake under horizontal polarization, vertical polarization, and without polarisers at 0°. The spectrum, when the lake is unpolarized, is the commonly seen water spectrum and its reflectivity is low throughout the entire wavelength range with a slight reflection peak at the blue-green band. Most remote sensors available today do surface sounding, and the low reflectivity of water bodies means that, in normal optical remote sensing images, water appears darker, thus making it difficult to use water spectral characteristics to identify remote sensing features and undertake parameter inversion of water quality data. At the vertical view direction, the water body's horizontal polarization and vertical polarization are close to zero, which is consistent with the earlier hypothesis that at the vertical direction of water bodies, there is no polarized reflection and the DOP is at zero.

Images 5.2b, c shows the reflectivity and the DOP spectral curve of the lake under horizontal polarization, vertical polarization, and no polarization when the viewing zenith angle is 30° and 45°, respectively. Both Fig. 5.2b, c show that the degree of reflectivity when there is no polarization in the visible bands is higher than that under

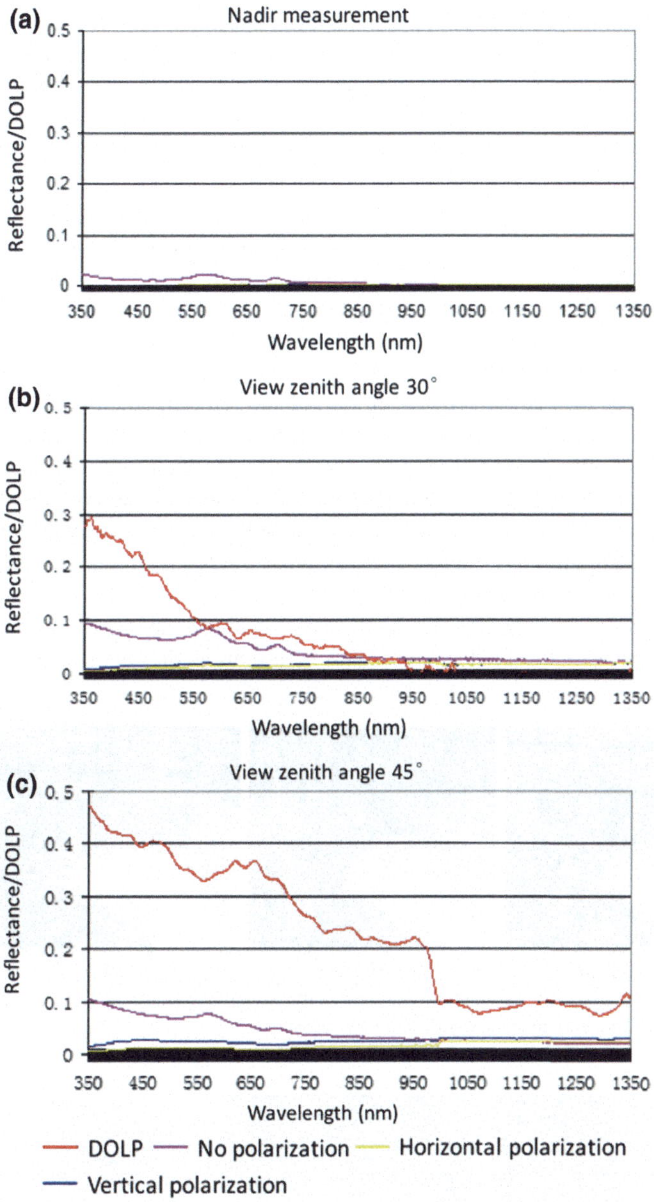

Fig. 5.2 Polarization spectrum of the water body at different angles of observation

vertical polarization. We surmises that this might be due to the water body surface's specular reflection or it might be due to one of the multi-angle spectral characteristic of the water. This means that data from multi-angle observations are important one-dimensional information for remote sensing inversion but more research is needed to establish solid reasons for why this is so. In Fig. 5.2b, the DOP spectra within the region of 350–900 nm are all larger than reflectivity values recorded without polarisers, while in Fig. 5.2c, the DOP spectra within the waveband range of 350–1350 nm are all larger than reflectivity values recorded without polarization. DOP, recorded at zenith angles of 30° and 45°, in some bands are several times larger than the reflectivity recorded without polarization at 0°.

One can also utilise satellite data from PARASOL to test the effectiveness of water's polarization hyperspectral characteristic. Figures 5.3 and 5.4 are PARASOL satellite images of a certain region in the Atlantic Ocean, taken on 29 November 2008: Fig. 5.3 is a high-intensity image of water which has no polarization at different wavelengths. Figure 5.3a–c respectively correspond to high-intensity images taken at 490, 670, and 865 nm. The white regions in the image are clouds while the black regions are parts of the ocean. Figure 5.4 shows DOP images taken by the PARASOL satellite at different wavelengths. Figure 5.4a–c respectively correspond to DOP images taken at 490, 670, and 865 nm.

From Figs. 5.3 and 5.4, it is clear that the region of the water body shows up much brighter in each of the DOP images seen here than in the high-intensity images.

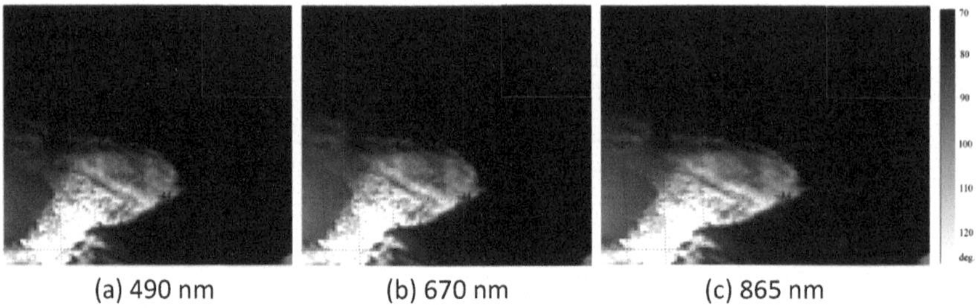

Fig. 5.3 High intensity images at different wavelengths

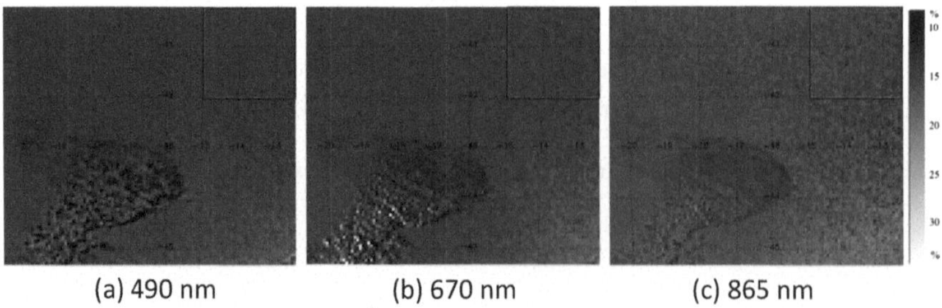

Fig. 5.4 DOP images at different wavelengths

Table 5.1 Comparison of average brightness of water in DOP and high-intensity images

	490 nm	670 nm	865 nm
Reflectivity image	14.07	3.47	1.97
DOP image	57.25	62.51	72.49
Brightness multiple	4.1	18.0	36.8

Table 5.1 lists statistics pertaining to the average brightness of the red-boxed regions (the water body) seen in Figs. 5.3 and 5.4. We get the brightness multiple through the average brightness of DOP images divided by the brightness multiple of high-intensity images. The brightness of DOP images of the same water body at 490 nm, 670 nm, and 865 nm correspond to 4.1 times, 18.0 times, and 36.8 times the brightness of high-intensity wavelength images, respectively. The multiple is smaller at 490 nm due to atmospheric polarization, water absorption, and other factors but generally, DOP images of water bodies tend to be brighter than high-intensity images thereof. This means that DOP images can successfully brighten regions in high-intensity images that appear darker and can, through luminance segmentation, identify and categorise features to improve accuracy of interpretation of remote sensing and inversion procedures.

The above analysis shows that multi-angle DOP spectra of water are larger than the spectrum of water in the same state without polarization. Using spectrometers to measure the features and patterns of water multi-angle polarization spectra is also applicable to space-borne multi-angle polarization remote sensors which are used in the identification and inversion of water bodies. This might be an important method of observation for remote sensing of objects with low reflectivity.

5.2 Strong Water Body Sun-Glint Polarization Separation (Bright Light Attenuation) and Measuring the Density of Water

This chapter focuses on sun-glint polarization separation and measurement of water density the better to understand polarization data collection from water bodies.

5.2.1 Sun-Glint Polarization Data in Reflection Data from Water Bodies

Calm water surfaces can be approximately regarded as mirrors. When sunlight strikes the surface, the strong reflected radiation that is formed on the direction of the line of reflection makes sun-glint. In windy conditions, an angle of inclination is generated on an otherwise horizontal water surface and bright spots are formed at the wave peaks, which cause sun-glint. Reflected radiation carry a minimal amount of data from the water itself and its intensity is related to the nature of the water surface as affected by the presence of water plankton, yellow substances, and foam; however, at the central region of influence of sun-glint, sun-glint is very intense and the amount of radiation could be up to several times greater than that from the water's upward radiation, which means that sensors could have reached their point of saturation, making it difficult to extract ocean colour data from the remote sensing information received, negatively affecting the accuracy of data collected with regards to water quality. With regards to the research outcome of the FY-1-02 batch of radiation simulation satellite images, Pan Delu et al. estimates that regions unavailable for ocean colour monitoring due to the presence of sun-glint account for up 20–25% of the entire region: such regions that can be modified to be rendered available are also available only if sensors are not saturated. Sun-glint is also hard to avoid in airborne remote sensing data. As such, finding out how to eliminate the effects of sun-glint is an important problem in water quality remote sensing monitoring.

The transmission of solar radiation from the exosphere sees it go through Rayleigh scattering and aerosol scattering in the atmosphere before a part of it returns to the sensors that satellites carry and a part of it undergoes diffuse reflection to reach the sea surface. When direct light strikes the sea surface, part of it might go through the atmosphere to reach sensors carried by satellites due to specular reflection, while the other part will be refracted and enter the water. At the sub-surface level of water, part of the radiation that enters will undergo scattering caused by ocean colour factors such as chlorophyll, yellow substances, suspended sediment, and other particles to be refracted once again when leaving the water surface, passing through the atmosphere, and reaching the sensors while another part of the radiation will continue downwards to reach the euphotic zone or deeper, undergoing partial reflection, refraction on the water surface, then reaching the sensor.

Extensive work has been undertaken on how to reduce and eliminate sun-glint. The outcomes of such research are already widely applied to all levels of remote sensing. Ji (1994) raised the idea of solar altitude range that causes specular reflection in water: he believed that a meticulous set-up of the relationships between the time when remote sensing was done, solar altitude, solar orientation, and course of navigation could effectively avoid specular reflection. Mao (1996) and Li (1997) discussed the factors that affect sun-glint. Anjum's team analysed POLDER data from 7 November 1996 recorded over the Indian subcontinent and found that, when the observation zenith angle was 55°, the water's reflection was markedly less intense than at other zenith angles.

Currently, there are two main methods available to avoid, or eliminate, sun-glint: to arrange the CET of satellites to be at twelve noon when possible to minimise the influence of sun-glint or to design the sensor such that it has three modes of scanning—vertically, with a forward inclination of 20° and with a backward inclination of 20°—then use image splicing to remote sense spots affected by sun-glint, such as the USA's "SeaStar" satellite's SeaWiFs sensor.

Although the use of polarization technology to remove solar reflection has no precedent in aerial remote sensing, it has long been used in photography. In photography, solar reflection is a common issue, especially when taking photographs of people through car windshields or of objects underwater. Photographers usually resolve this with the help of a polarising filter to remove solar reflection. Also, Cunningham et al. (2002) used a ship-borne polarization radiometer to observe the sea surface from a direction of 53° to minimise the impact of solar reflection and calculate water leaving radiance.

(1) **Sun-glint: formation mechanism and calculation model**

The reflection of all objects follows the laws of reflection: an incident, normal, and reflected beam all lie in the same plane and the angle of incidence is identical to the angle of reflection. Therefore, a smooth surface will induce specular reflection but a rough surface would not since there are various planes of reflection and the direction of the normal is different at different points, inducing diffuse reflection instead. Water surfaces, especially static water surfaces can be approximately regarded as smooth surfaces such that when a source of light strikes the surface, strong reflection radiation occurs in the direction of reflection symmetrical to the direction of incidence with respect to the normal thus forming a specular reflection of water. In windy conditions, the previously smooth surface now has many angles of inclination which create flares or bright spots at the peaks of the waves. These are also caused by the specular reflection of water.

The intensity of sun-glint radiation is not simply a function of wavelength, it is also the azimuth angle at which solar altitude and ocean colour scanners observe from and a function of sea surface wind speed. The following set of equations can be used to calculate radiation coefficient g, which is a result of direct sunlight striking a rough surface caused by wind speed V_w. The coefficient, g, can be written as follows.

$$g = [\rho(\omega)\cos\omega/\cos\theta_v\cos\theta_s]/\exp[(-\tan^2\beta/\sigma^2)/(4\pi\sigma^2\cos^4\beta)] \tag{5.4}$$

Each variable can be expressed as

$$\sigma = 0.003 + 0.00512V_w \tag{5.5}$$

$$\omega = 0.5\cos^{-1}[\cos\theta_v\cos\theta_s - \sin\theta_v\sin\theta_s\cos\phi] \tag{5.6}$$

$$\beta = \cos^{-1}[(\cos\theta_v + \cos\theta_s)/(2\cos\omega)] \tag{5.7}$$

From the equation above, $\rho(\omega)$ is the function of solar specular reflection; g is the coefficient of reflection of direct sunlight; θ_v is the altitude of ocean colour scanner; θ_s is solar altitude; ϕ is the difference between the azimuth angles of the sun and the sensor.

(2) **Factors affecting solar flare**

The factors that affect sun-glint primarily include the time when satellites pass through the equator, their date of transit, and wind velocity.

The relationship between satellites passing through the equator (CET) and sun-glint: when the CETs of satellites are different, the central position, scope, and radiation value of sun-glint also differ. As CET moves from morning to high noon, the central position of sun-glint also changes as it moves from the east side of the image to the centre of the image, the scope of the sun-glint also shrinks and the distribution changes from being almond-shaped to an oval, and the oval also shrinks until at noon, it is just a dot. Images with a launch window of twelve noon are less than one tenth as pixelated as images launched at eight in the morning due to the influence of sun-glint, thereby increasing the use of satellite data. When CET is pushed to beyond twelve noon, this trend is reversed: sun-glint centre starts moving to the west side of images and images are therefore more affected by the sun.

Relationship between transit date of satellites and sunlight: the central position of sun-glint changes between a latitude direction of North and South according to its date of transit, however, the date of transit has little effect on the distribution and size of a sun-glint. The relationship between the latitude of the sun-glint centre and a satellite's date of transit can generally be characterised as a sinusoidal function. During either the vernal or autumnal equinox, the centre is positioned on the equator; during winter and summer solstices, the centre lands at around 24° S and 22° N respectively.

Relationship between wind velocity and sunglint: different wind velocities give rise to different sea states which change the conditions for reflection of sunlight and thus affect the intensity and scope of sun-glint. It has to be noted that the altitudes and azimuth angles of the sun and sensor will affect the intensity and scope of sun-glint as well, even if the wind velocity remains constant; however, research has shown that, in a situation where altitude and azimuth angle of the sun and sensor remain unchanged, the intensity and scope of sun-glint increase with wind velocity.

(3) **Separation of sun-glint using polarization**

Natural light hits a smooth water surface at an angle of incidence that is not its Brewster angle: reflected and refracted light all become partially polarized and the reflected light's primary direction of vibration must be perpendicular to the incidence plane, while the refracted light's direction of vibration need only lie on the plane of incidence. If the angle of incidence is the Brewster angle then reflected light is fully linearly polarized and refracted light is a partially polarized wave.

Theoretically, an incoherent light beam that is emitted or reflected from any target surface at a given wavelength can be described using the Stokes vectors:

$$S = \begin{bmatrix} I = \langle E_x^2 \rangle + \langle E_y^2 \rangle \\ Q = \langle E_x^2 \rangle - \langle E_y^2 \rangle \\ U = \langle 2E_x E_y \cos\delta \rangle \\ V = \langle 2E_x E_y \sin\delta \rangle \end{bmatrix} \tag{5.8}$$

In the equation, I denotes the intensity of non-polarized light, which is the total radiance in remote sensing, as such the value is always positive; Q represents the difference in intensity between the linear polarized light in the x-direction and y-direction, its value can be positive, negative, or zero; U is the difference in intensity between the linear polarized light at 45° and −45°, its value can be positive, negative, or zero; V denotes the intensity of circular polarization, as geophysical subjects generally induce only linear polarization, so in remote sensing probes the assumption that $V = 0$ is often adopted, therefore, the utilisation of linear polarized light to analyse characteristics of objects is a very important part of remote sensing.

In passive remote sensing, the light source is natural light and when light hits a target surface and gets reflected, the target essentially acts as a polariser. When the reflected light enters the probe device, if a polarization filter is placed on a multi-angle sensor, then one can get a multi-angle polarization reflectance and the filter is now an analyser. If there is linear polarized light from reflected light produced after sunlight passes through a target object, then we can use the analyser to remove such reflected light.

According to the Fresnel equation, the incident beam's state of polarization is related to the amplitude of the reflected beam. One has to calculate the perpendicular and parallel components of the incident beam separately to predict reflectivity:

$$R_{\perp} = \frac{\sin^2(\theta_1 - \theta_2)}{\sin^2(\theta_1 + \theta_2)} \tag{5.9}$$

$$R_{=} = \frac{\tan^2(\theta_1 - \theta_2)}{\tan^2(\theta_1 + \theta_2)} \tag{5.10}$$

where, θ_1 is the angle of incidence, θ_2 is the angle of refraction, and when the sum of the two angles add to 90°, the parallel component ($R_{=}$) is equal to zero, and,

$$\tan\theta_1 = \frac{n_2}{n_1} \tag{5.11}$$

The above equation represents the law governing the Brewster angles (*i.e.*, those that satisfy the conditions imposed on θ_1 as shown above). When an incident beam enters a surface at a Brewster angle θ_1, the reflected light is completely polarized (n_1 is the refractivity of the incident beam's medium, n_2 is the refractivity of the incident beam's medium), its vibration is perpendicular to the incidence plane (the plane made up by the incident, reflected, and refracted beams). The refractivity of the water is 1.333, and the corresponding Brewster angle is thus 53.1°.

As the water scattering causes only very weak upward radiation while sun-glint cause very intense radiation, in central regions affected by sun-glint, radiance from sun-glint can be several times more intense than radiation from the water surface. When sun-glints are present, in sun-glint overpower-data collected from the water body, most pixels have reached saturation and that makes it difficult to extract ocean colour data from total radiance, decreasing the use of ocean colour images and information. Reflected light from water is partially polarized light and through the use of polarisers which can eliminate light, we can remove polarized light from sun-glint and decrease the intensity of non-polarized light by half, achieving our goal of separating water from sun-glint.

The incident beam that reaches the water surface is non-polarized light and after undergoing reflection, it becomes polarized. At this point, the water is essentially a polariser: when the light hits at the Brewster angle of 53° (i.e., for a solar altitude of 37°), the reflected light is completely polarized with an electrical field vector perpendicular to the incidence plane. Using an analyser before a sensor, we adjust the azimuth of the analyser such that the polarization azimuth and direction of polarization of reflected light are mutually perpendicular. The polarization filter will block out light and the reflected light would not be able to pass through the polarization filter and the data received by the sensor would be subjected to atmospheric scattering and water scattering, the degree of specular reflection is zero, and sun-glint has been successfully separated.

One can calculate solar altitude according to the formula $\sin h = \sin\phi \sin\delta + \cos\phi \cos\delta \cos t$. In this formula, ϕ represents geographic latitude, the value of solar declination angle δ can found by checking an Astronomical Almanac, and t represents time. According to this method of calculation, one can calculate the solar altitude at any given point on Earth at any given time (longitude, latitude). As such, one can calculate the optimal time, which is when the solar altitude is at 37°, for any part of the world, at which to conduct polarization remote sensing on water. Due to length constraints, only optimal times around the world during the vernal equinox are listed here. During the vernal equinox, $\delta = 0°$ (Table 5.2 shows local times).

From Table 5.2, one can tell that regions with northerly, or southerly, latitudes beyond 53° do not have a solar altitude of 37° even at noon, which means that these regions cannot take advantage of polarization reflection during the vernal equinox to completely remove sun-glint. Compiling the solar altitudes around the world at noon, areas with northerly, and southerly, latitudes of 0° to 30° have twelve months in a year when they are able to remove all sun-glint; areas with northerly, or southerly, latitudes of 40° have eight months in a year when they are able to remove all sun-glint (January, February, November, and December are non-optimal months); areas with northerly, and southerly, latitudes of 30° have half a year when they can completely remove sun-glint; accordingly, at the extremities of the globe, the solar altitude is

Table 5.2 Optimal times for polarimetric remote sensing of water around the world during Table 5.2 vernal equinox

Latitude (North)	0	N10	N20	N30	N40	N50	N60
Time	12 + 3:32	12 + 3:29	12 + 3:20	12 + 3:03	12 + 2:32	12 + 1:22	–
Latitude (South)	0	S10	S20	S30	S40	S50	S60

always below 37° so they cannot completely eliminate sun-glint throughout the year (Luo et al. 2007).

Those listed above are all situations involving the complete elimination of sun-glint, yet, it should be noted that sun-glint of water can contain certain data about the water itself, and can contribute significantly towards the inversion of certain indices. For example, specular reflection from water films, if completely separated, causes the loss of information regarding reflection. In reality, as scattering in water bodies tends to be very weak, sometimes, after complete separation of sun-glint, data about upwelling light streams from scattering is partially absorbed by polarization filters such that very little data is left when it eventually reaches the sensor. Also, to require that polarization remote sensing only be conducted at Brewster angles is an unnecessarily onerous requirement since there is no strong need to eliminate all sun-glint. As long as the radiation from sun-glint does not push the sensor to saturation, upward radiation from water bodies can still be successfully identified even if sun-glint is present. As long as the sunlight's angle of incidence is within a certain range of the Brewster angle, when reflected light formed by linear polarization light and natural light is partially polarized, one can modify the intensity of sun-glint by controlling the value of the polarization azimuth, so as to eliminate the effects of noise on otherwise useful data.

At any xoy plane, light intensity as observed by using a linear polariser in the direction of angle θ with respect to the x-axis is:

$$I(\theta) = [(1 - p) \cdot \frac{1}{2} + p \cdot \cos^2\theta] \cdot I_0 \tag{5.12}$$

p is the DOP, and through calculating its value at different solar altitudes, θ can be found by controlling the value of the polarization azimuth. When $\theta = 0°$,

$$I(\theta) = \frac{1}{2}(1 + p) \cdot I_0 \tag{5.13}$$

This means that polarized light in partially polarized light has been completely separated to leave only natural light. The light intensity of natural light is halved after going through a linear polariser, which is to say that the light intensity of partially polarized light after going through a polariser become 0–50% of that of incident light, significantly reducing the influence of sun-glint, therefore, according to sensor used, one can offer a fixed range of optimal values for solar incidence angle and polarization azimuth angle to conduct polarization remote sensing, ensuring that the intensity of sun-glint remains within a suitable range and subsequently offer a range of suitable times at which to conduct polarimetric remote sensing over water. These time periods will last longer than time periods where one aims for complete elimination of sun-glint and the sensors will be able to receive much more data.

To validate the aforementioned hypothesis, we simulated the specular reflection of water by using a polarization spectrometer. Figure 5.5 shows, respectively, the 0-degree polarization of the corresponding viewing angle, bidirectional reflectance

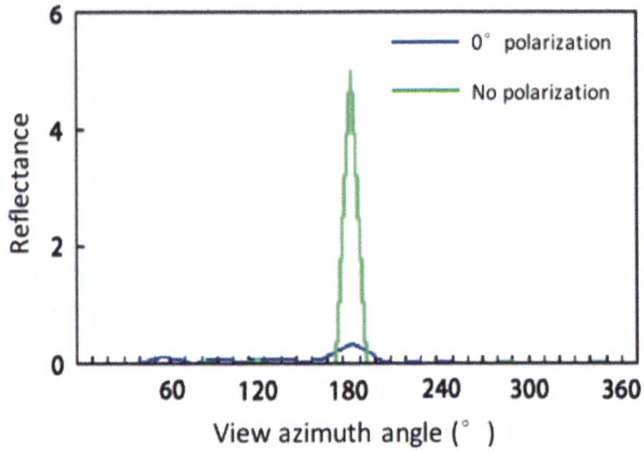

Fig. 5.5 Spectrum of a pure water body at an angle of incidence and viewing angle of 53°: 0-degree polarization and no polarization

(as detected without polarising lenses), and reflectance ratio spectrum of pure water body at incident angles of 53°, 50°, and 60°, for wavelengths of between 670 and 690 nm, and azimuth angles from 0° to 360°, with the *x*-coordinate as the azimuth and *y*-coordinate as the reflectance ratio in the opposite direction.

One can see from the above discussion that sun-glint can only be produced when the multi-angle probe lies within the plane of incidence, the values of reflectance ratio at any other orientation are all very small. The energy incident on a pure water surface, on a parallel plane of incidence in the direction where the viewing angle is 53°, is largely refracted or absorbed. At this point, reflection at 0-degree polarization is much weaker than that without polarization, which is to say that when the angle of incidence is the Brewster angle and the polarization azimuth of the sensor pre-polariser is 0°, the reflected light is completely polarized, is completely absorbed by the polariser, and completely separated from sun-glint.

One can tell, from Fig. 5.5, that the reflection ratio at 0° polarization exhibits a certain reflection peak and not at its theoretical value of zero: this is partly due to stray light disturbance effects and the difficulty of ensuring absolute accuracy when controlling the Brewster angle as the measuring device can only adjust angles to a certain precision (the Brewster angle for pure water bodies should be 53.1°). As such, the peak is most likely to be caused by the effects of non-polarized light.

From Figs. 5.6 and 5.7, one can tell that, when the angle of incidence is 50° or 60°, which is to say not at a Brewster angle, the value of the 0-degree polarization reflectance ratio at corresponding viewing angles remains much smaller than the value of the reflectance ratio when there is no polarization. In Fig. 5.6, the reflectance ratio at 0-degree polarization is 1.192 and the value of reflectance ratio without polarization is 4.571, and its separation efficiency is 73.92%; in Fig. 5.7, the value reflectance ratio at 0-degree polarization is 2.063 and the value of reflectance ratio without polarization is 5.871, and its separation efficiency is 65.32%. This means that separation efficiency improves as the angle of incidence becomes closer to the Brewster angle. This is because, the closer the angle of incidence is to the Brewster angle, the proportion of natural light in the reflected light beam decreases and that

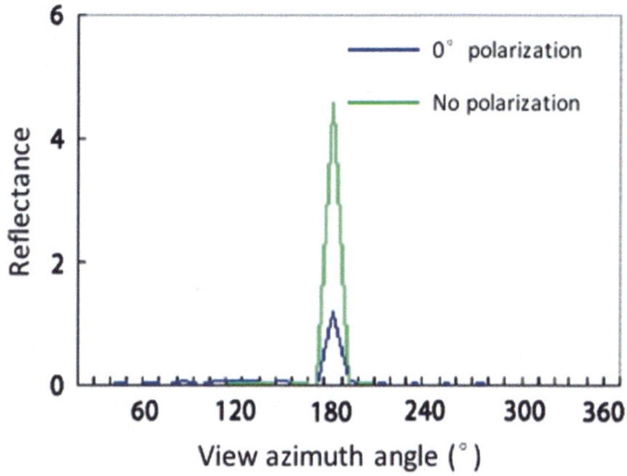

Fig. 5.6 Spectrum of pure water when the angle of incidence and viewing angle are both 50°: 0-degree polarization and no polarization

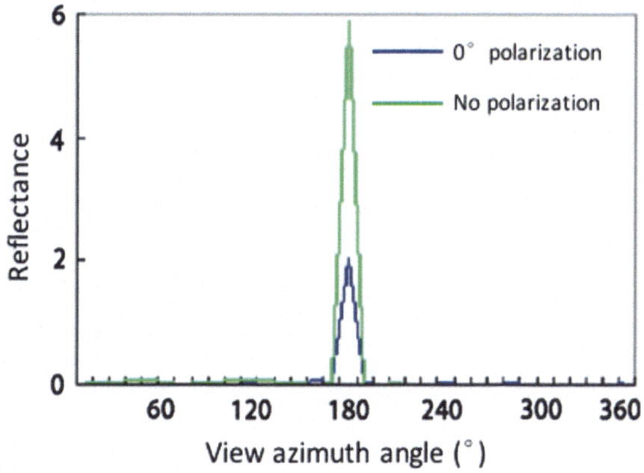

Fig. 5.7 Spectrum of pure water when the angle of incidence and viewing angle are both 60°: 0-degree polarization and no polarization

of polarized light increases. The polarized component of reflected light from water bodies can thus be completely separated.

5.2.2 Polarization Data and the Inversion of the Density of a Water Body

DOP is the physical measure used to describe polarized light and its fixed quantity shows the percentage of linear polarization light in all light, and is defined thus:

$$DOP = \left| \frac{I_{\perp} - I_{=}}{I_{\perp} + I_{=}} \right| \tag{5.14}$$

where, $I_=$ represents the light intensity of the reflected beam when its electrical field vector is parallel to the plane of incidence, $I_\perp$ denotes the light intensity of the reflected beam when its electrical field vector is perpendicular to the plane of incidence. $I_\perp$ corresponds to 90° polarization (I_{90°) and $I_=$ corresponds to 0° polarization (I_{0°). According to the Fresnel equation, with light intensity as the square of the electrical field vector amplitude:

$$P = \left| \frac{\left(-E_\perp \frac{\sin(\theta_1-\theta_2)}{\sin(\theta_1+\theta_2)}\right)^2 - \left(E_= \frac{\tan(\theta_1-\theta_2)}{\tan(\theta_1+\theta_2)}\right)^2}{\left(-E_\perp \frac{\sin(\theta_1-\theta_2)}{\sin(\theta_1+\theta_2)}\right)^2 + \left(E_= \frac{\tan(\theta_1-\theta_2)}{\tan(\theta_1+\theta_2)}\right)^2} \right| \tag{5.15}$$

where, $E_\perp$ represents the component of the incident beam's electrical field vector when it is perpendicular to the plane of incidence and $E_=$ represents the component of the incident beam's electrical field vector when it is parallel to the plane of incidence: θ_1 and θ_2 are the angle of incidence and the angle of refraction, respectively.

As for natural light, $E_\perp^2 = E_=^2$. According to Snell's law:

$$\frac{N}{Ns} = \frac{\sin\theta_1}{\sin\theta_2} \tag{5.16}$$

The refractivity of air is $N_s = 1$,

$$\frac{\sin\theta_1}{\sin\theta_2} = N \tag{5.17}$$

From the above three equations, we get

$$P = \frac{2\sin\theta_1 \tan\theta_1 \sqrt{N^2 - \sin^2\theta_1}}{N^2 - \sin^2\theta_1 + \sin^2\theta_1 \tan^2\theta_1} \tag{5.18}$$

The DOP of light reflected after it passes through the surface of a liquid body is decided by two factors: firstly, the angle of incidence of light and the refractivity of the liquid itself. Different bodies of liquid have different levels of refractivity as they are made up of different substances and structured and formed in different ways. The DOP of reflected light also changes according to the nature of the liquid through which the light passes.

With regards to the derivation of the function for p, when $p' = 0$, θ_1 =53.1°. At this point the function takes an extremely large value, meaning that the reflected beam is completely linearly polarized light and the angle of incidence is the Brewster. When the angle of incidence begins to increase gradually from 0°, the value of DOP of the reflected light also increases until reaching the Brewster angle, at which the DOP is at its greatest value, and decreases thereafter. This means that the closer the angle of incidence is to the Brewster angle, the better the linear polarization of the reflected light.

According to the discussion above, it can be concluded that: polarization is present in the reflected beam that passes through water. At this moment, the body of water itself is essentially a polariser. When the angle of incidence is at the Brewster angle of 53° (i.e., at a solar altitude of 37°), a calm water surface will fully polarise the reflected sunlight. Using an analyser before a sensor and adjusting the azimuth angle of the analyser such that the polarization azimuth and the reflected beam's polarization direction are mutually perpendicular, one will see that, due to the light-filtering effect of the polarization filter, not all of the reflected light can pass through the polarization filter and the data that the sensor receives is just that of atmospheric scattering and scattering of the water body and the intensity of the water's specular reflection can be neglected. This is how one can, without being affected by solar reflection, use the radiative transfer equation of the atmosphere, atmospheric water, and water to calculate water-related parameters.

While using the Gladstone-Dale relationship, Sobolev found that, in the visible light wave band, the K value of liquid is 0.340, N is its refractive index, and ρ is density, which gives

$$\frac{N-1}{\rho} = K \tag{5.19}$$

This can give us the relationship between density of liquids and the DOP:

$$\rho = \frac{\sin\theta_1\sqrt{\tan^2\theta_1(2 - P^2 + 2\sqrt{1-P^2}) + P^2} - P}{KP} \tag{5.20}$$

The refractive index of light in a medium varies according to wavelength λ. Such a phenomenon is known as dispersion. In a typical situation involving dispersion, the refractive index monotonically decreases as wavelength λ increases and the rate of change is larger at the shortwave end. In the visible region, seawater exhibits dispersion and different densities of seawater will give rise to different refractive indices. According to different angles of incidence, measuring the DOP of bodies of seawater with different densities, we can use inverse iteration to compare the calculated, and actual, densities, so as to identify the optimal wave band within which to measure the density of seawater, improving the accuracy of polarization technology used to measure seawater density.

The DOP of static water surfaces will increase gradually as the angle of incidence increases within the range of 10°–50°. This is because, when natural light enters a smooth liquid surface at an angle of incidence that is not the Brewster angle, reflected light becomes partially polarized and the primary vibrational direction of the reflected light is perpendicular to the plane of incidence, which is why light vibration perpendicular to the plane of incidence is greater than that parallel thereto. This is to say that the S-component increases as the angle increases and the P-component decreases as the angle decreases. At 50°, the DOP tends to a straight line, which is a result of the angle of incidence nearing the Brewster angle of seawater and

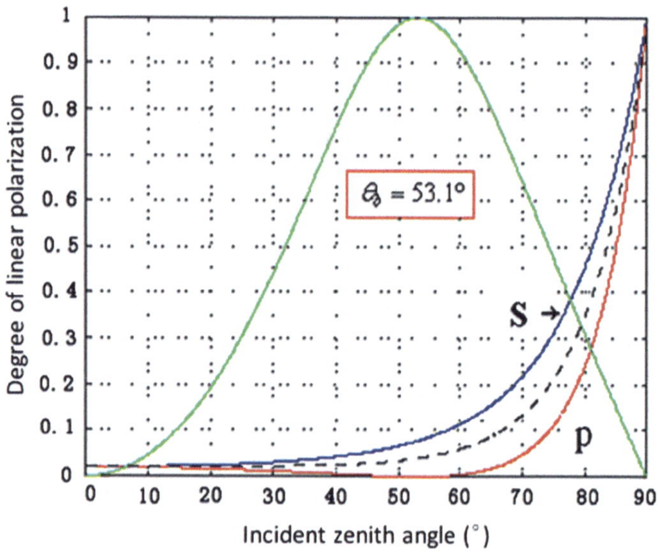

Fig. 5.8 DOP spectrum of water at different angles of incidence

the DOP of light in various colours approaching its maximum value. The changes in the S and P-component can be seen in Fig. 5.8.

When the angle of incidence coincides with the Brewster angle, P—which represents the DOP—reaches its maximum and as the angle continues to increases, the DOP will gradually decrease. As the DOP of the water is within the range of the angle of incidence (0°, 90°), it will monotonically increase until it reaches its maximum, at the Brewster angle of the water body, at which point it starts to decrease monotonically again, giving rise to the parabola shown.

Through the measurement of the DOP of 19 test samples, we used an inverse iteration method to calculate the density of seawater and compared that value to its actual density to determine the optimal waveband range for use in this calculation within a smaller range of deviation: the corresponding DOP within the spectral range will be its average value. We also used six test samples to compare the deviation between calculated and actual values (Table 5.3).

5.3 The Polarization Reflection Filtering Characteristics of Bare Soil

Simultaneous use of multi-waveband, multi-temporal, hyperspectral remote sensing data to improve our ability to identify and differentiate land surfaces, was how people began to notice the influence and contribution that angle information has made towards the differentiation and categorisation of remote sensing images: this

Table 5.3 Comparison of the calculated and actual values for seawater samples with different densities at the optimal waveband for different angles of incidence

Angle of incidence (°)	Wavelength λ /DOP *P*	Standard deviation in *P* δ	Calculated value (g/cm^3)	Actual density value ρ (g/cm^3)	Deviation ρ-ρ′ (g/cm^3)
10	651–671 nm/*P* = 0.04592927	0.00201022	1.011374	1.011650	0.000276
	650–670 nm/*P* = 0.04587868	0.001891164	1.015665	1.016138	0.000473
	649–670 nm/*P* = 0.04583371	0.002025185	1.019489	1.019542	0.000053
	649–671 nm/*P* = 0.04577823	0.002015186	1.024212	1.024167	−0.000045
	649–671 nm/*P* = 0.04510776	0.001948184	1.029973	1.029658	−0.000315
	648–670 nm/*P* = 0.04564836	0.002020110	1.035318	1.035342	0.000024
20	651–670 nm/*P* = 0.18983991	0.002019826	1.011189	1.011650	0.000461
	651–671 nm/*P* = 0.18961048	0.002017499	1.015744	1.016138	0.000394
	650–671 nm/*P* = 0.18938673	0.002022624	1.020197	1.019542	−0.000655
	651–672 nm/*P* = 0.18914669	0.002018622	1.029524	1.024167	−0.00082
	650–671 nm/*P* = 0.18891989	0.002137621	1.029524	1.029658	0.000134
	648–670 nm/*P* = 0.18866382	0.002008622	1.034662	1.035342	0.00068
30	650–670 nm/*P* = 0.43926218	0.001917146	1.011860	1.011650	−0.00021
	649–670 nm/*P* = 0.43872696	0.001902630	1.016485	1.016138	−0.000347
	649–670 nm/*P* = 0.43836763	0.001907396	1.019857	1.019542	−0.000315
	649–670 nm/*P* = 0.43787016	0.001922294	1.023914	1.024167	0.000253
	648–670 nm/*P* = 0.43726455	0.001907396	1.029182	1.029658	0.000476
	648–669 nm/*P* = 0.43658667	0.002019931	1.035096	1.035342	0.000246
40	652–671 nm/*P* = 0.75617286	0.002018868	1.011011	1.011650	0.000639

(continued)

Table 5.3 (continued)

Angle of incidence (°)	Wavelength λ /DOP P	Standard deviation in P δ	Calculated value (g/cm^3)	Actual density value ρ (g/cm^3)	Deviation ρ-ρ' (g/cm^3)
	652–672 nm/P = 0.75527228	0.002002117	1.016563	1.016138	−0.000425
	651–671 nm/P = 0.75480716	0.002013755	1.019434	1.019542	0.000108
	650–670 nm/P = 0.75400962	0.002010513	1.024365	1.024167	−0.000198
	650–671 nm/P = 0.75318482	0.002020868	1.029472	1.029658	0.000186
	648–670 nm/P = 0.75212196	0.002013191	1.036066	1.035342	−0.000724
50	650–671 nm/P = 0.98214361	0.001131510	1.011994	1.011650	−0.000344
	650–672 nm/P = 0.98183319	0.001260886	1.016465	1.016138	−0.000327
	650–670 nm/P = 0.98159895	0.001285502	1.019821	1.019542	−0.000279
	650–670 nm/P = 0.98126249	0.001135427	1.024615	1.024167	−0.000448
	649–669 nm/P = 0.98089620	0.001096511	1.029801	1.029658	−0.000143
	648–669 nm/P = 0.98051720	0.001112491	1.035131	1.035342	0.000211

includes the 3-dimensional spectral characteristics of land surfaces on a 2π-space. Early remote sensing primarily collected observational data from Earth along a perpendicular profile thereto. The distribution of features across the land surface was identified according to the ability of different features to absorb, reflect, and emit electromagnetic waves. This is based on an assumption that the distribution of the target feature's reflection spectrum is uniform in 2π-space (Lambertian reflectance). As remote sensing developed, the conclusions drawn as a result of this assumption were found to be quite different from actual distributions of land surfaces. Observing land surfaces and terrain from a single-angle perspective can easily lead to erroneous conclusions as there are many different terrestrial features that often have the same spectral reflectance at a single-angle perspective. With multi-angle observation, the chances of many different terrestrial features having the same spectral reflectance decreases, moreover, the substance composition, colour, form, and structure of Earth's features vary, as such both the state of polarization of its spectrum and its 3-dimensional spectral characteristics in 2π-space will differ. It is through these differences that we can more accurately identify different terrestrial features and terrains and contribute significantly towards the greater application of remote sensing.

The polarization reflectance of bare soil mostly comes from its surface. This is not to say that incident beams cannot pass through bare soil but that with regards to polarization, such polarization is negligible. Hapke's bidirectional reflectance model of bare soil has been widely applied in remote sensing inversion. This model is (Liang and Townshend 1996) as follows:

$$\rho(\theta_i, \theta_v, \varphi) = \frac{\omega}{4(\mu_i + \mu_v)}[(1 + B(\xi))P(g, \xi) + H(\mu_i, \omega)H(\mu_v, \omega) - 1] \quad (5.21)$$

where, $\mu_i = \cos\theta_i$, $\mu_v = \cos\theta_v$, ω is single-scattering albedo, g is asymmetric phase function. $B(\xi)$ is the backscattering function to calculate the hot-spot effect.

$$B(\xi) = \frac{S(0)}{\omega P(g, 0) \cdot [1 + (1/h)\tan(g/2)]} \quad (5.22)$$

Here $S(0)$ is defined as the size of the hot-spot, h is defined as the width of the hot-spot, and $p(g, \xi)$ is the particle phase function:

$$P(g, \xi) = \frac{1 - g^2}{(1 - 2g\cos(\pi - \xi) + g^2)^{3/2}} \quad (5.23)$$

$H(x, \omega)$ is the function used to calculate multiple scatterings:

$$H(x, \omega) = \frac{1 + 2x}{1 + 2x\sqrt{1 - \omega}} \quad (5.24)$$

Multi-angle polarimetric remote sensing uses the feature polarization data produced by the reflection, scattering, and transmission electromagnetic radiation process of subjects on land surfaces and in its atmosphere as the remote sensing data source for polarization dichroic reflection. Considering that a planar beam with a phase velocity equal to its speed of propagation along the *Z*-axis can always be broken down into two linearly polarized waves along the *X* and *Y*-axes, respectively, as $E_x(z, t)$ and $E_y(z, t)$:

$$\begin{cases} E_x(z, t) = E_x \sin(\omega t - \beta z - \phi_x) \\ E_y(z, t) = E_y \sin(\omega t - \beta z - \phi_y) \end{cases} \quad (5.25)$$

where, $E_x, E_y, \omega, \beta, \phi_x$, and ϕ_y respectively represent the harmonic wave amplitude, vibration frequency, phase constant, and initial phase. When the phase difference is $\phi = \phi_2 - \phi_1 = m\pi$ $(m = 0, \pm 1, \pm 2, \ldots)$, one is describing linearly polarized light. Similarly, when the amplitude and phase difference satisfy certain conditions, we can also get circular or elliptical polarization.

The Stokes parameters can be used to represent the state of polarization of any quasi-monochromatic plane wave from a target of the probe. The source-emitted light is non-polarized and after reflection from bare soil, according to the Fresnel equation, polarization is present in the reflection. The target feature is essentially a polariser at this point and two situations arise: one is that the light that reaching the sensor is a combination of non-polarized and polarized light, or partially polarized light; while the other is that the incident light is completely polarized and the reflected light is completely polarized light. As such, the intensity of the reflected light after it passes through the polarization filter can be expressed as:

$$I(\theta) = [(1 - p) \cdot \frac{1}{2} + p \cdot \cos^2 \theta] \cdot I \tag{5.26}$$

In the equation, p is DOP and θ is polarization azimuth. DOP is defined as the ratio of the intensity of the fully polarized component and the intensity of its light wave. Light passes through the polarization filter as partially polarized light, p is the intensity of the linearly polarized light component.

If we put a polariser before a sensor, then we would be able to measure the polarization bidirectional reflectance, therefore, we can set the bare soil base in Hapker's polarized direction reflectance model (BPDF) as shown below (Wu and Zhao 2005):

$$R_p = \frac{\omega}{8(\mu_0 + \mu)} \cdot [1 + (2\cos^2 \theta - 1) \cdot p] \cdot [(1 + B(g))P(g) + H(\mu_0)H(\mu) - 1] \tag{5.27}$$

In any XOY-plane, the light intensity observed in the direction of angle α with respect to the X-axis can be shown as:

$$I(\alpha) = \frac{1}{2} \cdot (I + Q \cdot \cos 2\alpha + U \cdot \sin 2\alpha) \tag{5.28}$$

Taking three viewing directions for which α is respectively equal to 0°, 60°, and 120° to get the I, Q, and U values of these simultaneous equations, we find the corresponding DOP P, and the polarization azimuth θ, which will in turn complete the set of model values.

To test the aforementioned model, one can design and conduct the following bare soil probe experiment:

1. **The production and measurement of bare soil samples**

Eight types of soil, namely black soil, latosol, brown soil, sand alluvial soil, peat soil, chernozem, dark loess soil, and a dark brown soil are used as soil samples. For the purposes of standardisation and comparison, we put the samples in sample cartridges and weighed them. Then, placing the soil on a clean glass panel, we added water to each soil sample with a burette, stirred it and allowed equilibration for 30 min. Then, during measurement, we collected the soil samples from their respective sample cartridges and reweighed them to determine their gravimetric moisture content. Each soil sample was treated as described above and it was determined that their moisture contents were, in order, 0 (oven-dried), 12, 18, and 22%. Using a glass plate applied at 45° to flatten the soil sample, we then measured their spectral characteristics.

The polarization spectrometer used for measured is produced by Changchun Institute of Optics, Fine Mechanics and Physics, Chinese Academy of Sciences. The spectrometer is composed of three parts: a light source system, a bidirectional polarimetric reflectance spectrometer system, and an automatic control system. The measurement is done by placing the sample on the sample stage in the centre of the dichroic photometer, adjusting the horizontal position and height, turning on the light source and setting the polarization filter in front of the light source at the angle required. Then, for each sample, we would measure the sample's reflectance spectrum in 2π-space without polarization, at 0° polarization, and 90° polarization at two wavelengths: A (690 nm to 760 nm) and B (760–1100 nm). Simultaneously, we would change the incident light source's elevation to measure the reflectance spectrum at different elevation angles to examine its effects and influence on soil in a 2π-space reflectance spectrum.

2. **Analysis of bare soil's polarimetric reflectance characteristics**

Figure 5.9 shows the reflectance spectral curve of black soil in the A waveband where the angle of incidence is respectively 50°, and 60°, with a 90° polarization filter.

From Fig. 5.9, we can tell that the peaks are, in order, at 60°, 50°, 40° and 20°. From the left-hand image, the 10° and 20° spectra almost overlap and in the right-hand image the curves of 10°, 20°, and 30° also almost overlap.

The above results show that: the angle of incidence of the light source has an extremely strong influence on the spectrum and at the direction of the angle of reflection, opposite to that of the angle of incidence, there is a clearly observable specular reflection in soil. This effect produces a conical spectral curve with respect a principal axis that is the reflected beam's optical axis. As such, there are certain peaks and troughs shown at different viewing angles and the spectral peak value at each viewing angle gradually reduce with respect to the cone's principal axis.

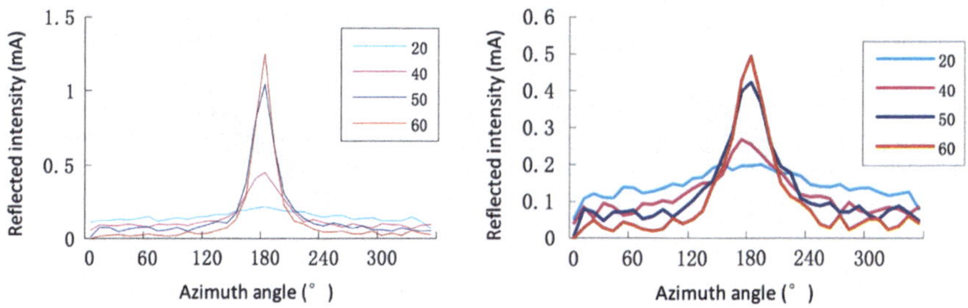

Fig. 5.9 Black soil spectra at an angle of incidence of 50° (left) and 60° (right) and polarization filter is at 90°

Therefore, as the angle of incidence increases, the spectral curve overlaps more frequently as it gets further from the angle of incidence.

Testing the soil model

By measuring the polarization direction reflectance ratio of brown soil surface when the incident zenith angle is 0°–60°, the viewing zenith angle is 0°–60°, and viewing azimuth angle is 0°–360° at the A-waveband (670–690 nm) and a polarization azimuth of 0°, 30°, 45°, and 90°, we use the measured value in comparison with the calculated values from the model.

Figure 5.10 shows the comparison between the measured values and the values calculated using the BPDF model: the measured values at 0°, 30°, 45°, and 90° are consistent with the values calculated using the BPDF model, showing that the BPDF model can simultaneously show the characteristics of polarization reflection and bidirectional reflection. Results of the experiment are based on the A-waveband (670–690 nm) of the polarization direction spectrometer. As the width of the waveband is 20 nm, we would need to use a hyper-spectrometer such as an ASD device next, to subdivide the spectral width and number of bands to test the BPDF model accuracy and applicability more precisely.

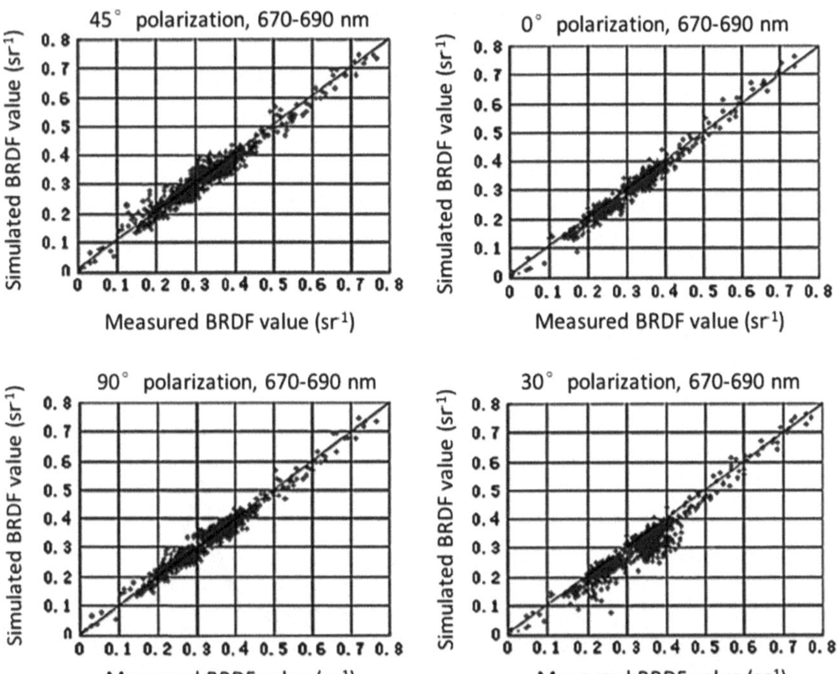

Fig. 5.10 Comparison of measured, and BPDF model, values

5.4 The High Signal-to-Background Ratio Filtering Relationship Between Polarization and Bare Soil Water Content

The soil water content, as the basic parameter in research into the formation, transformation, and exhaustion of terrestrial water resources, is the connection between surface water and groundwater and is also the basic element used in researching land surface energy exchange and has important effects on climate. The use of polarization technology to measure soil water content primarily relies on measuring the reflection from the bare soil surface or the electromagnetic radiation energy emitted therefrom. This process is subjected to, and associated with, soil surface roughness, its optical characteristics, and the observational angle. This chapter mainly expounds the relationship between soil water content and polarization.

In reality, to radiation wavelengths, most natural surfaces are rough: when the target surface is sufficiently rough such that, when irradiated with solar shortwave radiation, its reflected radiance intensity presents as a constant in 2π-space with the target object as its centre, which means that reflected radiance intensity does not change according to observational angle, we call such objects diffuse reflectors, or Lambertian reflectors. Diffuse reflection is also known as Lambertian reflection, which is also known as isotropic reflection. Theoretically, there are only a few objects close to those described as Lambertian reflectors in nature.

Reflection between diffuse and specular reflection is known as directional reflectance, also known as non-Lambertian reflectance. The bulk of objects found in nature produce directional reflectance, which is to say that they possess anisotropic properties when faced with solar shortwave radiation scattering: these objects are known as non-Lambertian objects. When the applications of remote sensing enter the stage of quantitative analysis, then we have to abandon the assumption that the target is a Lambertian reflector. The description of directional reflectance cannot entail simple use of the reflective index as the reflective index differs in each direction.

When a ray of a plane electromagnetic wave strikes a smooth surface at an angle of incidence that is the Brewster angle then the reflected wave is a linearly polarized wave and the refracted wave is partially polarized. If the incident wave is natural light, then it can be broken down into two equal parts: that perpendicular to the principal reflection planar vibration and that parallel thereto; meanwhile in the reflected wave, vibration parallel to the surface is non-existent since, if it did exists, then its electrical field vector's direction of vibration and that of its electromagnetic propagation would be the same: this violates the Maxwell principle. If the angle of incidence is not the Brewster angle, then the partially polarized characteristic of the reflected wave can be fully understood. If the medium is a completely rough surface then the wave emitted from different points in the medium has no fixed phase difference, so the reflected wave has, in general, no polarization. At this time, the surface presents as a Lambertian reflector.

We know that when the target feature presents as a non-Lambertian reflector and the angle of incidence is not the Brewster angle, then the reflection that occurred must include specular reflection, which means that there must be polarized light within the reflected light and the reflected light at this point is partially polarized. The extent of polarization can be measured using DOP, originally defined as:

$$p = \frac{I_p}{I_p + I_n} \quad (5.29)$$

where, I_p and I_n represent the intensity of polarized and natural light, respectively. Natural light has $p = 0$ while completely linearly polarized light has $p = 1$, so, for partially polarized light $0 \leq p \leq 1$. The greater the proportion of polarized light in reflected light, the greater the extent of polarization and the greater the DOP.

Studies have shown that, for bare soil, the higher the moisture content of the soil, the greater the DOP. This can be understood as liquid water filling the voids. This means that there must exist a threshold for the moisture content in a bare soil that can make the DOP equal to zero, such that when the moisture content exceeds the threshold, the DOP of the soil is greater than zero and the soil presents as a non-Lambertian reflector; below the threshold, the soil presents as a Lambertian reflector.

1. **Polarimetric reflection ratio and bare soil moisture content threshold**

To measure polarization of soil with a certain amount of moisture, we fill the sample cartridge with soil samples (three replicates), weigh each sample then put each dry soil sample on a clean glass panel, then routinely add water and stir each sample using a burette, leaving it to equilibrate overnight. The moisture content of black soil samples were: 16.2, 18.0, 20.2, 24.6, 25.2, 26.1, 28, 34.8, and 36.1%. We placed each of the aforementioned soil samples back into its cartridge, using a glass piece at 45° to flatten the soil sample before measuring the reflectance ratio at different moisture contents. After measuring the reflectance ratio of each soil sample, we immediately measure its whiteboard value. All tests were conducted under dark-room conditions.

In Fig. 5.11, the trend is such that the DOP of black soil with a particle size of 1 mm at an angle of incidence of 40°, wavelength of 760 to 1100 nm, azimuth angle of 180°, and viewing angle of 50° increases as the moisture content increases.

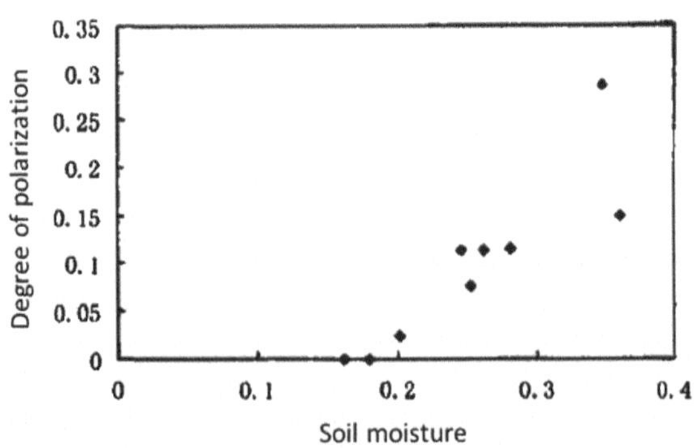

Fig. 5.11 Polarization spectrum of black soil (angle of incidence of 40° at different moisture contents)

Using SPSS statistical analysis software to conduct regression analysis on the data obtained, we find that:

$$Y = -0.4204 + 2.3629X - 4.6888X^3 \tag{5.30}$$

where Y is the DOP of the soil, X is the moisture content of the soil, and multiple correlation coefficients $R = 0.82563$ and $R^2 = 0.68166$. Conducting an F-test on the above model, we found that that there is a functional relationship between soil moisture content and its DOP. When DOP is zero, the soil moisture content was 19.3%.

In Fig. 5.12, the trend that the DOP of black soil with a particle size of 1 mm, at a an angle of incidence of 50°, wavelength of 760–1100 nm, azimuth angle of 180°, and viewing angle of 50° increases as the moisture content increases. Again, we get the following regression model:

$$Y = -0.2843 + 1.8526X - 3.239X^3 \tag{5.31}$$

where Y is the DOP of the soil, X is the moisture content of the soil, and multiple correlation coefficients $R = 0.91747$ and $R^2 = 0.84175$. Conducting an F-test on the above model, we found that that there is a functional relationship between soil moisture content and its DOP. When the DOP is zero, the soil moisture content was 16.0%.

In Fig. 5.13, the trend that the DOP of black soil with a particle size of 1 mm, at a an angle of incidence of 60°, wavelength of 760 to 1100 nm, azimuth angle of 180°, and viewing angle of 60° increases as the moisture content increases. Again, we get the following regression model:

$$Y = -0.2144 + 1.4974X - 0.5840X^2 \tag{5.32}$$

where Y is the DOP of the soil, X is the moisture content of the soil, and multiple correlation coefficients $R = 0.95206$ and $R^2 = 0.90642$. Conducting an F-test on the above model, we found that that there is a functional relationship between soil

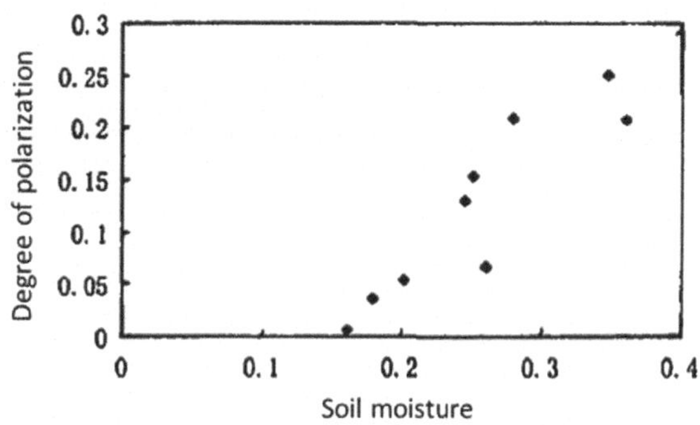

Fig. 5.12 DOP spectrum of black soil (angle of incidence of 50° at different moisture contents)

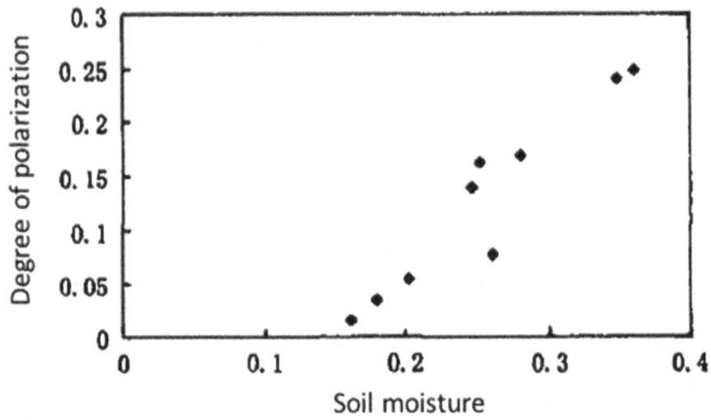

Fig. 5.13 DOP spectrum of black soil (angle of incidence of 60° at different moisture contents)

moisture content and its DOP. When the DOP is zero, the soil moisture content was 15.1%.

From experiments at different angles of incidence, one can tell that the soil moisture content threshold differs. Combining multi-angle remote sensing and polarization remote sensing such that each manages to exploit its respective advantages, we can improve the diversity of the methods we currently have for bare soil remote sensing monitoring and this can also contribute towards the classification of soil by remote sensing, and to the remote sensing of soil moisture content. This will in turn provide a theoretical basis for the determination of whether, or not, current remote sensing platforms can be applied to multi-angle remote sensing of soil and the design of future sensors.

2. **The minimum soil polarization intensity and its moisture content**

There are three absorption peaks caused by O–H on the soil spectrum. The spectral absorption peaks close to 1450 and 1940 nm are caused by a difference in water content and the spectral minimum absorption peak are caused by organic compounds. This portion of the chapter will mainly focus on explaining the changes caused by water content in the vicinity of 1450 and 1940 nm.

From Figs. 5.14 and 5.15, as the soil moisture content increases, the soil's multi-angle polarization hyperspectral reflective index first decreases but at a certain moisture content is reached, the soil's multi-angle polarization hyperspectral reflective index begins to increase. Neema et al. attributes this inflection in reflective index to the fact that soil particles have most likely maximised their ability to absorb water. When more water is added to the soil, then the specular reflection therefrom increases and the soil's hyperspectral reflective index at this point begins to increase as its moisture content increases. Through experimentation, we also found that, in other states of polarization the inflection occurs at a moisture content of around 30%.

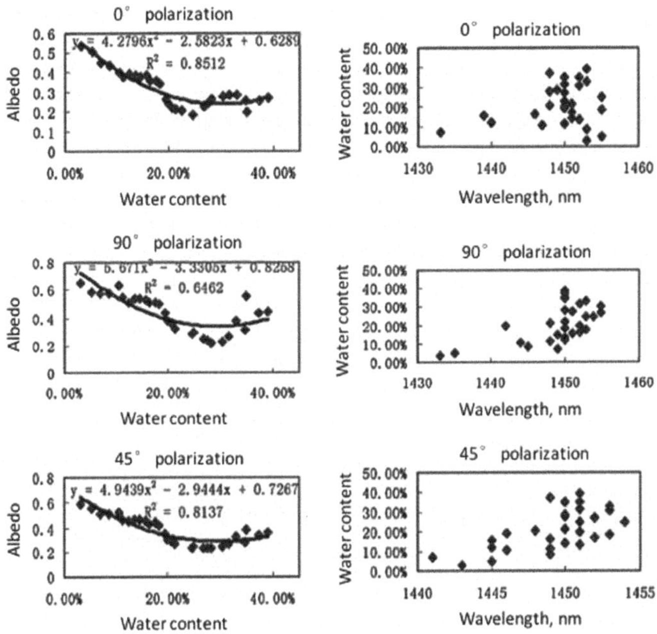

Fig. 5.14 Soil spectral minima and their relationship with water content at around 1450 nm

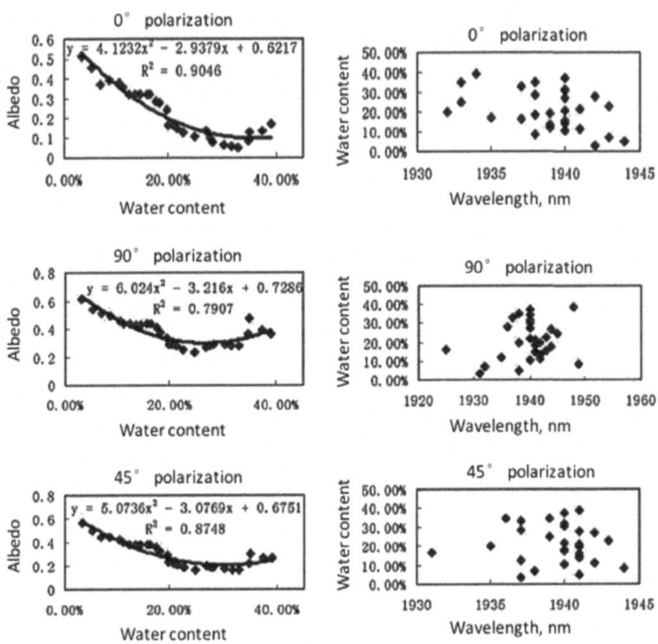

Fig. 5.15 Soil spectral minima and their relationship with water content at around 1940 nm

References

Cunningham A, Wood P, McKee D (2002) Brewster-angle measurements of sea-surface reflectance using a high resolution spectroradiometer. J Opt A: Pure Appl Opt 4(4):S29

Liang S, Townshend J (1996) A modified Hapke model for soil bidirectional reflectance. Remote Sens Environ 55(1):1–10

Luo Y, Zhao Y, Li X et al (2007) Research and application of multi-angle polarization characteristics of water body mirror reflection. Sci China, Ser D Earth Sci 50(6):946–952

Wu TX, Zhao YS (2005) The bidirectional polarized reflectance model of soil. IEEE Trans Geosci Remote Sens 43(12):2854–2859

Chapter 6
Remote Sensing of Land Surfaces 5: Characteristics of Radiative Transfer on Surface Polarization

This chapter focuses on four typical types of land surfaces (rock, water, soil, and vegetation) as targets of observation to introduce the fifth essential feature of remote sensing of ground objects: the characteristics of radiative transfer. These include using the polarization reflectance of the vegetation canopy to demonstrate energy transfer by radiation; using the model illustrating the degree of polarization of the vegetation canopy's reflected light to derive and characterise the pattern of polarized radiative transfer; using the relationship between a single leaf of vegetation and polarization to create a method for the characterisation of single leaf polarization; and using a few types of crops and their impacts on polarization.

6.1 Characteristics of Polarimetric Reflectance of the Vegetation Canopy

Vegetation is the most typical of ground objects. This section focuses on the characteristics of polarization reflectance of the vegetation canopy to demonstrate the properties of radiative transfer.

When observation is conducted by human eye or sensor at an oblique angle, it can be found that part of the leaf surface appears white rather than green (Fig. 6.1). A majority of the photons are specularly reflected directly from the surface and due to the high intensity of such radiation, it tends to be cause dazzling. In addition, the part with a white colour illustrates the absence of wavelength selectivity, meaning that the reflectivity in the visible range is uniform. Based on the Fresnel principle, this part of the light is partly polarized (Vanderbilt et al. 1985).

After incident radiation interacts with the leaf, some radiation will be directly reflected from the leaf surface and thus does not contain any information about the leaf blade as it did not enter it; while the remainder of the radiation will enter the leaf and interacts with its internal structure before being partly absorbed while the rest of the radiation escapes. This part of the radiation also contains internal

L. Yan et al., *Polarization Remote Sensing Physics*, Springer Remote Sensing/Photogrammetry, https://doi.org/10.1007/978-981-15-2886-6_6

Fig. 6.1 Oblique observations of corn

information about the leaf (chlorophyll content, nitrogen content, etc.) (Vanderbilt 1980). The part with surface reflection is only related to the characteristics of the leaf and the corresponding relationship between incidence and observation, specifically the relative orientation of incidence and observation directions, the refractive index of the wax layer, and the roughness of the leaf surface.

To determine the intensity of photosynthesis of crops in the vegetation canopy, the structure of the crop canopy has been listed as a priority in earlier studies. The angle of the leaf is the most important factor to describe the canopy geometry: however, field measurements are very difficult. With the development of oblique sensors and 3-d digital technologies, the measurement of leaf gaps is now possible. The disadvantage of these techniques is that they are unable to acquire useful information at large-scale in a short period of time, especially during agricultural and environmental remote sensing, long-distance detection of canopy structure is an effective means of distinguishing between different types of crops and plants. This allows researchers to focus on the applications of remote sensing to the discrimination of vegetation canopy types. Using polarization information to establish a connection with the angle of a leaf has been shown to be feasible by both theoretical and practical research.

The relationship between canopy structure and the characteristics of crop canopy polarization reflectance is mainly evident in the relationship between the angle and the polarization of the leaf. Researchers found that different crops, at different growth stages, have different canopy structures, so that the characteristics of polarization reflectance vary: this provides a basis for the association between the polarization information and canopy structure. Figure 6.2 shows the relationship between polarization and the other factors (including leaf area index (LAI), mean tilt angle (MTA), plant height (PH), and the degree of greenness of a leaf).

Subsequently, a detailed study of the relationship linking polarization and the tilt angle is conducted: there is a linear relationship between them, as shown in

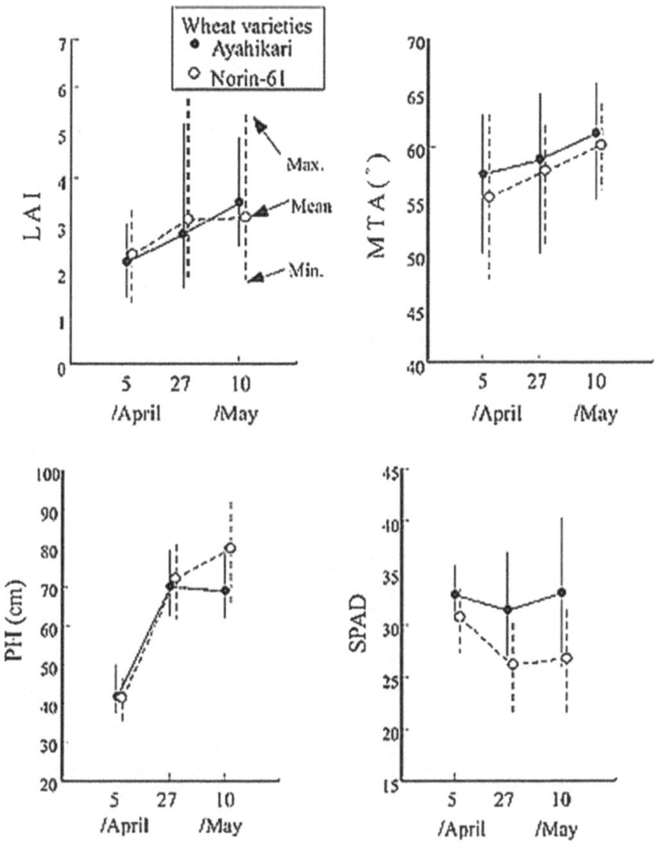

(a) Leaf area index LAI, (b) mean tilt angle MTA, (c) plant height PH, (d) degree of greenness

Fig. 6.2 Variations of two crop canopies under different states of polarization

Fig. 6.3. From the comparison of measured, and estimated, values, it is found that the utilisation of an empirical model obtained from regression can predict the tilt angle of the blades from crop canopy data.

Although the tilt angle of a leaf can be calculated using the regression model, more field experiments should be conducted in future studies to improve the stability and provide calibration of actual measurements.

In addition, Roudeaux also studied the effect of LTA within the vegetation canopy on the polarization reflectance observed, and built the famous Rondeaux & Herman model based thereon (a detailed description follows in Sect. 6.2.1). Different distributions of leaf tilt angle presented the canopy with more corresponding G function data, while Fig. 6.4 shows the relationship between the solar zenith angle and the G function values of vertical, homogeneous, and horizontal type vegetation canopies,

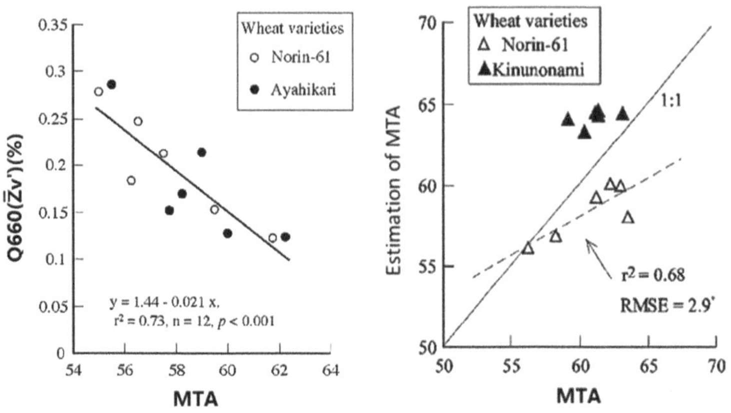

Fig. 6.3 Correlation between polarization and tilt angle (left), Comparison between estimated, and actual, values

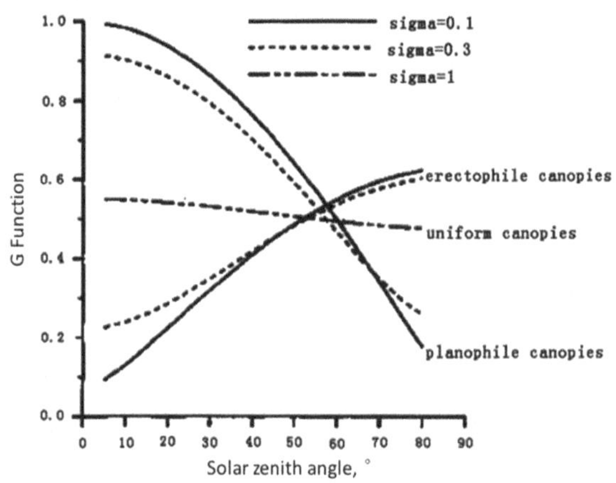

Fig. 6.4 Relationship between the solar zenith angle and the G function values of vertical, homogeneous, and horizontal vegetation canopies

in which Rondeaux used a Gaussian distribution to simulate the LTA distribution and σ represents the variance of the Gaussian distribution function (the observed σ value of a homogeneous canopy is 1). In the Gaussian distribution function, the LTA of a horizontal canopy is set to 0°, while that of a vertical canopy is 90°.

Figure 6.5 shows the polarization reflectance difference for different LTA distributions in Ronudeaux' stimulation of the soybean canopy under two different sets of conditions. Figure 6.5a shows the distribution of LTA centred on 35°, the variation of its polarimetric reflectance at the main plane that varies with observation angles when variances are set to 0.3, 0.4, and 0.5 respectively (especially at larger observation angles). Figure 6.5b shows dehydrated soybean data, whose LTA distribution is centred on 30°. A similar conclusion can be drawn to that arising from Fig. 6.5a, but under these circumstances the difference caused by changing the variance is smaller.

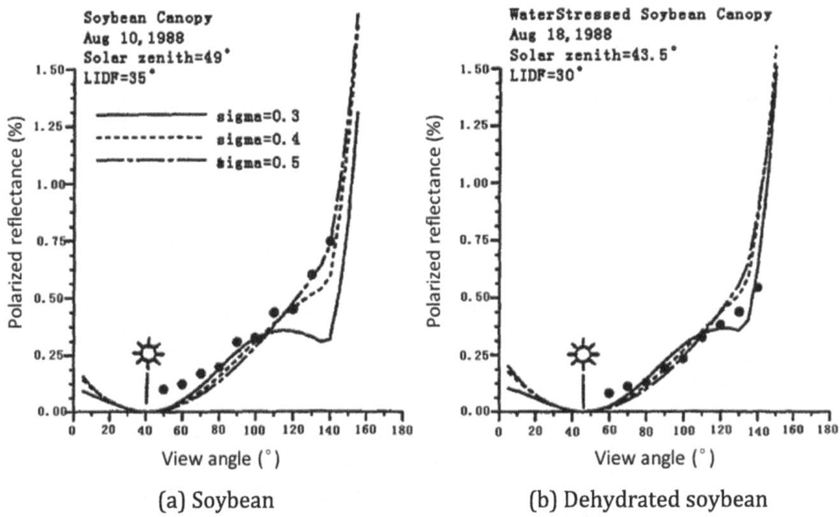

Fig. 6.5 Polarization reflectance and LTA distribution of a canopy under two different conditions (with, and without, water limitation)

6.2 DOP Models of the Reflected Light from the Vegetation Canopy

The DOP models of the reflected light from vegetation canopies, including the Rondeaux & Herman model, Bréon model, Nadal & Bréon model, and the Maignan single-parameter model, are analysed.

6.2.1 Rondeaux & Herman Model

The earliest, and most cited, model of the DOP of reflected light of vegetation canopies was proposed by Rondeaux in 1991 (Rondeaux and Herman 1991). Figure 6.5 shows the spatial geometric relationship governing the interaction of incident radiation and the canopy (human eye or sensor data). Through previous studies, it is known that only specular reflection on the surface contributes to the DOP, while diffuse scattering within the leaf does not, therefore, the radiation flux density, with polarization characteristics caught by the sensor is:

$$d\Phi_r(z) = dA \cdot S(z)dz \cdot f(\theta_n)d\omega_n \cdot E(z) \cdot \cos\omega \cdot R(\omega) \tag{6.1}$$

where, dAdz is the volume metadata, S(z) represents leaf area density at height z, $f(\theta_n)$ gives the probability that the leaf orientation is aligned with θ_n, and $d\omega_n$ is the micro-solid angle normal to the leaf. Thus $dA \cdot S(z)dz \cdot f(\theta_n)d\omega_n$ represents the sum of

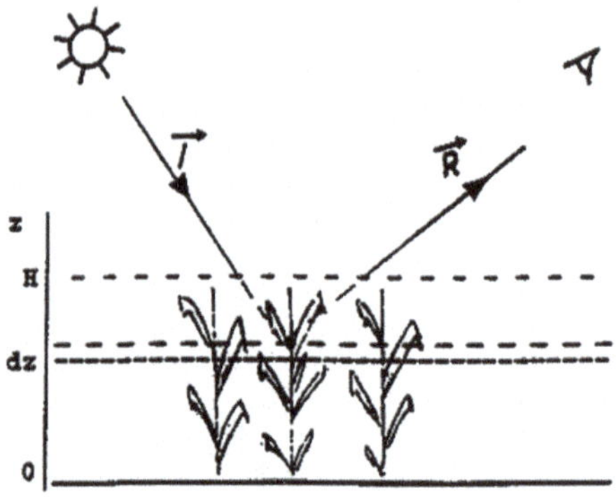

Fig. 6.6 Geometric relationships of incident radiation, vegetation canopy and observations

the one-sided leaf area of all leaves within micro-solid angle $d\omega_n$ centred around θ_n. cos w is the cosine value of the phase angle, E(z) is the radiant flux density projected onto the horizontal surface, and R(ω) refers to the specular reflectivity (Fig. 6.6).

So, radiance received by sensor is:

$$dL_r(z, \theta_r, \varphi_r) = \frac{d\phi_r(z)}{dA \cdot \cos\theta_r \cdot d\omega_r} = \frac{S(z)dzf(\theta_n)R(\omega)E(z)}{4\cos\theta_r} \tag{6.2}$$

Assuming the leaves of canopy interior are multi-layered (Fig. 6.7), then the probability of the reflected incident radiation (after passing through height z) being observed is equal to the product of the extinction coefficient of incidence τ_i and the extinction coefficient of reflection τ_r, and the total radiance received by the sensor is:

$$L_r(\theta_r, \varphi_r) = \int_0^H dL(z, \theta_r, \varphi_r)\tau_i\tau_r \tag{6.3}$$

And,

$$\tau = \exp\{[-G(\mu)/\mu] \int_z^H S(z)dz\} \tag{6.4}$$

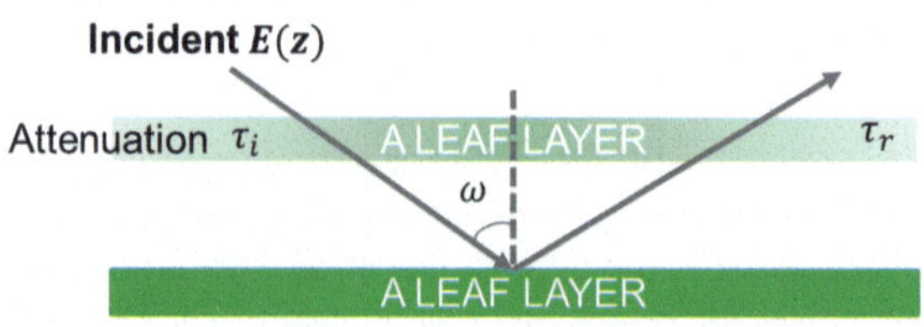

Fig. 6.7 The effect of a multi-layered leaf on the incident radiation

In which $\mu = \cos\theta$ (solving for τ_i, $\theta = \theta_i$; solving for τ_r, $\theta = \theta_r$), gives

$$G(\theta) = \int\int f(\theta_n) \cos \omega d\omega_n \tag{6.5}$$

Then, Eq. (6.3) can be rearranged to give:

$$L_r(\theta_r, \varphi_r) = \frac{f(\theta_n)}{4\cos\theta_r} R(\omega) E_0 \cdot \int_0^H \exp\{-[\frac{G(\theta_i)}{\mu_i} + \frac{G(\theta_r)}{\mu_r}] \int_z^H S(z')dz'\} \cdot S(z)dz \tag{6.6}$$

For uniform vegetation, the leaf area density is a constant, and the material is thick enough to cover the ground surface, then

$$L_r(\theta_r, \varphi_r) = \frac{f(\theta_n)}{4\cos\theta_r} R(\omega) E_0 / [\frac{G(\theta_i)}{\mu_i} + \frac{G(\theta_r)}{\mu_r}] \tag{6.7}$$

Figure 6.8 shows a comparison of theoretical, and measured, values using this model, suggesting that the model is accurate.

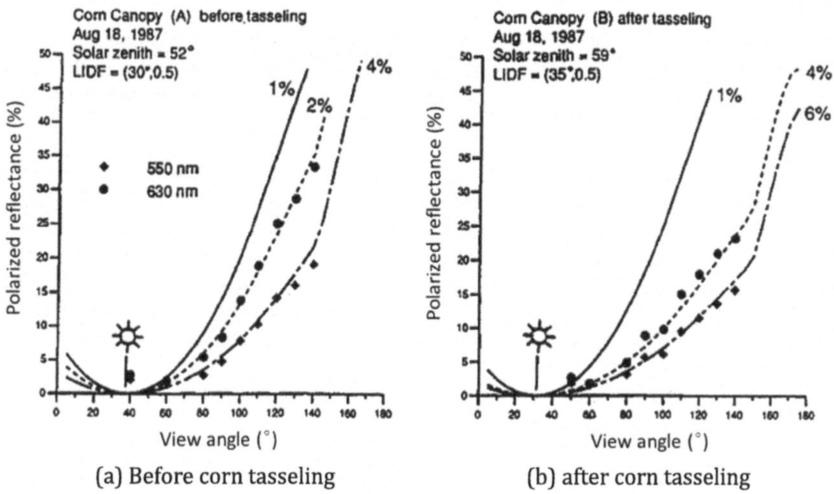

Fig. 6.8 Comparison of theoretical, and measured, values using the Rondeaux & Herman model. Here, LIDF is the leaf inclination distribution function

6.2.2 Bréon Model

Based on the Rondeaux model, Bréon hypothesises that LTA follows a uniform distribution, then the polarized reflectance, ρ_P, can be transformed into (Bréon et al. 1995):

$$\rho_P(\theta_s, \theta_r, \varphi) = \frac{\pi L_r(\theta_s, \theta_r, \varphi)}{\cos\theta_s E_0} = \frac{R(\omega)}{4(\cos\theta_s + \cos\theta_r)} \tag{6.8}$$

Bréon went further by considering the bare soil model as it is impossible for vegetation to be dense enough to always satisfy the necessary conditions, and thus the ground surface is considered as a series of small specular reflections along a fixed direction to obtain the bare soil model:

$$\rho_P(\theta_s, \theta_r, \varphi) = \frac{\pi L_r(\theta_s, \theta_r, \varphi)}{\cos\theta_s E_0} = \frac{R(\omega)}{4\cos\theta_s \cos\theta_r} \tag{6.9}$$

In reality, the reflectivity of any ground object can be written as the weighted combination of those of the vegetation and bare soil. This model reflects the properties of polarization reflectance of land surfaces and is seen as a physical model related only to geometric conditions without including other unknown parameters. The disadvantage is that it only considers uniformly distributed crops such as vegetation and bare soil, and it lacks accuracy for other complex types of crops. Figure 6.9 shows a comparison of simulated, and measured, data form vegetation and the bare soil model.

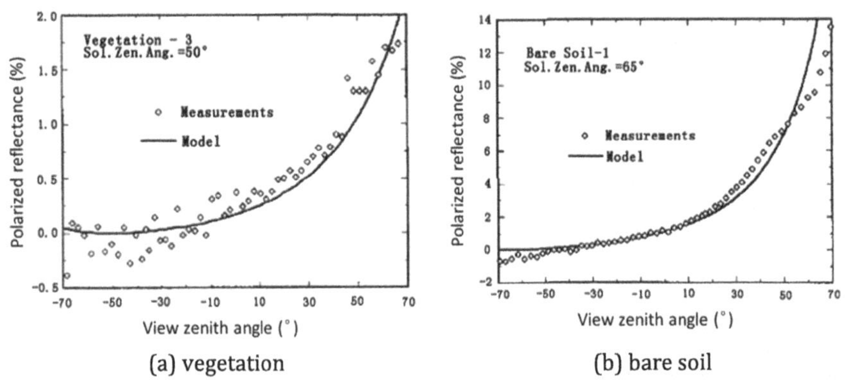

Fig. 6.9 Comparison of simulated, and measured, polarimetric reflectance in the Bréon model

6.2.3 Nadal & Bréon Model

When Nadal was using POLDER data to observe the reflectance of land surfaces, he discovered that a stable relationship exists between the polarimetric reflectance $R_p(\theta_s, \theta_r, \varphi)$ and $\frac{F_p(\alpha)}{\mu_s+\mu_r}$, and polarimetric reflectance at smaller scattering angles tends to saturate instead of increasing (Nadal and Breon 1999). Based on this, Nadal suggested a semi-empirical polarization reflectance model:

$$R_P(\theta_s, \theta_r, \varphi) = \rho[1 - \exp(-\beta \frac{F_p(\alpha)}{\mu_s + \mu_r})] \tag{6.10}$$

Here, ρ and β are values fitted by using measured data, where ρ represents the saturation value, and $\rho\beta$ represents the gradient when the scattering angle is relatively large. Figure 6.10 depicts the relationship between this linearity and saturation.

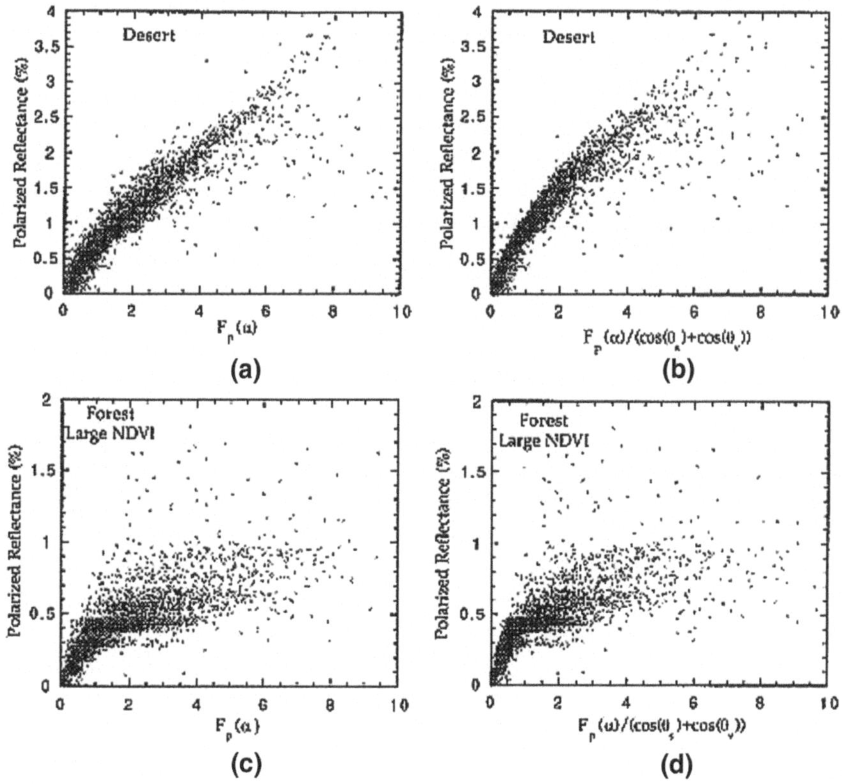

Fig. 6.10 Linear relationship in the Nadal & Bréon model

6.2.4 Maignan's One-Parameter Model

The Rondeaux model is obtained based on the interaction of photons and the canopy, while the atmosphere affects satellite remote sensing observations. POLDER is the longest serving polarization imaging instrument in orbit, which supplies polarimetric reflectance information in three bands (443, 670, and 865 nm), including information on surface roughness, water content, and LTA of the ground surface although it is mainly used to measure the polarization of the atmosphere. In other words, polarization has become an effective means with which to measure ground surface targets. Simultaneously, the expression of bidirectional polarized reflectance distribution functions of ground surface should also be an essential method in representing target polarization distributions. POLDER also provided theoretical and actual support to the establishment of BPDF.

Figure 6.11 shows the polarimetric reflectance distribution of five major types of ground objects at four solar zenith angles: the polarimetric reflectance is closely associated with the geometric orientation of the observation, especially when the heterogeneous reflection of polarization by reflection is more obvious. Eventually, the scattering angle tends to zero and increases with the phase angle.

After comparing the Rondeaux–Bréon polarization model for vegetation, Bréon's polarization model for soil, Nadal's model, and the Waqute model, Maignan discovered that the newly established one-parameter model of polarimetric reflectance is suitable for most studies of the characteristics of polarization reflectance with higher accuracy (Maignan et al. 2009). The comparison shows that:

$$R_p^{surf}(\Omega_s, \Omega_v) = \frac{C\exp(-\tan\alpha_1)\exp(-v)F_p(\alpha_1, n)}{4(\mu_s + \mu_v)} \tag{6.11}$$

The new model, as proposed by Maignan, will be used to estimate the polarimetric reflectance information of aerosols over land surfaces (Fig. 6.12).

Litvinov proposed a reflection model targeting soil and vegetation using polarising filters, and the acquisition of polarization data using polarization scanners (Litivinov et al. 2010). The basic theory underpinning this model is such that the inherent bidirectional reflectance matrix is used to describe surface reflection of objects, in which the Stokes parameters of scattering is closely associated with the incident radiation field. To describe the polarization characteristics of scattered radiation and the surface reflection intensity separately, the bidirectional reflectance function and bidirectional polarization distribution function are combined and used to replace the bidirectional reflectance matrix, which mainly congregate in the region of visible and near infrared bands in this model.

The left-hand part of Fig. 6.14 represents the reflection intensity ratio of the soil (proportion of the first component of Stokes parameter), while the right-hand part represents proportion of polarization reflectance. 1, $1'$ and $1''$ represent scattering angles formed under different flight paths (the same applies to 2 and 3). Also, 1, 2, and 3 represent the combinations (670 nm and 1588.86 nm), (1588.86 nm and

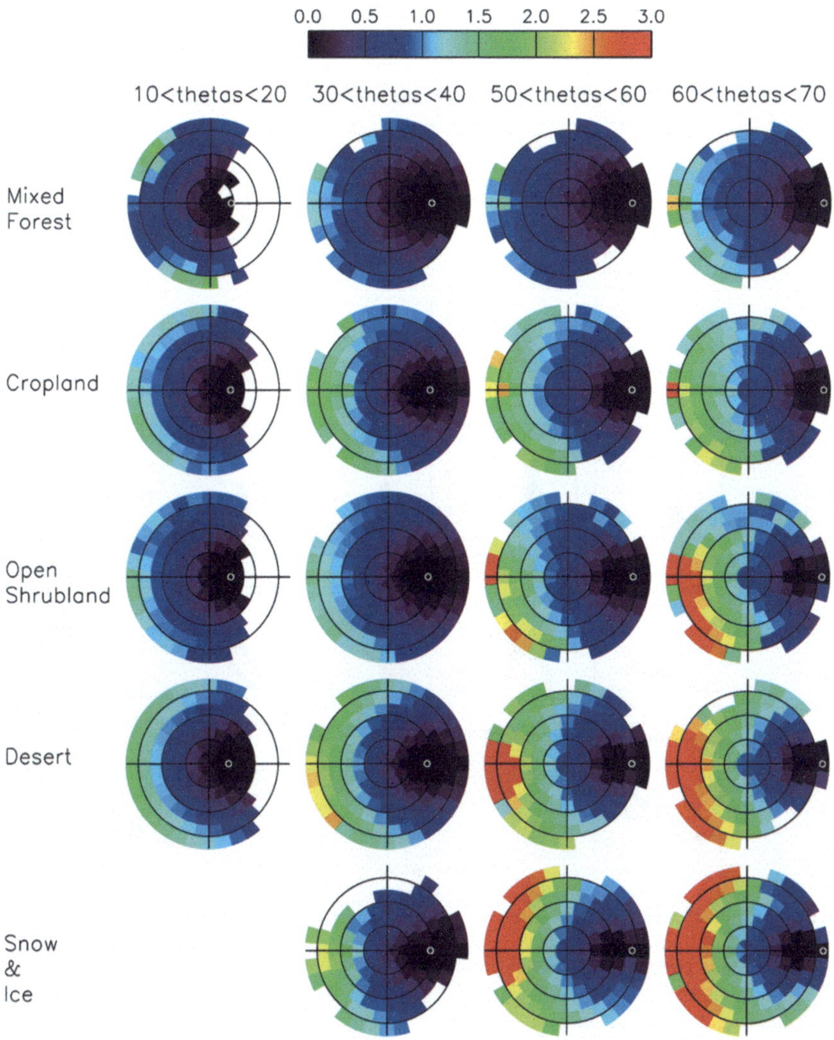

Fig. 6.11 Polarized reflectance distribution: polar diagram of different ground objects. The black circles represent constant viewing angles every twenty degrees, the horizontal straight line represents the principle plane, with the backscatter direction to the right. The small white circles is the averaged sun position

2264.38 nm), and (670 nm and 2264.38 nm), respectively. Figures 6.13 and 6.14 show different vegetation measurements.

By comparing the reflection intensity ratio of soil and vegetation, it is found that the rate of change in the scattering geometry with incident radiation angle is constant in the visible and near infrared regions, in other words, the ratio of the reflected intensity is independent of the scattering angle and the scattering geometry.

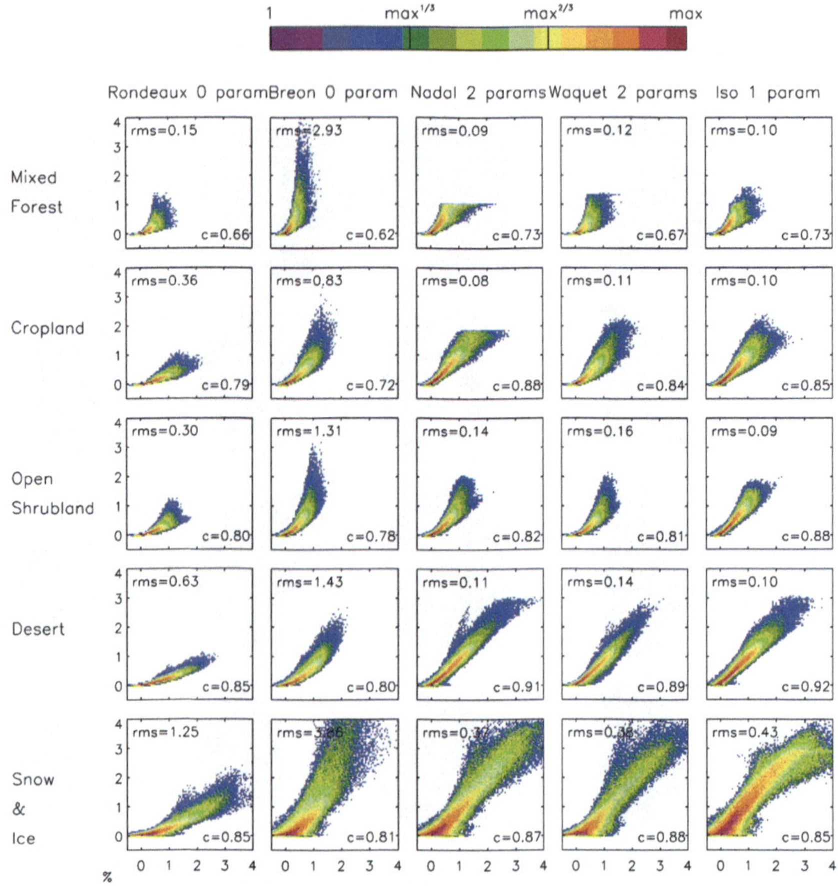

(a) Rondeaux-Bréon model, (b) Bréon model, (c) Nadal model, (d) Waqute model, (e) Maignan model

Fig. 6.12 Fitting effect diagram of five ground objects

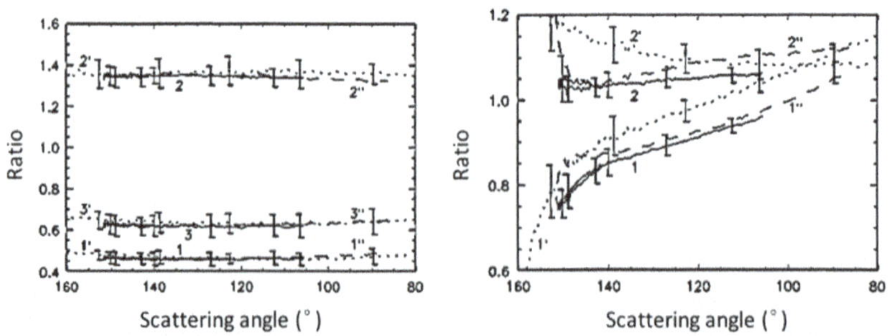

(a) Reflection intensity ratio of the soil (b) The proportion of polarization reflectance

Fig. 6.13 Measured polarimetric reflectance of soil

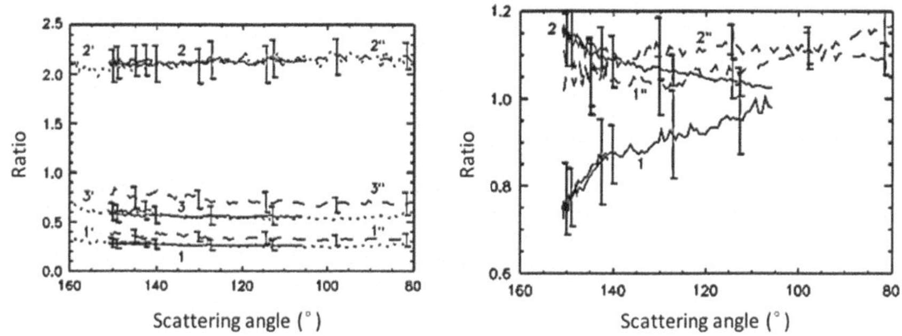

Fig. 6.14 Measured polarimetric reflectance of vegetation

Through the study of the characteristics of polarimetric reflectance and the review of polarized reflectance models used for vegetation and soil, it is discovered that, for polarization remote sensing, atmospheric polarization and ground object polarization are mutual restraints to each other, therefore examination of both areas is essential to improving the overall accuracy.

6.3 Polarimetric Properties of a Single-Leaf

The polarization property of ground objects is mainly related to the complex refractive index, surface roughness, and the orientation of incidence and observation. Among them, the complex refractive index and the surface roughness are associated with plant species (discussed in Sect. 6.3.1), while the orientation of incidence and observation will be analysed in Sect. 6.3.2.

6.3.1 The Relationship Between Polarization by Reflection and Plant Species

Plants vary widely in nature, while their leaves have distinctive shapes, thickness, smoothness, moisture content, and so on, do they have different reflectances? Single leaves of Nasturtium, Pittosporum, and the rubber tree are used to exemplify the comparison of their reflectance ratio. Tested samples of Nasturtium leaf appear greyish-green, smooth, and moderately concave with a radius of curvature of around 50 mm; Pittosporum is streamlined, glossy, green in colour, smooth, and concave along the main veins with a length of around 140 mm and maximum breadth of around 75 mm; a rubber tree leaf is long, oval-shaped, dark green, concave along the main veins (the cross-section is an obtuse V-shape) with a major axis of 140 mm and

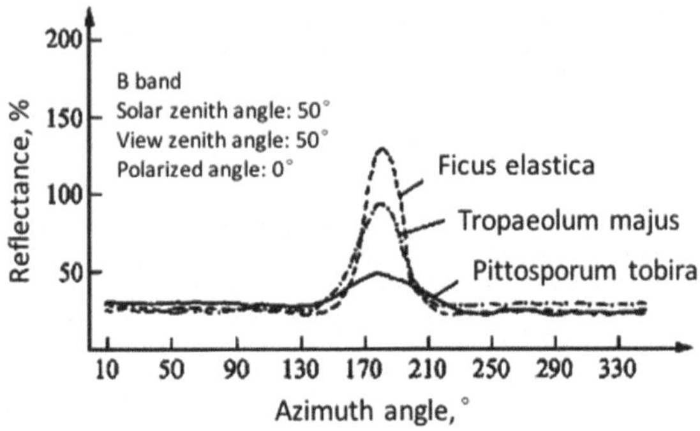

Fig. 6.15 Polarized reflectance of leaves

minor axis of 75 mm. The photometer has A (630–690 nm), and B (760–1100 nm), bands.

As shown in Fig. 6.15, the polarimetric reflectance values of the single leaf of different plants are distinctive. The reflectance spectrum of the single leaves of Nasturtium, Pittosporum, and a rubber tree at an incidence angle of 50°, viewing angle of 50°, azimuth angle of 10°–350°, and a polarization angle of 0° in the B-band is shown in Fig. 6.15. Their shared property is that curves between azimuth angle of 150°–220° show significant changes and the crest appears at an azimuth of 180°, while for the other azimuthal positions the curve is stable. Therefore, the polarization reflectance characteristics in the azimuthal range of 140°–220° are the main focus of the present study.

Using a bidirectional reflectance spectrophotometer to measure the polarization reflectance data of five types of deciduous tree leaves from the forests in Northeast China in early June, the data were analysed to obtain the polarimetric reflectance of a single-leaf of different types to provide a scientific basis for the study and application of polarization and multi-angle remote sensing: this represented a new direction for vegetation remote sensing, especially forest remote sensing (Song et al. 2004).

There were five species of samples acquired on 5 June 2002 from the Changchun Moon Lake experimental district. The poplar leaf is green, with a smooth, glossy surface, large leaf area, and clear veins in the same plane as the mesophyll. The wych elm leaf is long, oval-shaped, light green in colour, with a hairy, but relatively, glossy surface that is convex along the veins. The white birch leaf is light-green in colour, its shape is similar to the leaf of wych elm but is smaller, with a smooth, glossy surface, and unclear veins. The oak leaf is green, with a smooth surface and unclear veins. The purple willow leaf is green, long, oval-shaped with little hair, and its surface is relatively rough with clear concave veins.

From Fig. 6.16, different deciduous leaves have different polarization reflectance ratios. The polarization reflectance curves at azimuthal angle ω in regions 0°–140° and 220°–350° are more stable with smaller differences and diffuse reflection properties; from 140° to 220°, the differences in polarization reflectance characteristics among these leaves are significant, and their reflectance ratio in descending order was: poplar, wych elm, white birch, oak, and purple willow, with all their curves

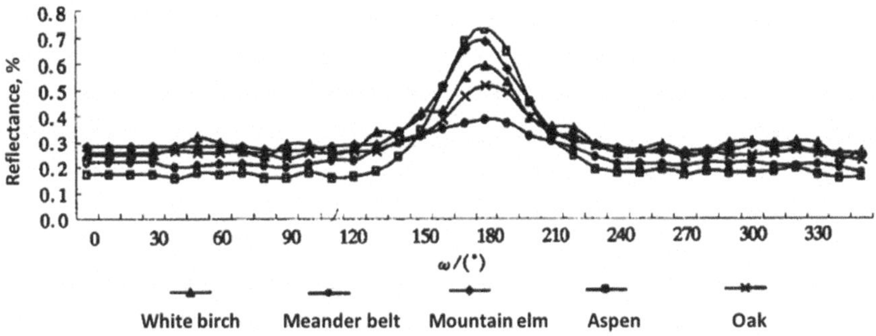

Fig. 6.16 Polarimetric reflectance of different deciduous leaves in the A-band

peaking at an azimuth angle of 180°. When changing the band, and the angle of incidence, the same conclusion was obtained.

Through previous studies, and the analytical processing of data in this experiment, the differences in polarization reflectance ratio at azimuth angles of 140°–220° is smaller, thus insignificant to target recognition and radiation angle correction: in the following analysis, the main focus will be on the polarization reflectance characteristics at azimuth angles of 120°–240° (extending 20° on each side to reflect continuity).

6.3.2 How Polarization Reflectance Is Related to Incidence and Observation Orientations

Figures 6.17 and 6.18 show the reflectance of Nasturtium at incident zenith angles of 30°, 40°, 50°, and 60°, at a viewing angle of 60°, under polarization angles of 0°, 90°, and non-polarising conditions in the A and B bands: at the same viewing angle and different zenith angles of the incident beam, polarimetric reflectance ratio increases with the increase in zenith angle of the incident beam, and reaches its maximum when the polarization angle is 0° and azimuth angle is 180° (i.e., facing the light); however, a peak value also appears at 90° polarization but with a smaller peak amplitude. The peak value of bidirectional polarization reflectance is the arithmetic mean of peak values at 0°, and 90°, polarizations.

Figures 6.19 and 6.20 depict the reflectance curves of a single nasturtium leaf at a zenith angle of the incidence beam of 60°, at viewing angles of 30°, 40°, 50°, and 60°, under polarization angles of 0° and 90°, and non-polarising conditions in different bands. From Figs. 6.19 and 6.20, at the same zenith angle of incident beam, polarimetric reflectance increases with the increase in viewing zenith angle, and reaches its maximum when the azimuth angle is 180°. Also, the polarimetric reflectance value of a single nasturtium leaf changed most significantly between 140° and 220° regardless of whether the viewing angle was fixed while the zenith

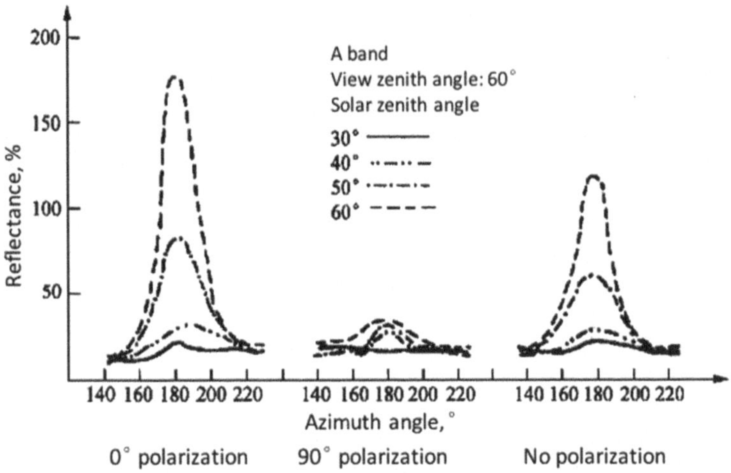

Fig. 6.17 Effect of different zenith angles of the incident beam on the reflectance of Nasturtium leaves (A band)

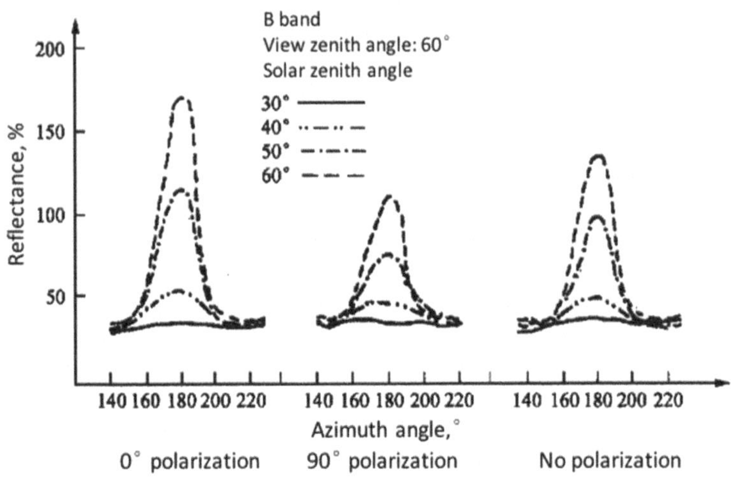

Fig. 6.18 Effect of different zenith angles of the incident beam on the reflectance of Nasturtium leaves (B band)

angle of incidence was changed or vice versa. The peak value occurs at an azimuth angle of 180°, otherwise the values are smaller although not significantly so: a similar trend can be seen in the other curves.

The above conclusion is derived from collated data pertaining to the polarimetric reflectance characteristics of the single nasturtium, but the trends observed are applicable to other plants.

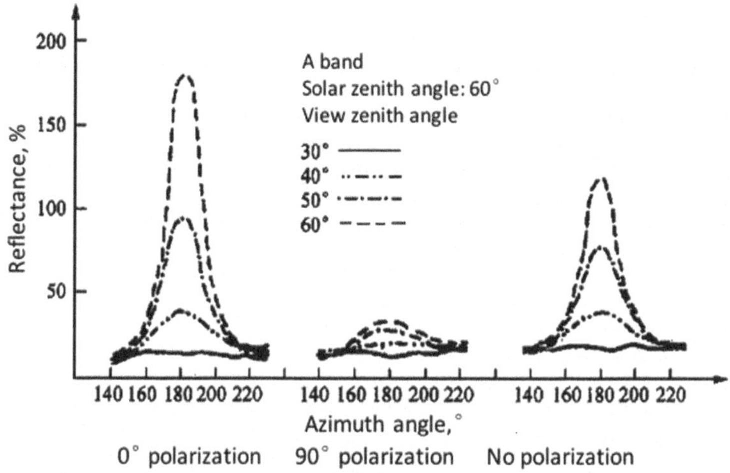

Fig. 6.19 The effect of different viewing angles on the reflectance of nasturtium leaf

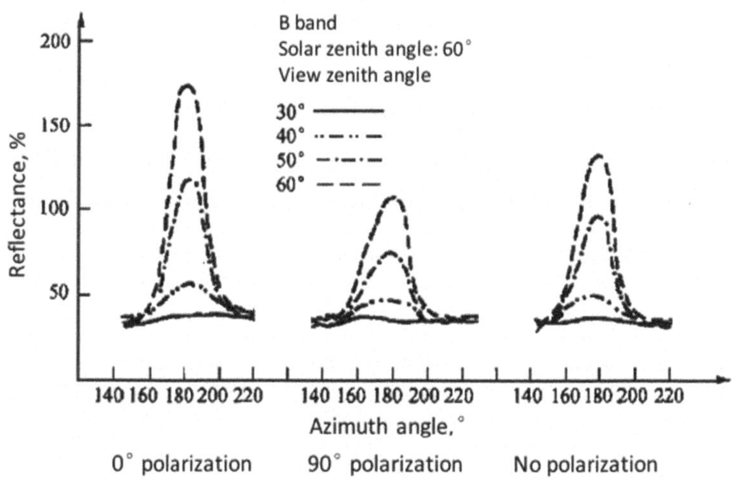

Fig. 6.20 The effect of different viewing angles on the reflectance of pittosporum leaf

From Figs. 6.17, 6.18, 6.19 and 6.20, the polarimetric reflectance measured is correlated to the relative orientations of incident and viewing angles. For the same plant, different relative orientations lead to different polarimetric reflectance (Fig. 6.21).

Figure 6.22 shows the polarization by reflection curve and bidirectional reflection curves of different single leaves of deciduous at angles of incident beam λ of 50° and 60°, a viewing angle α of 50°, and polarization angles of 0° and 90° in the A-band. Through horizontal comparison, the polarimetric reflectance of a single-leaf, in ascending order, is at 90° polarization, bidirectional reflectance, and 0° polarization,

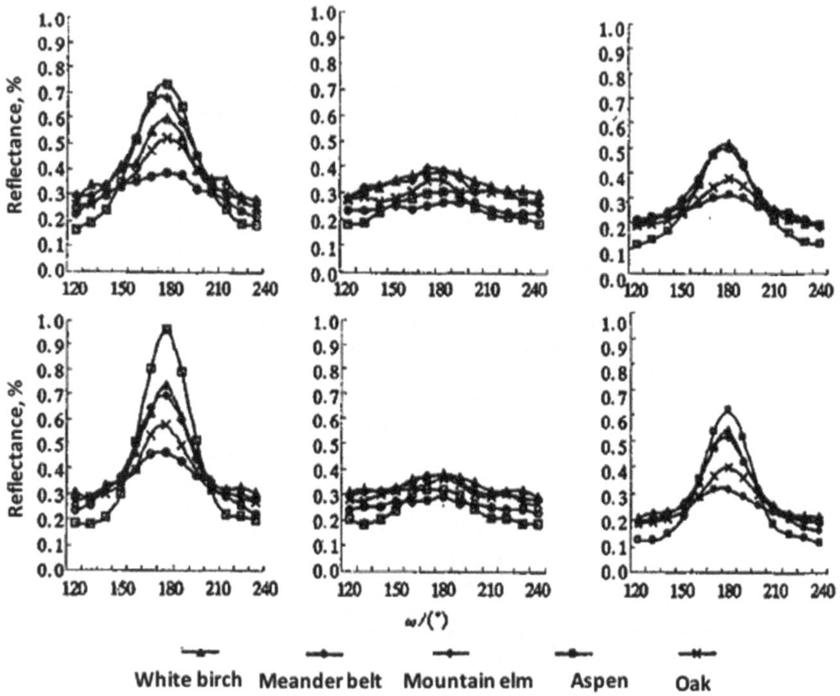

(a) polarimetric reflectance at 0° polarization, (b) polarimetric reflectance at 90° polarization, (c) bidirectional reflectance

Fig. 6.21 Polarization reflectance of different deciduous single-leaf samples at incidence angle of 50° (up) and 60° (down) in the A-band

their relationship being: the value of bidirectional reflectance is half the sum of the reflectance under 0° and 90° polarization plus a small additive constant. Through vertical comparisons, the polarimetric reflectance of different deciduous single-leaves at 90° and bidirectional reflectance increase with the incident angle. When the incident angle exceeds 50°, the polarization reflectance of poplar at 90° polarization exceeds that of wych elm and white birth; for other deciduous single-leaves the order of their polarimetric reflectance does not change. For bidirectional reflectance, at an incidence angle of 50°, the three do not vary to any significant extent; when the incident angle reaches 60°, the reflectance of poplar exceeds that of wych elm and white birch has the largest reflectance. With increases in incident angles, the reflectance of wych elm and white birch are similar and are higher than the polarization reflectance of oak. Purple willow has a low polarization reflectance. At 0° polarization, the polarimetric reflectance of different deciduous single-leaf samples increases with increasing incident angle while the polarimetric reflectance characteristics are subtle and little different is shown among different species (only poplar has a larger curvature). A

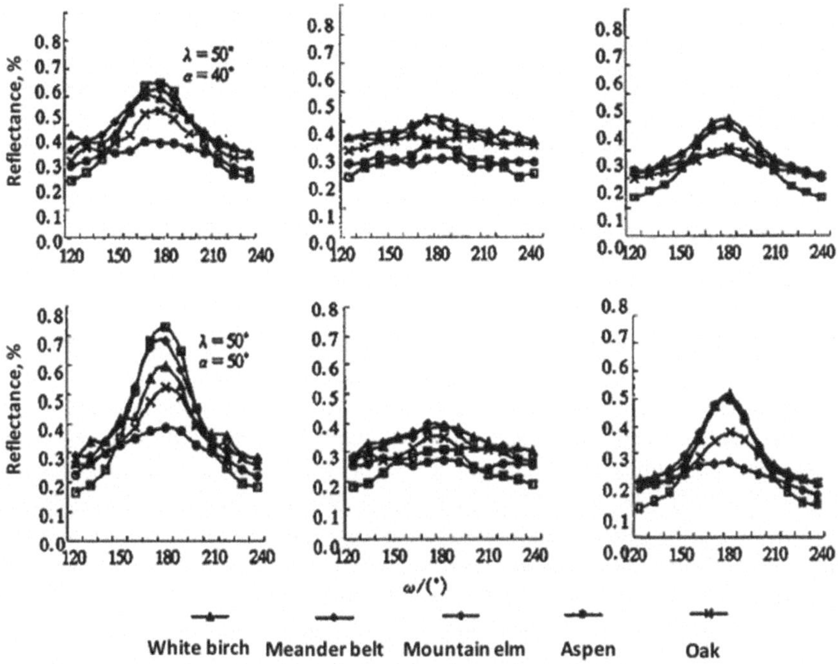

(a) polarimetric reflectance at 0° polarization, (b) polarimetric reflectance at 90° polarization, (c) bidirectional reflectance

Fig. 6.22 Polarimetric reflectance of different deciduous single-leaf samples at incidence angles of 40° (up) and 50° (down) in the A-band

combined view in both horizontal and vertical comparisons shows that peaks appear at an azimuthal angle of 180°.

Figure 6.22 shows the bidirectional reflectance and polarimetric reflectance of different deciduous single-leaf samples at 0° and 90° for polarization angle θ. Through horizontal comparisons, under a fixed condition of the zenith angle of the incident beam, their polarimetric reflectance in ascending order is at 90° polarization, bidirectional reflectance, and 0° polarization. Through vertical comparisons, also under a fixed condition of the zenith angle of the incident beam, their reflectance increases when the viewing angle increases but the rate of change varied among the three. Poplar leaf's reflectance increases the fastest at the biggest curvature, and its polarimetric reflectance exceeds that of the wych elm and white birch when the viewing angle is greater than 40°. For bidirectional reflectance, the polarimetric reflectances of the three are similar and only distinguishable at viewing angles of 40° and 50°. Similar to the situation when the viewing angle is fixed and the incidence angle is changing, at 0° polarization, the polarimetric reflectance of different leaves increases with the viewing angle, but the changes in polarimetric reflectance characteristics are insignificant. The above data analysis indicates that the surface structure of leaves, especially their smoothness, is essential to the values of polarimetric, and bidirectional, reflectances. Poplar leaf is smooth, thus it has a high polarimetric reflectance, and while purple willow leaf is rough and has a lower reflectance: the other two samples lay between these two.

6.4 Polarimetric Reflectance Characteristics of Different Crops

This section focuses on the nature of the polarization to prove, experimentally, the characteristics of radiative transfer. During the tilted observation of corn, sorghum, and wheat canopies, a white colour appears due to specular reflection and not a pure green colour as specular reflection is different from diffuse reflection, and light therefrom does not penetrate to the interior of the leaves where it would otherwise react with the cytochrome, cell wall, and water. Instead, it is due to the reflection from the leaf surface that it can be used as the basis for identifying species. Surface characteristics are related to factors such as: stage of growth, water composition, and temperature (Grant et al. 1987a, b; Goel and Strebel 1984).

Under microscopic examination, surface cuticles are not smooth but full of microscopic features: each cuticle is multi-layered and comprises a wax layer. Under the electron microscope four optical phenomena can be observed, all contributing to the understanding of how the cuticle layer affects light-scattering: the seemingly smooth cuticle is able to perform specular reflection, and the scattering characteristics of acicular structures on the cuticle layer is related to the particle sizes as determined by the Rayleigh and Mie scattering theory. Such scattered light undergoes linear polarization; incident light on the leaf surface and the light transmitted into the leaf are largely scattered by a majority of the leaves due to the different refractive indices of different leaves (Shibayama and Watanabe 2007, 2008).

Leaf blade, and the polarization portion of canopy scattering, are measured to obtain polarization information in the laboratory and from an aircraft: Egan, and his group, found that using airborne sensors to obtain the degree of linear polarization can provide additional information from remote sensing to distinguish between types of measurement. Egan reported the potentially important conclusion that dried leaves usually play a depolarising effect.

Vanderbilt and his group conducted studies on polarization by way of the reflection characteristics of a wheat canopy, and focussed on the relationship between the specular reflection, diffuse reflection, and polarization (Fig. 6.23).

The main region of this study is between 0.48 and 0.72 mm. The degree of linear polarization is calculated using the following formula,

$$R_I = \frac{R_{\max} + R_{\min}}{2} \tag{6.12}$$

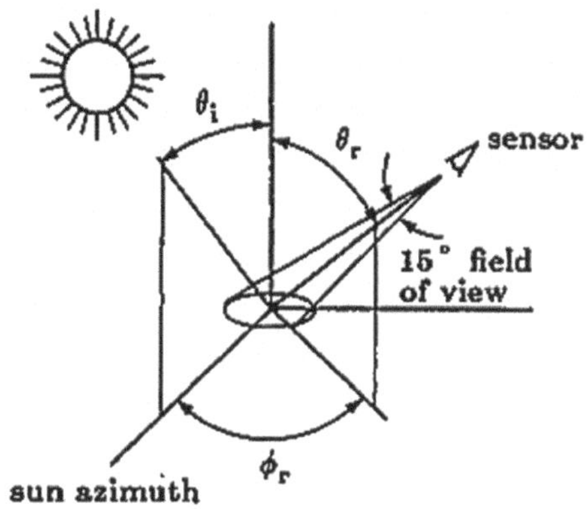

Fig. 6.23 Geometric relationship between measurements

$$R_Q = \frac{R_{\max} - R_{\min}}{2} \tag{6.13}$$

$$Degree\ of\ Linear\ Polarization = \frac{R_Q}{R_I} \cdot 100\% \tag{6.14}$$

From left to right (Fig. 6.24): polarimetric reflectance of canopy; reflection and polarization; diffuse reflection and specular reflection; degree of linear polarization. The third diagram in Fig. 6.24 represents the proportion of polarization in the specular reflection, which is related to the incidence angle and refractivity as depicted in the diagram. For specular reflection, the ratio of specular reflection of polarization and the specular reflection light is equal to the ratio of Stoke's first, and second, parameters. This value is related to the incidence angle and the refractivity of the cuticle of the wheat blade (Fig. 6.25).

A dichroic photometer was used to measure the polarimetric reflectance and bidirectional reflection data of a single poplar leaf in four phenophases: this was later analysed to obtain the relationship between them and their response to different phenophases. This provides a basis for the study of polarization remote sensing and multi-angle remote sensing and initiates a new direction for forest remote sensing.

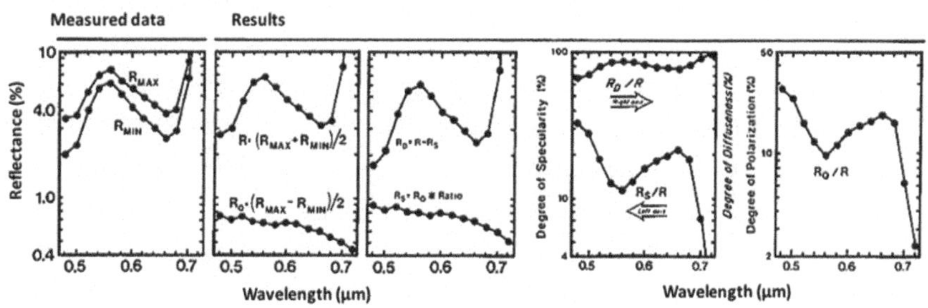

Fig. 6.24 Measured, and calculated, results

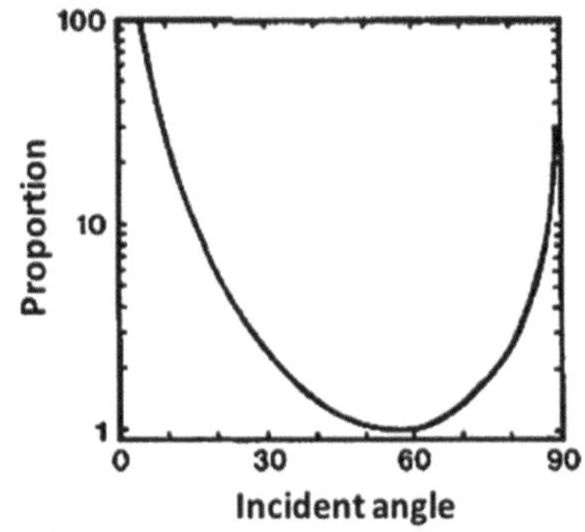

Fig. 6.25 Relationship between incidence angle and the proportion of polarimetric reflection in the entire reflection

The single poplar leaf specimens were collected in the Forest Park in the Moon Lake remote sensing experimental station. Data acquisition occurred on 28 May, 5 July, 12 August, and 16 September 2002. At around 9:30 in the morning, typical leaves of four healthy poplar trees are collected and labelled in the sampling area, and then wrapped with a damp towel before being sealed in a plastic bag and sent to the laboratory an hour later. The colour of the poplar leaf in its four phenophases is yellow-green, green, dark-green, and dark-green: the thickness of the leaves increases with phenophase progression. Leaves of the last two phenophases were similar, having glossy, smooth surfaces with clear veins that are on the same horizontal level as the mesophyll.

Figure 6.26 shows the polarization reflectance curve at a 50° zenith angle of incidence and a 50° viewing zenith angle: the polarimetric reflectance of poplar leaves in different phenophases varies. For azimuthal angles from 0° to 140° and 220° to 350°, the differences are small and diffuse reflection dominates: from 140° to 220°, the differences are more significant and their polarimetric reflectances in descending order are the: fourth, second, first, and third phenophases. Maxima occur at an azimuth angle of 180°. The same conclusion can be obtained by changing the band.

Figure 6.27 depicts the variations in the reflectance of a single poplar leaf, in dif-

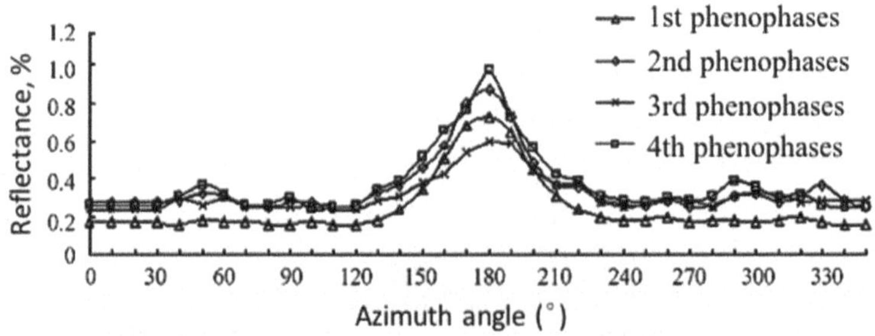

Fig. 6.26 Polarimetric reflectance of poplar leaves in four phenophases at different azimuth angles, under 90° polarization, using the A-band

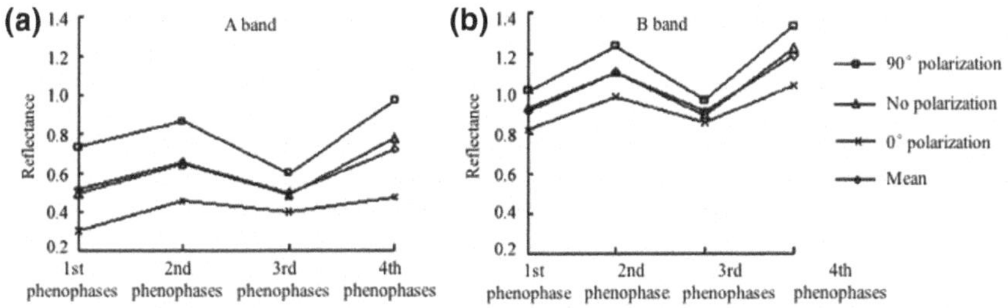

Fig. 6.27 Relationship between polarized reflectance and reflectance in the A (**a**) and B (**b**) bands

ferent phenophases, at an incident zenith angle of 50°, a viewing zenith angle of 50°, and azimuth angles of 180° in both A and B bands, featuring 90° and 0° polarization, bidirectional reflection, and mean values. After comparison, it was found that, in the first three phenophases, the goodness of fit of the mean values of bidirectional reflectance and polarization at 90° and 0° was high, thus under a certain accuracy, polarimetric reflectance data can be used to deduce the bidirectional reflectance data by inversion. For the fourth phenophase, the bidirectional reflectance differs slightly from the mean value due to changes in its internal structure and biochemical composition caused by aging, although the polarimetric and bidirectional reflectance characteristics are decided by surface structure, yet the multiple scattering caused by internal structural features is also a factor contributing to the differences between the four indicators.

A similar experiment is conducted on corn samples at different growth periods. They are divided into four phenophases and samples were collected from Qian Jin farm on the outskirts of Changchun City. The fifth corn leaf is taken from the plant at around 07:00 in the morning, and it is then wrapped in a wet towel and sealed in a plastic bag for preservation. The central vein portion is cut in the laboratory and taped onto the stage, facing upwards and flattened by hand. For the sound seedlings, the leaf is yellowish-green, thin, and glossy; the leaf in the tasselling stage is green, thick, hairy, and rough; the leaf in the milk stage is dark green, rough, and thicker than that in the tasselling stage; the leaf in its mature stage is yellowish-brown and glossy. Measurement is conducted in bands A (630 to 690 nm) and B (760 to 1100 nm) under two conditions: 0° (direction of the transmission axis) and 90° (direction of the extinction axis) polarization.

Figure 6.28 shows the reflectance (ϕ) of corn leaves in the sound seedling stage, tasselling stage, and the mature stage at an incident angle β of 50°, viewing angle γ of 50°, and azimuth angles α from 0° to 350° in the A-band at 0° polarization. Their common attribute is the significant variations shown at azimuthal angles from 140° to 220° and peak values at an azimuth angle of 180°, but the peak value for leaves in the sound seedling stage is much higher than that mature specimens.

The figure shows the reflectance at an incident angle β of 50°, viewing angles γ of 30°, 40°, 50°, and 60°, at 0° polarization in the A-band. Significant changes may be seen in the azimuthal range from 140° to 220° and peak values occur at an

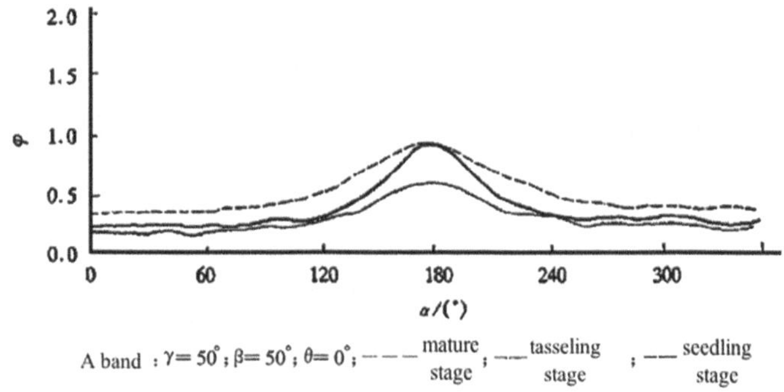

Fig. 6.28 Polarimetric reflectance of corn leaves in different phenophases

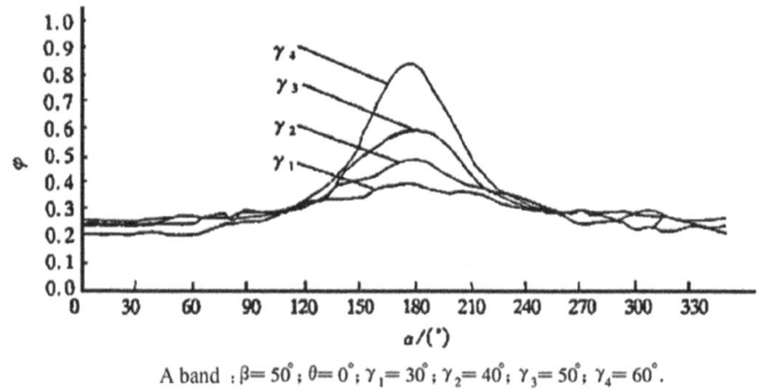

Fig. 6.29 Polarimetric reflectance of corn leaf blades at different azimuth angles in the A-band

azimuth angle of 180° (in all such situations); however in other azimuthal regions the reflectance values are smaller and show little change (Fig. 6.29).

From the studies of reflection, polarimetric reflection, band, and viewing directions of wheat, it is found that the spectral characteristics at a reflectivity of less than 0.5 μm and the red light region of 0.66 μm is caused by absorption of chlorophyll. The maximum value appears at an azimuth angle of 180° (the backscatter direction), while the minimum value of reflectivity appears in the nadir and forward directions (at an azimuth angle of 0°). Differing from reflectivity, a characteristic spectral shape does not appear because of chlorophyll absorption in the information pertaining to polarization by reflection, as shown in Fig. 6.30: its maximum value appears in forward directions (at an azimuth angle of 0°) and negative values appears at the nadir instead. This reflects the fact that polarization reflection did not enter the leaf interior due to specular reflection on the leaf surface.

The measured polarimetric reflectance of wheat before its tasselling stage is greater than that during the tasseling stage, and the polarimetric reflection increases with larger observation angles. This is due to hindrance of the leaf reflection by the

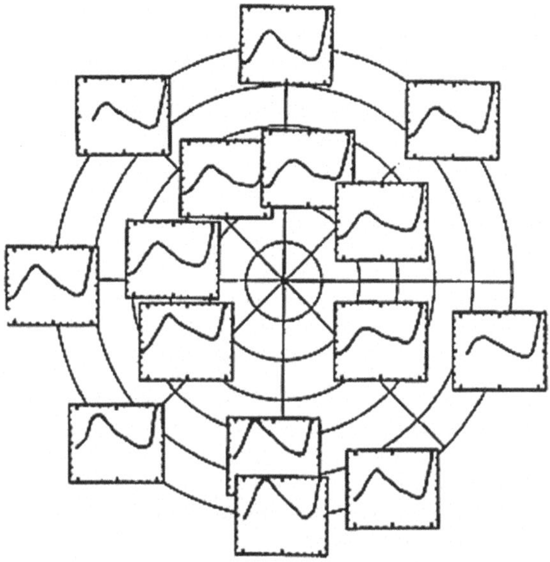

Fig. 6.30 Reflectivity at different azimuth angles, the higher part represents forward directions while the lower part represents backwards directions

tip of the blades entering the specular reflection part of the sensor, thus reducing the amount of polarization reflection information (Fig. 6.31).

By measuring the polarimetric reflectance of the wheat canopy, it was found that the polarimetric reflectance information thereof comes mainly from the blade tip. Specular reflection does not arise or interact with the interior organisation or

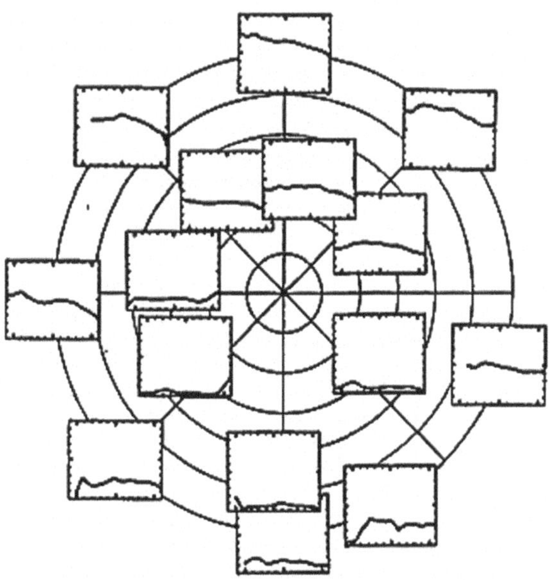

Fig. 6.31 Polarimetric reflectance of different azimuth angle, the higher part represents forward directions while the lower part represents backwards directions

cytochrome. When conducting observations of the canopy when tilted towards the solar incident direction, the polarimetric reflectance information during the tasselling stage is much lower than before it: this shows that polarization measurements can be used as an indication of heading information. In particular, polarization measurement can be an effective method for determining specular and diffuse reflection.

Figure 6.32 shows the degree of linear polarization (DOLP) of needle and shoot samples. Radiation specularly reflected by *Picea koraiensis* and *Pinus koraiensis* needle samples exhibit similar tendencies: the DOLP increases from back to forward-scattering directions, and decreases from strongly (650 nm) to weakly (820 nm) absorbing wavelengths. Its contribution to the radiation reflected near specular directions (VZA of arouns 40°) varies between 71% at 650 nm and 13% at 820 nm. The DOLP of shoots also follows these trends although their magnitudes are reduced.

Radiation specularly reflected from the needle samples exhibits a weak spectral dependency (Fig. 6.33a), as predicted by theory. Their polarized directional-conical

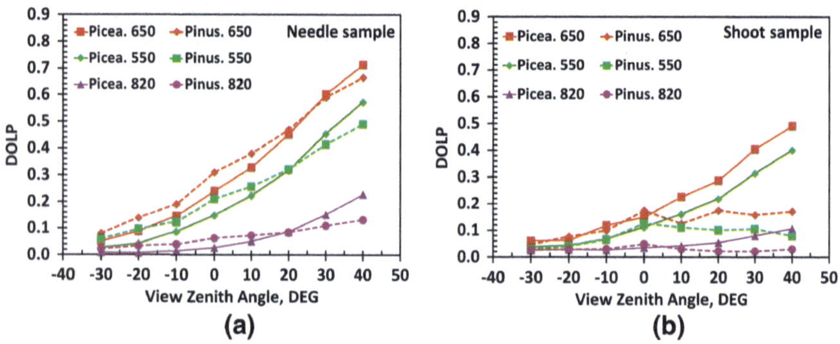

Fig. 6.32 Degree of linear polarization (DOLP) of *Picea koraiensis* (solid lines) and *Pinus koraiensis* (dashed lines) needles (**a**) and shoots (**b**) at green (550 nm), red (650 nm) and near infrared (820 nm) as a function of VZA

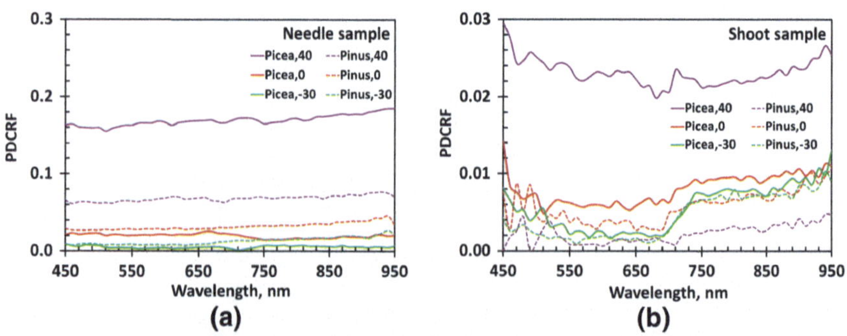

Fig. 6.33 Polarized directional conical reflectance factor (PDCRF) of *Picea koraiensis* (solid lines) and *Pinus koraiensis* (dashed lines) needle samples (**a**) and shoots (**b**) in the spectral interval between 450 and 950 nm for view zenith angles (VZA) of −30°, 0°, and 40°

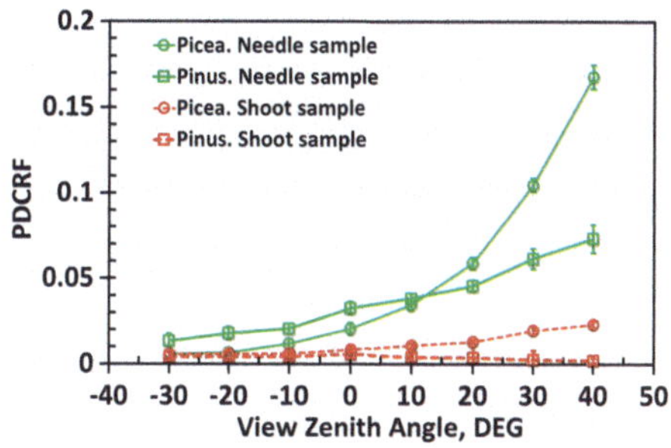

Fig. 6.34 Angular distribution of average PDCRF of needle samples (solid lines) and shoot samples (dashed lines) averaged over 450 to 950 nm. Vertical bars denote ±1 standard deviation

reflectance factor (PDCRF)_ increases from almost negligible values in backscattering directions (VZA < 0°) to about 0.17 when VZA = 40° (Figs. 6.33a and 6.34). The PDCRF of the shoot reflected radiation displays similar behaviour (Figs. 6.33b and 6.34), however, its magnitude is reduced by a factor of about 10. Note that the weak spectral dependency of the surface reflected radiation S_L explains the DOLP spectral behaviour: the contribution of the specularly reflected radiation to the total reflected radiation is small when the diffuse component is large, as in the near-infrared NIR region, and is large when the diffuse component is small, as in the pigment-absorbing blue and red spectral bands.

References

Bréon FM, Tanre D, Lecomte P et al (1995) Polarized reflectance of bare soils and vegetation: measurements and models. IEEE Trans Geosci Remote Sens 33(2):487–99

Goel NS, Strebel DE (1984) Simple beta distribution representation of leaf orientation in vegetation canopies. J Agron Crop Sci 76:800–802

Grant L, Daughtry C, Vanderbiit V (1987a) Polarized and non-polarized leaf reflectances of *Coleus blumei*. Environ Experim Botany 27:139–145

Grant L, Daughtry C, Vanderbiit V (1987b) Variations in the polarized leaf reflectance of *Sorghum bicolor*. Remote Sens Environ 21:333–339

Litivinov P, Hasekamp O, Cairns B et al (2010) Reflection models for soil and vegetation surfaces from multiple-viewing angle photopolarimetric measurements. J Quant Spectrosc Radiat Transfer 111:529–539

Maignan F, Bréon FM, Fédèle E et al (2009) Polarized reflectances of natural surfaces: spaceborne measurements and analytical modeling. Remote Sens Environ 113:2642–50

Nadal F, Breon FM (1999) Parameterization of surface polarized reflectance derived from POLDER spaceborne measurements. IEEE Trans Geosci Remote Sens 37:1709–1718

Rondeaux G, Herman M (1991) Polarization of light reflected by crop canopies. Remote Sens Environ 38:63–75

Shibayama M, Watanabe Y (2007) Estimating the mean leaf inclination angle of wheat canopies using reflected polarized light. Plant Prod Sci 10(3):329–42

Shibayama M, Watanabe Y (2008) Testing polarization measurements with adjusted view zenith angles in varying illumination conditions for detecting leaf orientation in wheat canopy. Plant Prod Sci 11(4):498–506

Song KS, Zhang B, Zhao YS (2004) Study on polarized characteristics of deciduous tree leaves in Northeast of China. J Northeast Normal Univ (Nat Sci Ed) 36(1):88–94 (in Chinese)
Vanderbllt VC (1980) A model of vegetation canopy polarization response. LARS Tech Rep 54
Vanderbilt VC, Grant L, Biehl L et al (1985) Specuar, diffuse and polarized light scattered by two wheat canopies. Appl Opt 24:2408–2418

Part III
The Physical Part of Atmospheric Polarized Remote Sensing

Atmospheric attenuation effects are the largest source of error in remote sensing, and are often experimentally measured but lacking a theoretical basis for their precise characterisation. Atmospheric observations used polarization as a means with which to explore the theory of atmospheric polarization remote sensing, and to improve the establishment of a new theoretical foundation for the atmospheric window. Thus, the important connotations of atmospheric polarized remote sensing are: the nature and physical characteristics of the full-sky polarization pattern (Chap. 7), regional characteristics of atmospheric polarization neutral points and decoupling of ground-atmospheric parameters (Chap. 8), and multi-angle stereoscopic observation chromatography of atmospheric particles under full-sky polarization vector field (Chap. 9).

Chapter 7
Atmospheric Remote Sensing 1: The Nature and Physical Characteristics of the Full-Sky Polarization Pattern

This chapter presents the physical phenomenon of the full-sky polarization effect generated by atmospheric effects, which is measurable and stable, and has become a feasible theoretical basis for quantitative analysis of atmospheric remote sensing defined as the sky polarization pattern, therefore, the systematic observations of, and research into, sky polarization patterns provided possibilities for the systematic, quantitative observation of atmospheric effects. These include: using the vector field pattern to demonstrate the physical characteristics of sky polarization pattern theory; using experimentally refined vector field theory to illustrate the measurement and parameter distribution of the sky polarization pattern; analysis of the factors affecting the sky polarization pattern to deduce its formation conditions; and using the stability of, and changes in, sky polarization patterns to prove the repeatability of this theory.

7.1 Theory of Sky Polarization Patterns

Sunlight could be absorbed and scattered by molecules and aerosol particles in the atmosphere during transmission: this is expressed in terms of the propagation direction, frequency, phase angle, amplitude, polarization, and so on. At a certain time and location, the sky polarization pattern is relatively stable which provides navigation information for some insects such as sand ants and crickets (Jiang et al. 2001; Hartmann and Wehner 1995; Collett et al. 1998) Other than a stable sky polarization pattern, another reason that insects such as sand ants and crickets are able to navigate by the polarization pattern is that they have visual nerve systems that are extremely sensitive to solar polarization distributions. Measurement and analysis of the sky polarization parameter distribution is an important part of this chapter.

L. Yan et al., *Polarization Remote Sensing Physics*, Springer Remote Sensing/Photogrammetry, https://doi.org/10.1007/978-981-15-2886-6_7

7.1.1 Theoretical Analysis of Skylight Polarization

Aerosol particles and molecules are the main components of the atmosphere, in which molecules include: 78% of N_2, 21% of O_2, and 1% of O_3, CO_2, H_2O and others (N_2O, CH_4, NH_3, and so on).

Aerosol particles mainly include: smoke, dust, fog, and small water droplets. Aerosol is a solid-liquid suspension with a solid core (such as dust, pollen, microorganisms, sea salt, etc.) and a liquid exterior, located at a height of 5 km or less. Aerosols, as condensation nuclei for droplets and ice crystals, are an absorber of solar radiation, a body for its scattering, and participate in various chemical recycling processes: as such it is an important part of the atmosphere. Atmospheric aerosols take various sizes with a broad distribution of particles smaller than the radius 0.1 μm which are called the Aitken nuclei, particles between 0.1 and 1 μm are called big particles, and those particles larger than 1 μm are called giant particles. Liquid particles are generally visible and have larger scales, and are called cloud droplets, fog droplets, rain droplets, ice crystals, or hail according to appearances. Their atmospheric concentration is generally several to several hundred thousand per cubic metre, the mass concentration is a few microgrammes per cubic metre, and all such gaseous components constitute the atmosphere.

To discuss the polarization characteristics affecting the radiative transfer process, it is necessary to start with Rayleigh scattering. The Rayleigh scattering model assumes that the wavelength of particles in the physical dimension is much smaller than the wavelength of light ($(x = 2\pi r/\lambda) << 0.1$). The dimension of oxygen and nitrogen molecules is $x < 0.1$, which satisfies Rayleigh's assumptions, while the other floating particles in the atmosphere, such as cloud and other non-gaseous particles, are not included in Rayleigh's assumption, therefore in cloudless conditions the scattering of sunlight is mainly Rayleigh scattering, and the scattered light is mainly linearly polarized, as described by the degree and azimuth angle of polarization. Natural light emitted by the sun can be decomposed into two orthogonal linear polarizations in the horizontal and vertical directions (Fig. 7.1).

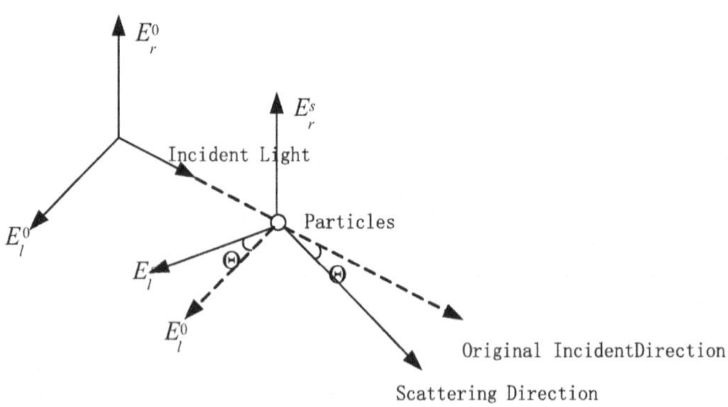

Fig. 7.1 Rayleigh particle scattering model

The scattering process altered the polarization of the light, which can be quantitatively described. In Rayleigh particle scattering, the degree of polarization when an incident beam is natural light is

$$P(\Theta) = \frac{1 - \cos^2\Theta}{1 + \cos^2\Theta} = \frac{\sin^2\Theta}{1 + \cos^2\Theta} \tag{7.1}$$

In Eq. (7.1), when scattering angle Θ is 90° or 270°, the maximum degree of polarization is $P_{\max}$. Ideally the maximum degree of polarization is 100% where the scattered beam is completely linearly polarized. At scattering angles of 0° and 180°, the scattered beam is natural and non-polarized (all other positions represent partially polarized light): however, as multiple scattering, and surface scattering, of molecules and particles have many neutral points in a cloudless sky, a 100% degree of polarization is unlikely to arise.

The radiation transfer equation describes the electromagnetic wave propagation in the medium and its distribution (Guonan 1985; Yao 2006), and the monochromatic radiation transfer equation in a plane-parallel atmosphere is written as:

$$\begin{aligned}\Delta L_\lambda = &-L_\lambda(z,\theta,\varphi)k_{ex,\lambda}\Delta l + \pi F_{\lambda,0}e^{-\sec\theta_0\int_0^\infty k_{ex,\lambda}(z')dz'}\beta_\lambda(z,\theta,\varphi,\theta_0,\varphi_0)\Delta l\\ &+\int_0^{2\pi}\int_0^{\pi} L_\lambda(z,\theta',\varphi')\beta_\lambda(z,\theta,\varphi,\theta',\varphi')\sin\theta' d\theta' d\varphi'\Delta l\\ &+B_\lambda[T(z)]k_{ab,\lambda}(z)\Delta l\end{aligned} \tag{7.2}$$

In the above equation, the incident direction of sunlight is (θ_0, φ_0), while $\pi F_{\lambda,0}$ is the monochromatic irradiance outside the atmosphere, the direction of observation is (θ, φ), the changes Δl_λ of the brightness of a certain wavelength $L_\lambda(z, \theta, \varphi)$ after a certain column is decided by the four items on the right-hand side of the equation, i.e., the reduction, single-scattering contribution, the contribution of multiple scattering, and heat radiation of l_λ passing through column Δl. For the shortwave band, heat radiation is not considered.

Equation (7.2) can be altered to:

$$\mu\frac{dL_\lambda}{d\delta} = L_\lambda - J_\lambda \tag{7.3}$$

In which source function

$$\begin{aligned}J_\lambda = &\frac{\widetilde{\omega}_{0,\lambda}}{4\pi}\left[\pi F_{\lambda,0}e^{-\delta/\mu_0}p_\lambda(\delta,\Theta_0) + \int_0^{2\pi}\int_{-1}^{1} L_\lambda(\delta,\mu',\varphi')p_\lambda(\delta,\Theta')d\mu' d\varphi'\right]\\ &+(1-\widetilde{\omega}_{0\lambda})B_\lambda[T(\delta)]\end{aligned} \tag{7.4}$$

where Θ_0 and Θ' represent the angles between the observation directions from (θ_0, φ_0) to (θ', φ') and then to (θ, φ) respectively, where δ is the optical thickness, $\mu = \cos\theta$ and $p_\lambda = (\delta, \Theta)$ are phase functions (Θ is the angle between the direction of incident sunlight and the direction of scattered light, or the scattering angle), and $\widetilde{\omega}_{0\lambda}$ is the single scattering albedo.

Equation (7.3) is the simplified equation for radiative transfer in a plane-parallel atmosphere. The source function J_λ includes single scattering, multiple scattering, and heat radiation.

As the polarization process alters the polarization state of light, scattered light in the sky possess certain polarization distribution. The following will only consider the air molecules and particularly the single scattering to derive the law of polarization distribution of sky scattered light.

If only single scattering is considered, then

$$J_\lambda = \frac{\widetilde{\omega}_{0,\lambda}}{4\pi}\pi F_{\lambda,0}e^{-\delta/\mu_0}p_\lambda(\delta,\Theta_0) \tag{7.5}$$

Using the radiation transfer function we solve

$$L_\lambda(\delta,\mu,\varphi) = \int_0^\delta J_\lambda(\delta')e^{-(\delta-\delta')/\mu}\frac{d\delta'}{\mu} \tag{7.6}$$

Let phase function p_λ be invariant with height, and solve the single scattering of radiation transfer equation

$$L_\lambda(\delta,\mu,\varphi) = \frac{\widetilde{\omega}_{0,\lambda}}{4\pi}F_{\lambda,0}p_\lambda(\Theta_0)\frac{\mu_0}{\mu_0-\mu}\left(e^{-\delta/\mu_0}-e^{-\delta/\mu}\right) \tag{7.7}$$

The atmospheric molecular scattering phase function is

$$p(\Theta) = \frac{3}{4}\left(1+\cos^2\Theta\right) \tag{7.8}$$

Two polarization components of the scattered light can be written as

$$L_r(\delta,\mu,\varphi) = \frac{\widetilde{\omega}_0}{4}F_0\frac{3}{4}\frac{\mu_0}{\mu_0-\mu}\left(e^{-\delta/\mu_0}-e^{-\delta/\mu}\right) \tag{7.9}$$

$$L_l(\delta,\mu,\varphi) = \frac{\widetilde{\omega}_0}{4}F_0\frac{3}{4}\cos^2\Theta\frac{\mu_0}{\mu_0-\mu}\left(e^{-\delta/\mu_0}-e^{-\delta/\mu}\right) \tag{7.10}$$

Defined by the degree of polarization, the degree of polarization distribution of the sky is:

$$P(\Theta) = \frac{1-\cos^2\Theta}{1+\cos^2\Theta} = \frac{\sin^2\Theta}{1+\cos^2\Theta} \tag{7.11}$$

When $\Theta = 0$, the degree of polarization of scattered light $P = 0$, which is natural light; when $\Theta = \pi/2$, $P = 1$ that is linearly polarized light, while it is partially polarized light for all other angles.

Figure 7.2 shows the angular distribution diagram of the degree of linear polarization when incident light is non-polarized. The diagram is symmetrical corresponding to the direction of incident beam, while the scattered light in the forward and backward directions remains non-polarized, yet at a 90° scattering angle the scattered light is completely polarized. In other directions, scattered light is partially polarized with a degree of polarization between 0 and 100%.

Equation (7.8) demonstrates that the degree of polarization distribution of scattered skylight is the same as that of the scattered light per unit volume of air. (In Mie-scattering, the DOP only relates to degree) Under the assumption of single scattering, it is easy to solve and thus easier to qualitatively analyse some problems, but as it overlooks multiple scattering, the reality is different.

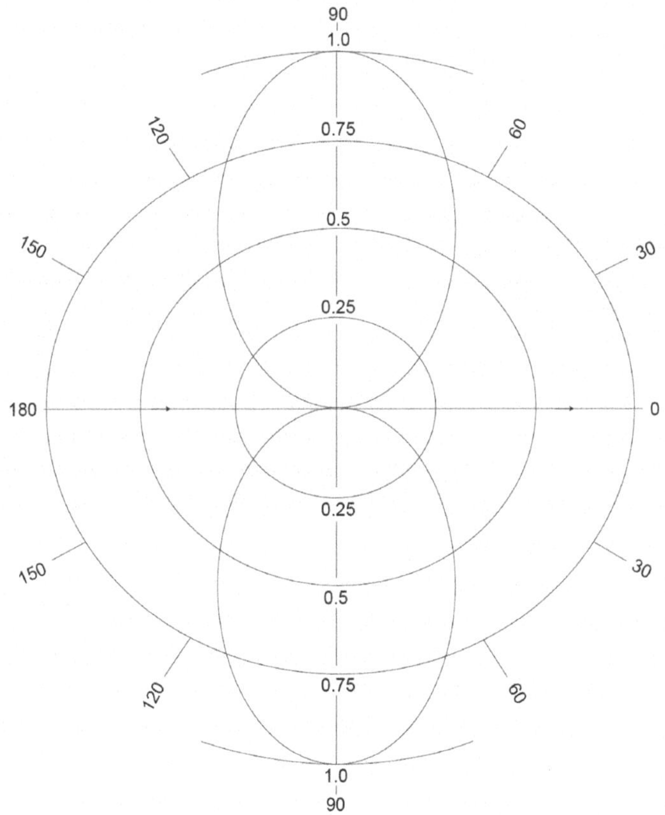

Fig. 7.2 Distribution of the linear degree of polarization (single scattering)

7.1.2 Mathematical Model for the Sky Polarization Distribution

In astronomy, the celestial sphere is a geocentric, imaginary sphere of infinite radius: all celestial bodies regardless of their distance from Earth will be projected onto the celestial sphere. Moons and stars are moving along their paths on the celestial sphere. To determine the position of celestial bodies is to determine the positions of celestial bodies on the celestial sphere. In addition to using the well-known Cartesian coordinate system to locate the Sun, the Moon, and other celestial bodies, the following coordinate systems: the ecliptic coordinate system, the first equatorial coordinate system, equatorial coordinate system, and horizontal coordinate system (Yang and Zhang 1988; Sheng et al. 2003) are used.

This book establishes the sky polarization model diagram using a horizontal coordinate system. In the horizontal coordinate system, the horizontal surface where the observer is located extends infinitely to the big circle intersecting with the celestial sphere and becomes a horizon circle, and the numerous great circles perpendicular to the horizon circle passing through zenith and nadir form the Horizon Meridional, while small circles parallel to the horizon circle are horizon parallels. Azimuth and elevation angle are used to indicate the position of celestial bodies in the horizontal coordinate system. The azimuth angle is the angle between the Horizon Meridional and the upper branch of the meridian passing through the celestial bodies, letting the upper branch of the meridian (south direction) be the starting point, moving towards the east and then north as negative, and moving towards west and then north as positive; if the sun is in the east, its azimuth angle is $-90°$, and $\pm 180°$ when it is in the north. The altitude angle is the angle between the line connecting the celestial centre, the observer, and the horizon circle. Taking the horizon circle as a starting point, the direction towards the zenith is positive while that towards the nadir is negative. When the centre of a celestial body is directly incident on the meridian, its azimuth angle and the hour angle are both 0°, and that is also the celestial transit time.

In the horizontal coordinate system illustrated in Fig. 7.3, S represents the Sun, OP represents the direction of observation in the sky, and Z refers to the zenith. From Rayleigh scattering theory it is known that the test beam's E-vector direction of vibration is perpendicular to the plane formed by the sun S, the ground observation point O, and the observation direction OP. The polarization azimuth angle φ is defined as the angle between the sky observation point P's E-vector vibration direction and the meridian (arc ZP) after passing point P. Point P's E-vector vibration direction is perpendicular to PS, and $\varphi = 90° - \angle ZPS$.

In spherical triangle ZPS, it is known by the law of cosines of spherical triangles, that

$$\cos\theta = \cos\left(\frac{\pi}{2} - h_p\right)\cos\left(\frac{\pi}{2} - h_s\right) + \sin\left(\frac{\pi}{2} - h_p\right)\sin\left(\frac{\pi}{2} - h_s\right)\cos(A_s - A_p)$$

This can be simplified to:

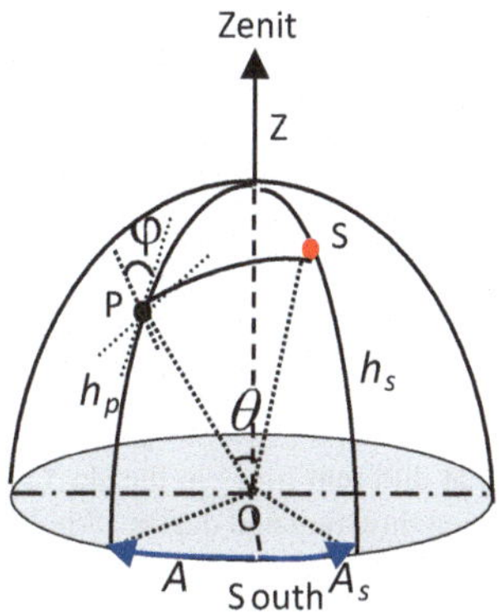

Fig. 7.3 E vector polarization information in P

$$\cos\theta = \sin h_p \sin h_s + \cos h_p \cos h_s \cos(A_s - A_p) \tag{7.12}$$

where θ is the angle between direction of incident beam OS and the observation direction OP, where h_s is the solar altitude angle, A_s is the solar azimuth angle, h_p is the altitude angle of the observation point P in a spherical sky, and A_p is the azimuth angle of observation point P. The southerly direction represents an azimuth angle of 0° while the westerly direction represents an azimuth angle of 90°.

In spherical triangle ZPS, it is known by the law of cosines of spherical triangles, that

$$\frac{\sin\angle ZPS}{\sin(90^\circ - h_s)} = \frac{\sin(A_s - A_p)}{\sin\theta}$$

Then:

$$\frac{\sin(90^\circ - \varphi)}{\sin(90^\circ - h_s)} = \frac{\sin(A_s - A_p)}{\sin\theta}$$

This can be simplified to:

$$\cos\varphi = \frac{\sin(A_s - A_p)}{\sin\theta}\cos h_s \tag{7.13}$$

From Eqs. (7.11) and (7.13), the degree of polarization P of the observed beam and the polarization azimuth angle φ can be calculated as shown in Fig. 7.4.

Figure 7.4 shows the polarization pattern of the sky at different solar altitude angles, where the direction and width of black dotted lines represent the polarization direction and degree of polarization, respectively, and the shaded part represents the observations of the sky visible to the naked eye. It can be inferred that the polarization pattern of the sky is closely associated with the solar altitude angle (or time), and

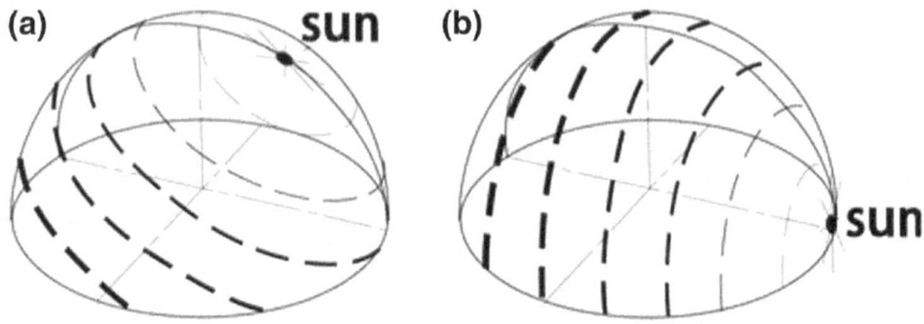

Fig. 7.4 The polarization pattern at different times in the sky (The direction and width of black dotted lines represent the polarization direction and degree of polarization, respectively)

under different solar altitude angles there are different polarization patterns, similar differences are also shown in the intensity of polarized light and its direction. At a lower solar altitude angle, the intensity of polarization is comparatively large, in other words the intensity of polarization in the morning or at dusk is stronger than that in the afternoon. Also, from the figure, there are two obvious lines of symmetry on the polarization pattern, one being the big circle that is at a 90° angular distance from the anti-sun or sun spot, which has the largest degree of polarization; another being the connection between the anti-sun or sun spot, which has a degree of polarization that is maximised at 90°.

As compared to the degree of polarization, polarization angle is also an important component of the polarization pattern, and biological creatures mainly depend on the distribution of polarization angles to obtain directional information. In cloudy weather, scattering and polarization by clouds and cloud atmospheric particles on sunlight cause differences between the polarization patterns obtained in cloudy and sunny weather. These polarization patterns are decided by the properties (dielectric attributes and structural characteristics) of the atmosphere, therefore, it is concluded that the physical characteristics of the distribution of sky-polarized light can be demonstrated with a vector field pattern to reflect its changes in time, space, and in other regards.

7.2 Measurement of the Sky Polarization Pattern and Its Parameter Distribution

The sky polarization pattern needs to be characterised by actual measurement, thus this section will describe how to conduct the static and dynamic measurement of full-sky polarization patterns, and ascertain the distribution of the full-sky polarization pattern by actual measurements.

7.2.1 Stocks' Parameter Distribution and Measurement Studies

Figure 7.5 shows the original image of the same region in the sky viewed with the aid of polarisers. During the photography, a clockwise rotation of 45° was applied to the polariser after each frame-capture.

Figure 7.6 represents the images obtained after intensity-density slicing of the original image.

The brightness value of any part of the images after adding the polariser can be written as $0.5I_g + I_{\max}\cos^2\alpha$ (where α is the angle between the polarising direction of linearly polarized light and the direction of the polariser). Then the difference between a point on any two images is $I_{\max}(\cos^2\alpha - \cos^2\beta)$, which explains the existence of $I_{\max}$. Huge differences in the images also imply that $I_{\max}$ is quite large, therefore, in the scope of full-sky measurement, the degree of linear polarization cannot be neglected.

	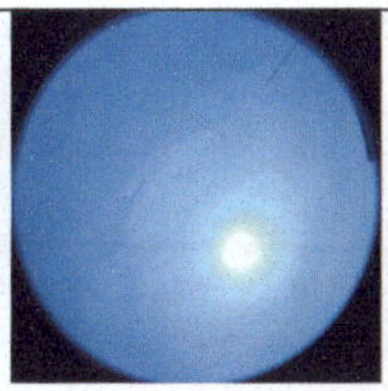	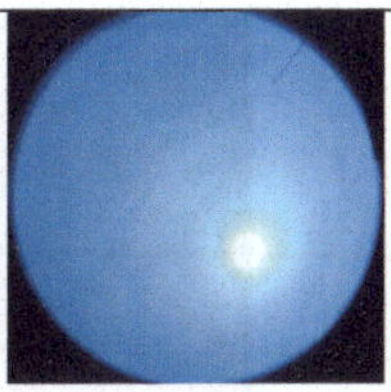	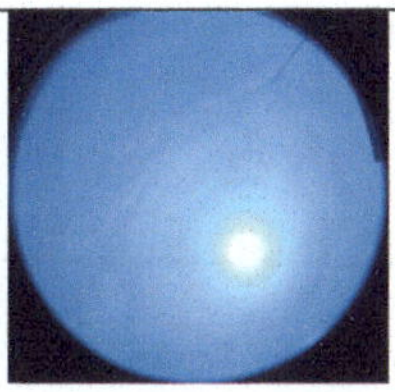
Original image	Image after 45° clockwise rotation	Image after 90° clockwise rotation	Image after 135° clockwise rotation

Fig. 7.5 The original image

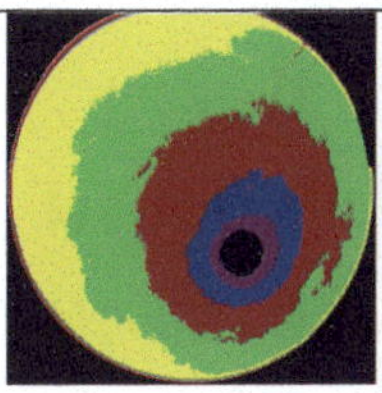	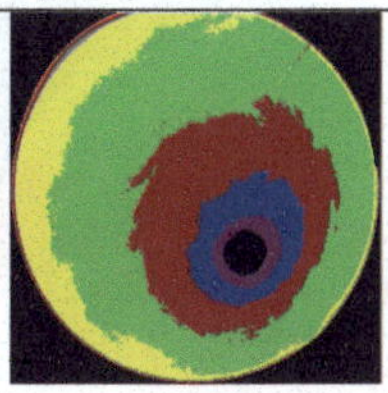	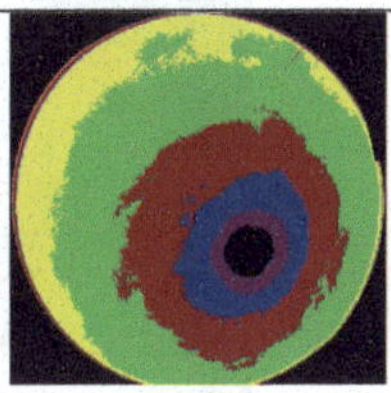	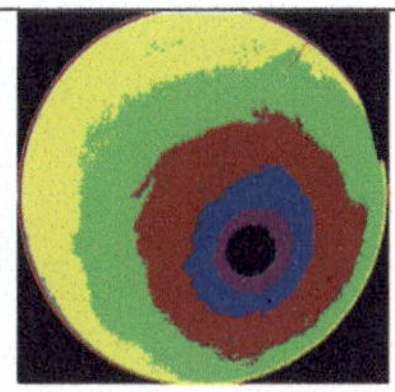
0° intensity-density slice	45° intensity-density slice	90° intensity-density slice	135° intensity-density slice

Fig. 7.6 Intensity-density slice images

Figures 7.7, 7.8, and 7.9 represent the distribution of Stocks' parameters I, Q, and U calculated from the above four images. After comparison it can be discovered that the distribution of the I-component is largely similar to the distribution of the intensity-density slice of the original image.

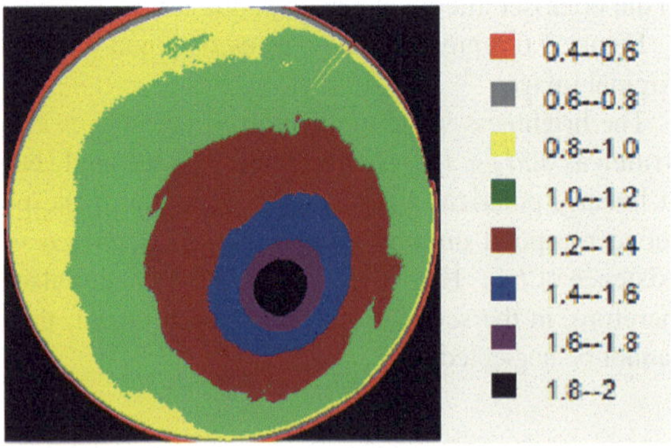

Fig. 7.7 Distribution image (I-component)

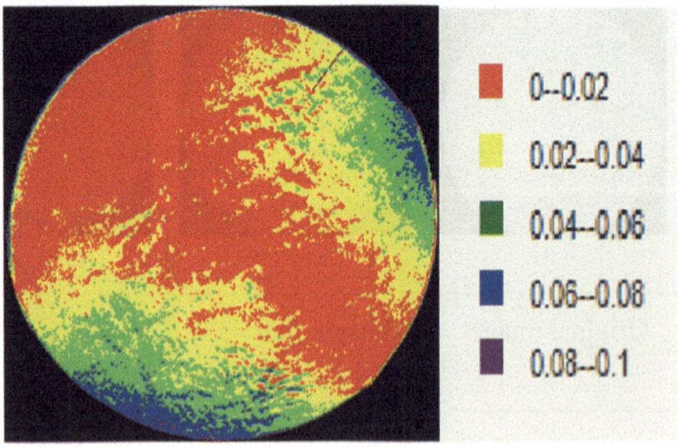

Fig. 7.8 Distribution image (|Q| component)

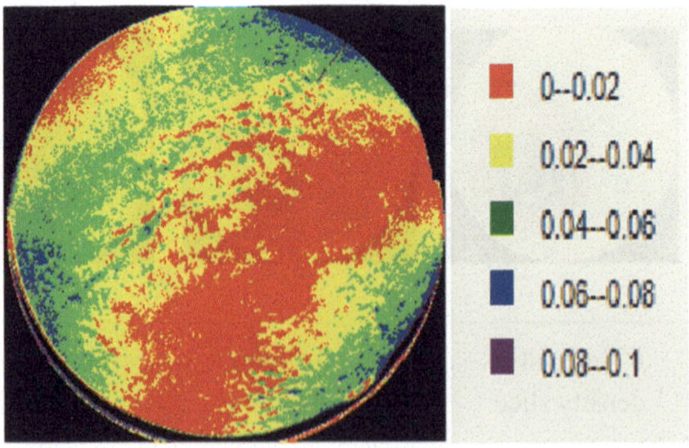

Fig. 7.9 Distribution image (|U| component)

The total light intensity is $I = I_g + I_{\max}$, and this explains why the distribution of the I-component is similar to the distribution of the intensity-density slice of the original image. After analysing the distribution of Stocks' parameters Q and U, it is found that the distributions of these two components have no obvious pattern. To further analyse this, more measurements were undertaken.

Figures 7.10 and 7.11 show the processed distribution of components Q and U captured on 7 May 2008 at 4 pm on the fifth floor of the Remote Sensing Building of Peking University. General comparisons reflect random distributions. The only known fact is that the component values of Q and U near the sun are relatively small.

Based on components Q and U:

$$Q = I(0°) - I(90°), U = I(45°) - I(135°)$$

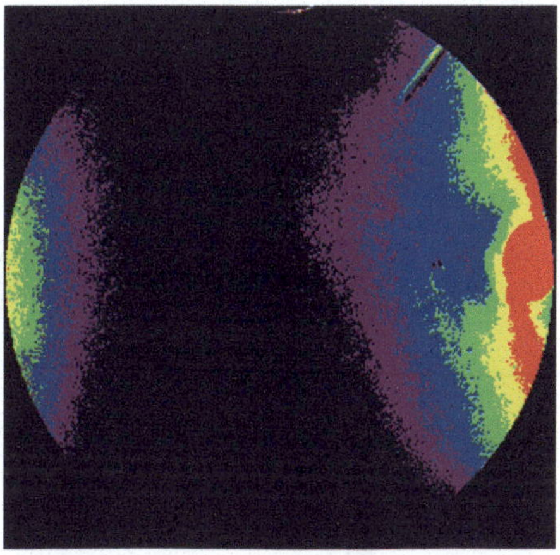

Fig. 7.10 Distribution image (|Q| component) simulation

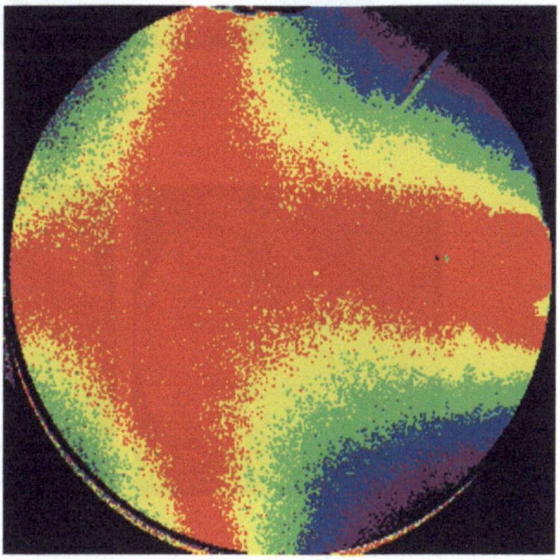

Fig. 7.11 Distribution image (|U| component) simulation

The values of Q and U are interchangeable with the variation of the original orientation of the polariser, thus there is no pattern to their distributions. At the spot near the sun, the images captured at different angles have similar luminance values due to over-exposure, thus the component values of Q and U are smaller.

7.2.2 *Study of the Measurement of Degree of Polarization and the Polarization Azimuth Angles*

Figure 7.12 shows the distribution diagram of full-sky degree of polarization. The red region shows over-exposure of the sun and its surroundings, and the degree of polarization generally follows an annular distribution. Theoretically, the full-sky polarization pattern is symmetrical. The symmetrical distribution can also be found on the images taken at other times.

Figure 7.13 shows a processed distribution of degree of polarization captured on 21 May 2008 at 1 pm on the fifth floor of the Remote Sensing Building. The two images are generally symmetrical, which is consistent with prevailing theory.

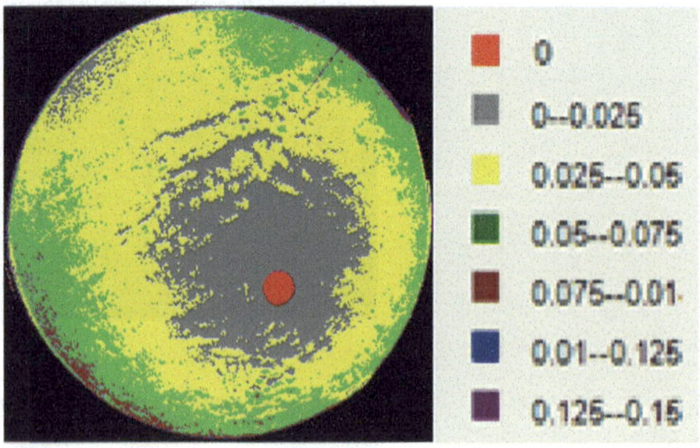

Fig. 7.12 Degree of polarization distribution

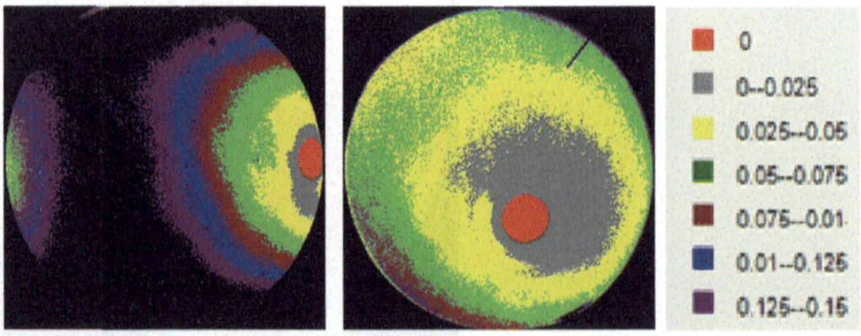

Fig. 7.13 Degree of polarization distribution and image

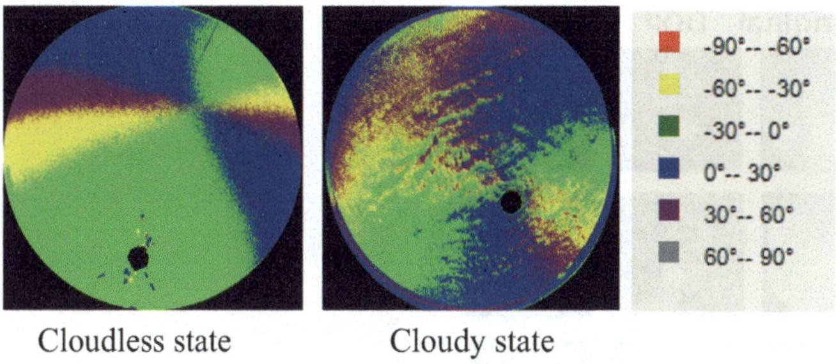

Fig. 7.14 Polarization azimuth angle distribution

The polarization azimuth angle of a point refers to the angle between the polarising direction of the point's linearly polarized beam and the initial direction of the polariser. Figure 7.14 shows the distribution diagrams of polarization azimuth angles when the sky is both clear and cloudy.

As shown in the figure, the distribution of polarization angles under clear sky condition is more regular, while under cloudy conditions the polarization angle is affected by multiple scattering due to the cloudy weather and thus its distribution has changed, but in general the symmetry remains the same.

To obtain the distribution pattern of full-sky polarization light within a certain time frame, continuous observation experiments were conducted on 30 May 2008 from 11 am to 5 pm on the fifth floor of the Remote Sensing Building. The full sky is captured once every hour. Figure 7.15 shows a comparison of the original image, degree of polarization, and polarization azimuth angle of the images taken.

The sky is clear with nearly no clouds on the day of this experiment. Minor processing was required for the degree of polarization distribution at 17:00 to remove the impact of clouds on the left-hand part of the image.

From the experimental results the following law of full-sky polarization distribution can be concluded.

Firstly, the full-sky degree of polarization distribution is not unalterable but changes with the changing position of the Sun, as inferred from the figures.

Also, the degree of polarization follows an annular distribution. When the solar altitude angle is higher when nearing noon, an annular distribution can be observed from the sky, and in time another annular distribution can be discovered, as a result the full-sky degree of polarization distribution is the superposition of the two annular distributions.

In general, the distributions of polarization azimuth angles also follow a certain distribution as inferred from the figures, and this is related to the position of the Sun. The experimental results in this chapter can further demonstrate the feasibility and effectiveness of using vector field theory to express sky polarization distribution theory.

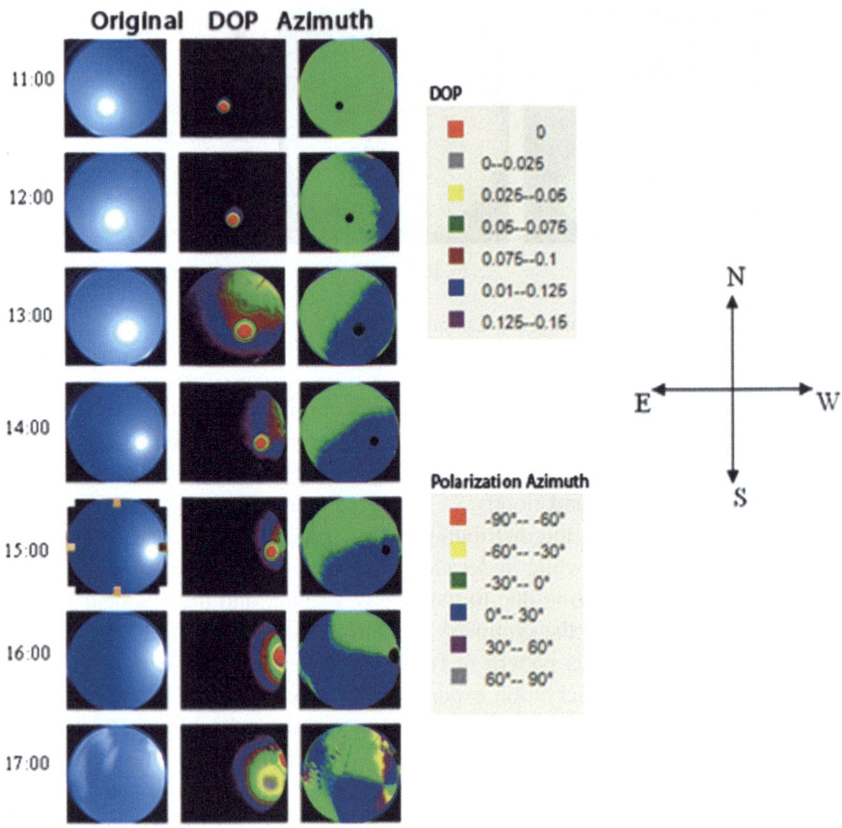

Fig. 7.15 Continuous experimental renderings

7.3 Impacts on the Degree of Polarization of the Sky Under Different Conditions

Polarization navigation is a relatively new research topic: biological creatures are able to navigate using sky polarization distribution because they have a visual nerve system that is extremely sensitive to changes in sky polarization direction, and also because of the existence of a relatively stable atmospheric polarization pattern in the sky, however, visible wavelengths, weather conditions, and the position of the Sun all have certain impacts on the sky polarization distribution (Waterman 1954, 1955; Ivanoff 1958; Ivanoff and Waterman 1985; Fang and Ning 2006; Waterman 2006). The applicability and impacts of such conditions on navigation are the main focus of this section. As the sky polarization distribution is mainly described by the parameters of the degree of polarization and polarization azimuth angle, thus using experimental methods to analyse the impact of these two parameters is the main focus in this section.

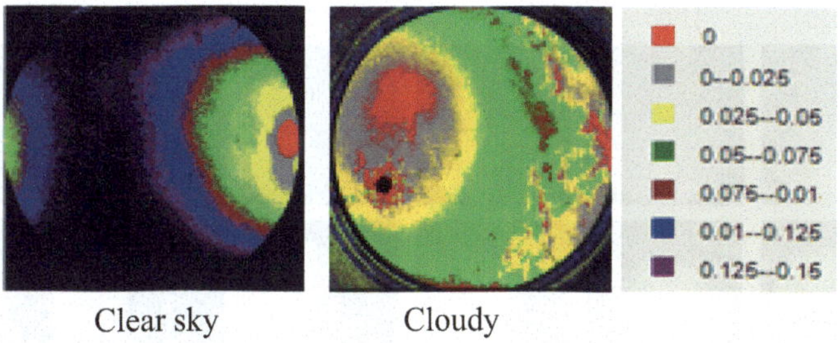

Fig. 7.16 The distribution of the degree of polarization under different weather conditions

7.3.1 Measured Results of Different Weather Conditions

As the sky polarization distribution is closely associated with atmospheric particles scattering sunlight, a comparison was conducted between the data collected on both clear, and cloudy, days. Figure 7.16 (left) shows the degree of polarization image measured under clear and cloudless weather conditions, while on the right can be seen the image taken during cloudy weather. The images are created using the sky polarization pattern measurement model introduced in the previous section.

The red region in the figure is the point where the atmospheric degree of polarization of the sky is zero. From the image taken under cloudless condition, it can be seen that the atmospheric degree of polarization demonstrates a regular annular distribution around this point, and its degree of polarization increases when expanding outwards, reaching its maximum at the black region and declining thereafter. The image on the right (captured in cloudy conditions) is similar to the image captured under cloudless conditions with an annular distribution, but its annularity is irregular with a green region indicating a lower degree of polarization on the bottom-right-hand corner; the main colours in the image are green and yellow, suggesting that its degree of polarization is much lower than that under cloudless conditions due to the depolarization effects caused by multiple-scattering in the atmosphere, which reduced the degree of polarization of the image.

7.3.2 Measurement Results of Different Solar Altitude Angles

To obtain the distribution of the full-sky degree of polarization and polarization angles within a period of time, continuous observations were conducted. Figure 7.17 shows images taken on 1 September 2008 from 10:30 to 15:00 on the full-sky degree of polarization of the Zhongguancun area once every hour. The location of experiment is on the fifth floor of the Remote Sensing Building. The shape and size of the polarization distribution at different solar altitude angles are varied, when the solar altitude angle is high, nearing noon, two annular distributions can be observed in the

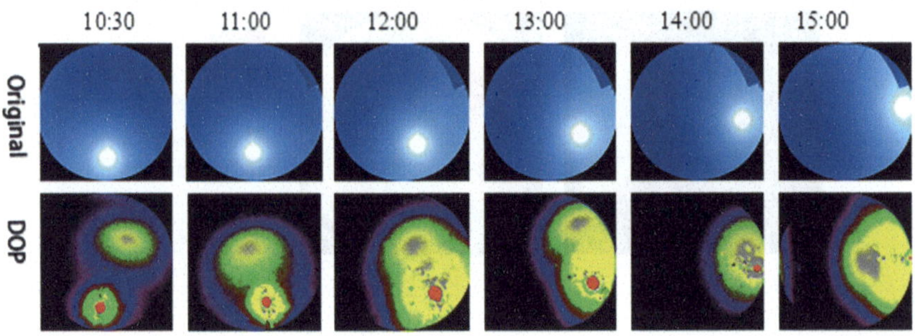

Fig. 7.17 Variation of degree of polarization with the change in solar elevation angle in a clear sky

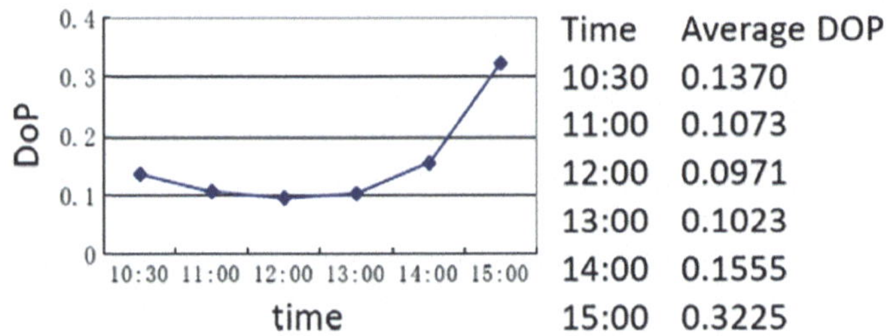

Fig. 7.18 Relationship between the average degree of polarization and the solar elevation angles

sky, and the degree of polarization of the entire sky is the superposition of the two annular distributions. In time, the other annular shape disappears.

Brightness statistics are collected from the degree of polarization images in Fig. 7.17 and the mean values are recorded in Fig. 7.18.

In Fig. 7.18, the mean value of the degree of polarization distribution decreases from 10:30 to 12:00 and reaches its minimum value at noon and the mean value increases thereafter. From the results in Figs. 9.17 and 9.18, it is concluded that the shape and size of the sky's degree of polarization distribution changes with the variation of solar elevation angle. The lower the solar elevation angle, the higher the mean value of the degree of polarization, and vice versa (this result is consistent with prevailing theory).

7.3.3 Measurement Results of Different Observation Bands

Figure 7.19 shows the sky's degree of polarization distributions at the violet, blue, and red bands under sunny and clear weather conditions. The three images taken in each row are taken at around the same time of the day, with a difference of less than 1 min between them.

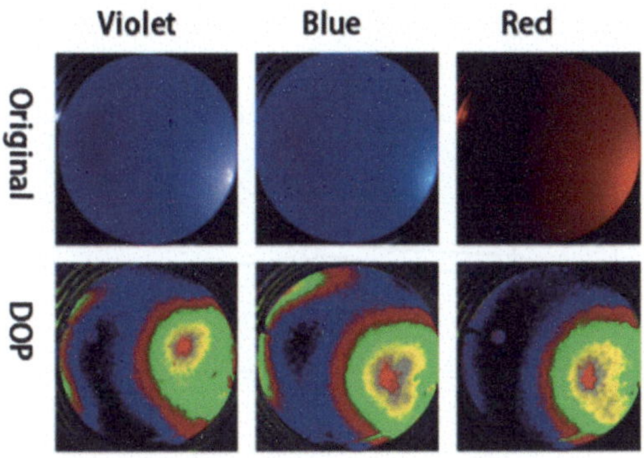

Wavelength (nm)	Mean value of the degree of polarization
400	0.136
450	0.152
650	**0.360**

Fig. 7.19 The degree of polarization distribution of the sky under clear weather conditions in different bands

In Fig. 7.19, the sky's degree of polarization distributions in the purple, blue, and red bands under clear sky conditions is slightly different: the maximum values of the degree of polarization occur in different regions, which means that the visible light spectrum affects the degree of polarization distribution. Through the calculation of mean values of the degree of polarization at different bands, it can be discovered that the degree of polarization increases with the increase in visible light wavelength under clear weather conditions, and the maximum value occurs in the red band.

Figure 7.20 shows the sky's degree of polarization distributions in the purple, blue, and red bands under cloudy and rainy weather conditions. The three images taken in each row are captured at around the same time of the day, with a difference of less than 1 min between them. Through the calculation of mean values of the degree of polarization at different bands, it can be discovered that the degree of polarization under cloudy and rainy condition is smaller than that in clear sky conditions, and the mean value of the degree of polarization in the red band decreases significantly, suggesting that the longer the wavelength the greater the depolarization effect.

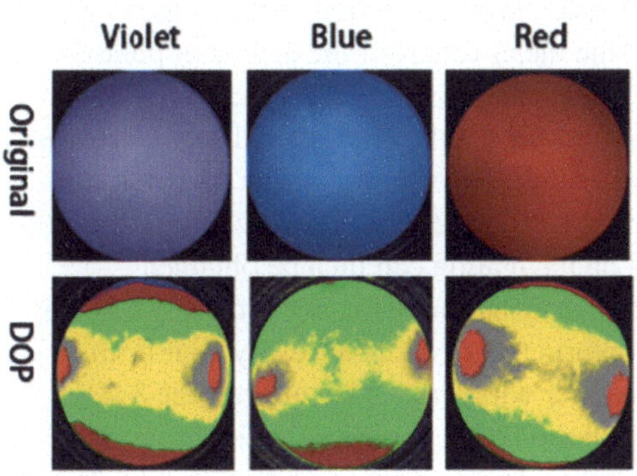

Wavelength (nm)	Mean value of degree of polarization
400	0.109
450	0.136
650	**0.102**

Fig. 7.20 The degree of polarization distribution of the sky under cloudy and rainy weather conditions in different bands

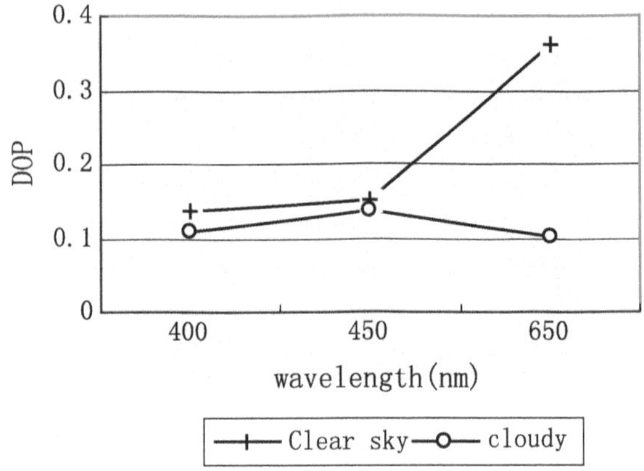

Fig. 7.21 Mean value of the image's degree of polarization in different bands

To illustrate the intensity of the sky's degree of polarization observed in different bands, the mean values of the degree of polarization images of Figs. 7.19 and 7.20, in different bands, are retrieved. The statistical results are as follows: the mean values of the degree of polarization of the clear sky in the violet, blue, and red bands are 0.136, 0.152, and 0.360 respectively, while the mean values of the degree of polarization of cloudy and rainy sky in the three bands are 0.109, 0.136, and 0.102. As shown in Fig. 7.21, the degree of polarization of clear sky increases with longer wavelengths, while the degree of polarization of cloudy and rainy sky has its maximum value in the blue band and is lower in the violet and red bands.

In Fig. 7.21 the average degree of polarization values of the three bands under cloudy weather conditions is smaller than the values under clear weather conditions, which means that in cloudy weather, the intensive depolarization effects of the multiple scattering induced by atmospheric particles caused the degree of polarization of the entire image to be lowered. As for animals navigating with polarized light, they can employ the distribution of sky polarization as their navigation information source as long as the degree of polarization exceeds the lower limits of their polarization sensitivities. Studies have proven that successful navigation requires the degree of polarization to be at least 5–10%, and the mean values of the degree of polarization of both clear and cloudy days in the experiments above are larger than this value, suggesting that the sky polarization distribution under the above conditions can be used as an information source for polarization-based navigation. Thus, if a polarization angle sensor were designed to be able to detect a degree of polarization distribution change of more than 5%, information on navigation can then be obtained

In addition, the degree of polarization distribution of the sky is closely associated with the observation bands. Full sky capturing was conducted every hour from 10 am to 4 pm on 22 January 2010 (sunny weather with wind at Beaufort force 3–4) in the Zhongguancun area to observe the distribution of sky polarization in different bands. Brightness statistics of all the polarization distribution images are then collected and their means are calculated (Table 7.1). For a more intuitive view of the relationship

Table 7.1 Relationship between the mean values of degree of polarization and the observation band

Time	Degree of polarization (red)	Degree of polarization (blue)	Degree of polarization (violet)
10:00	0.1439569	0.12530330	0.07400164
11:00	0.09539123	0.07921814	0.03204104
12:00	0.09601002	0.08403128	0.02421901
13:00	0.09074238	0.08914048	0.03130787
14:00	0.10416720	0.08769692	0.01189412
15:00	0.08474455	0.07150745	0.010632205
16:00	0.09664912	0.08455271	0.04264618

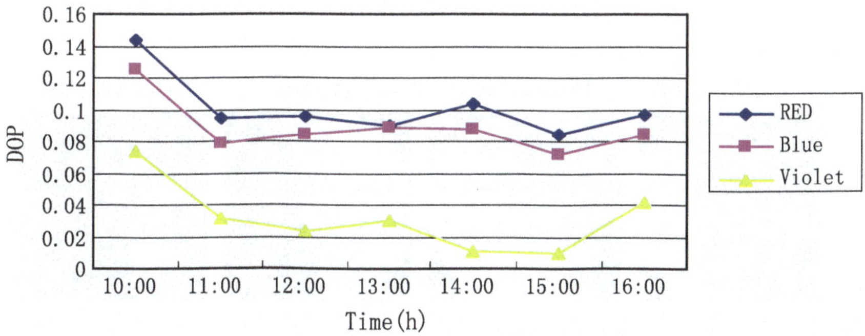

Fig. 7.22 Relationship between the mean values of degree of polarization and the solar altitude angle

between the degree of polarization and the observation bands, data from Table 7.1 are transformed into the diagram shown in Fig. 7.22.

The degree of polarization distribution of the sky is closely associated with the observation bands (Fig. 7.22). The longer the wavelength of visible light, the higher the mean value of the degree of polarization and vice versa. Also, the mean value of the degree of polarization at the same observation bands decreases gradually from 10 am to 12 pm, and reaches its minimum value at the largest solar elevation angle (12 pm), and, in time, the degree of polarization increases gradually, decreasing again between 2 pm and 3 pm: this is slightly different from the theoretical prediction possibly due to the influence of atmospheric turbulence induced by strong winds. From the data in Table 7.1, atmospheric turbulence exerts a strong influence on short waves and a weaker influence on long waves, implying that, under clear and sunny weather conditions, long waves are more suitable for navigation.

The above measurements suggest that the degree of polarization on clear days will decline with decreasing wavelength, but on windy and cloudy days the depolarization effects of short waves are obvious and the degree of polarization will decrease as a consequence, while shorter wavelengths will not be affected. Considering the factors above, the blue band is the most suitable for navigation.

7.4 The Applicability of Different Polarization Angles

How are different polarization angles applicable to navigation? The distribution of skylight polarization under different weather conditions is measured to obtain the distribution of skylight polarization under four weather conditions: sunny and clear, shaded by trees, slightly cloudy, and cloudy (Fig. 7.23).

A mean value is obtained for each of the category in Fig. 7.23 to form Table 7.2, in which, although weather conditions have significant effects on the degree of polarization in the sky, the mean values of the degree of polarization are all greater than

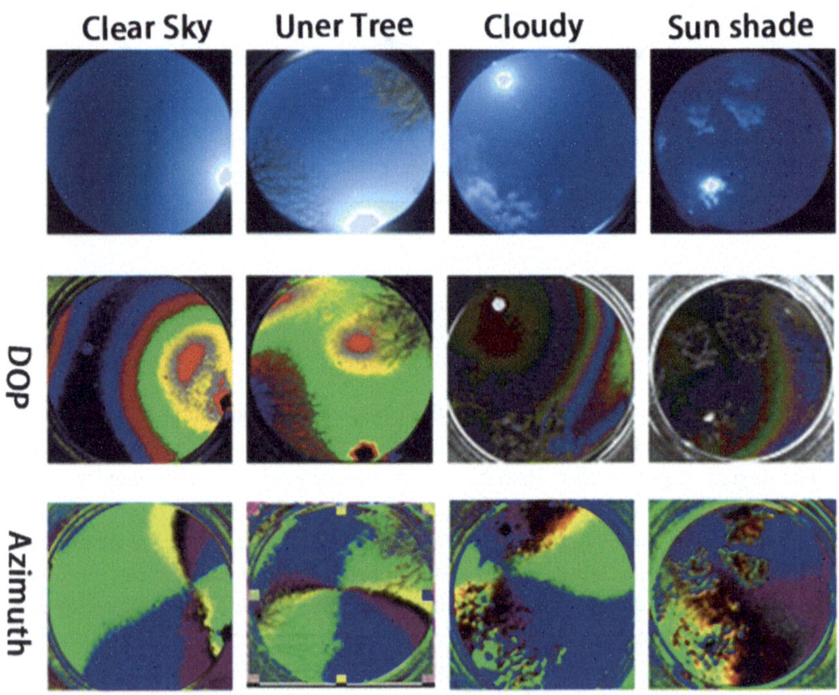

Fig. 7.23 Sky polarization distribution under different conditions

Table 7.2 Mean DOLP of different weather

Weather condition	Sunny and clear	Shaded by trees	Slightly cloudy	Cloudy
Mean degree of polarization	0.2087	0.1257	0.1320	0.1562

10% satisfying the requirements for navigation. Also, the distributions under different weather conditions follow a certain rule, thus navigation information can be retrieved from them.

7.4.1 Measurement Results in Different Weather Conditions

Biological creatures can use skylight polarization to navigate, relying on their sensitive polarization visual system and ability to process information regarding polarization angles. Thus, the analysis of skylight polarization angle under different conditions is extremely important. Figure 7.24 shows the skylight polarization distribution observed under different weather conditions. As the skylight polarization distribution is inseparable from the scattering of sunlight by atmospheric particles, comparative analysis is conducted of their distribution on sunny and cloudy days. On the left of Fig. 7.24 is the image of the full-sky polarization angle in sunny weather, while on the right is that during cloudy weather. The full-sky polarization image is created by using the aforementioned model of sky polarization patterns.

As shown in the image, in sunny weather the skylight polarization angles surrounding the point of convergence (Fig. 7.24) follow a regular symmetrical distribution, similar to the image in cloudy weather in which the sky polarization angle distribution is irregularly symmetrical, the angles do not follow a gradient with cruciform appearances of uneven sizes (unclear boundaries and overlapping at the yellow and purple regions). The image is mainly blue or green, which means that the sizes of its polarization angles differ from those in clear and sunny weather due to depolarization effects caused by atmospheric multiple scattering that reduces all polarization angles. Yet, the characteristic polarization angle distribution being symmetrical to the solar meridian does not change, so that information for purposes of navigation can be obtained from it.

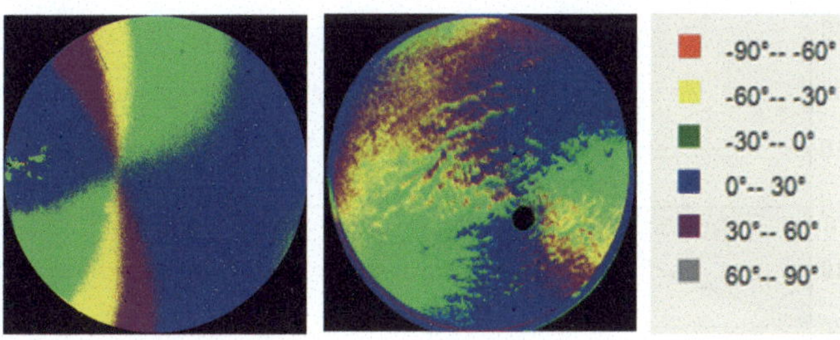

Sunny and clear Cloudy

Fig. 7.24 The distribution of polarization angle under different weather conditions

7.4.2 Measurement Results of Different Solar Elevation Angles

To measure the distribution of the full-sky polarization angle, continuous observational experiments were conducted. Figure 7.25 illustrates the captured full-sky polarization distribution in the Zhongguancun area at hourly intervals from 11 am to 4 pm on 1 September 2008, the venue of the experiment being the fifth floor of the Remote Sensing Building, Peking University. Individual examination of the images of polarization angle distribution reveals that this distribution is similar to that in Fig. 7.24, but the shape and size of the distributions differ at different solar elevation angles. In conclusion, with the change in solar elevation angles, the full-sky polarization angle distribution rotates around the zenith.

Extracting data along the solar meridian direction of each image in Fig. 7.25, it is discovered that, after processing, there is an ascending and then descending change in polarization angle at a certain point on the image as shown in Fig. 7.26, and that satisfies the theory in that the sky polarization pattern revolves around the zenith.

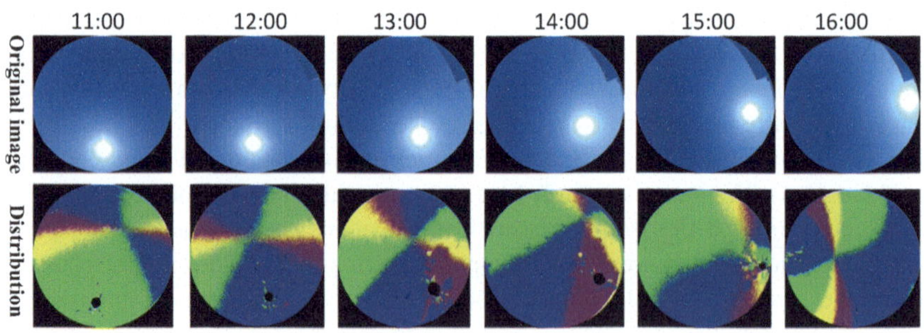

Fig. 7.25 The change in solar elevation angle with polarization angle in clear sky

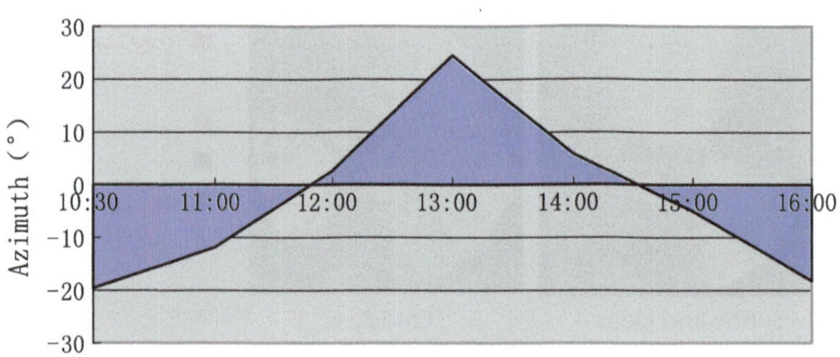

Fig. 7.26 The change in polarization angle at a certain point in a clear sky

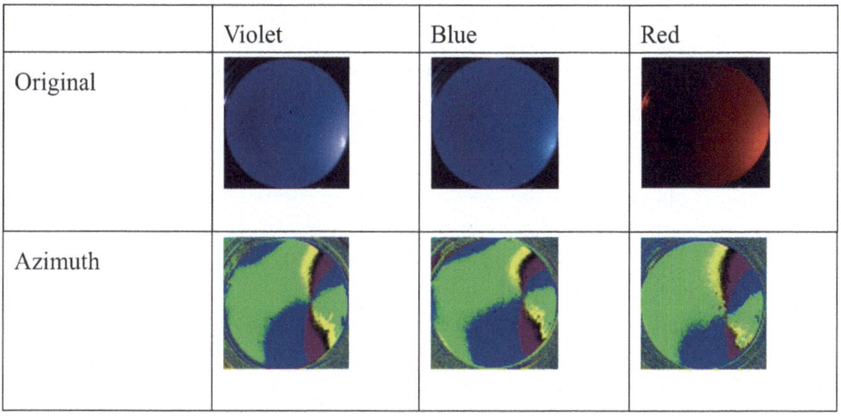

Fig. 7.27 Sky polarization azimuth distribution under clear sky in different bands

7.4.3 Measurement Results of Different Observation Bands

Figure 7.27 depicts the sky polarization distribution in three bands (violet, blue, and red) under sunny and clear weather conditions. The three images in each row are captured at about the same time, with less than 1 min between them.

All three images have almost the same polarization angle distribution that revolves around the same point in a regular pattern as shown in Fig. 7.27. This is consistent with the result of the model.

The distribution in the solar meridian direction is extracted as shown in Fig. 7.28, and the direction of the solar meridian in different bands is generally the same and is stable. Combined with the distribution of the degree of polarization data, navigation information can be obtained from the blue band under clear weather conditions.

Figure 7.29 illustrates the distribution of sky polarization angles in three bands (violet, blue, and red) under shading from trees in clear weather. The three images in each row are captured at about the same time, with less than 1 min between them.

From Fig. 7.29, it can be found that the polarization angle distributions in the violet, blue, and red bands under shading from trees in clear weather are generally the same, revolving around a point in a regular pattern that is similar to the result predicted by the model. In the violet and red bands, the distribution of polarization angles has a larger green region, demonstrating that the polarization angles are smaller in these two bands. Albeit so, the position with the largest linear polarization does not change, thus stable information for navigation can be obtained under shading from trees.

The distribution in the solar meridian direction of Fig. 7.29 is extracted as shown in Fig. 7.30, and the direction of the solar meridian of different bands is generally the same and stable. Combined with the distribution of the degree of polarization, navigation information can be obtained from the blue band under clear weather conditions.

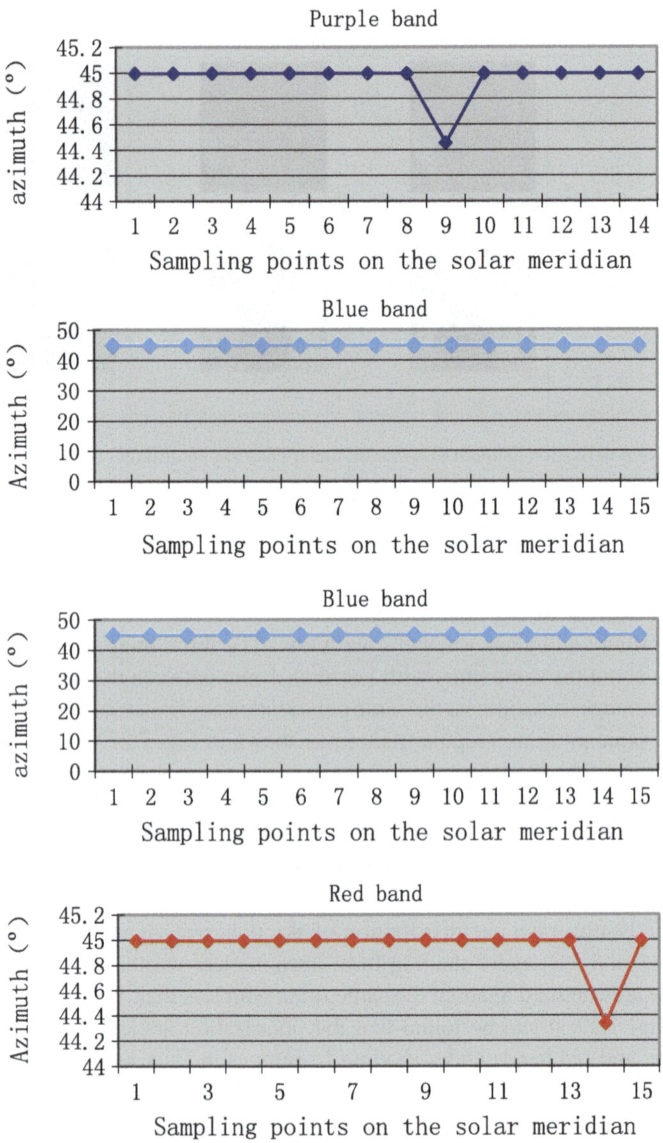

Fig. 7.28 Relationship between polarization angle and bands under clear weather conditions

From Fig. 7.29, it can be found that the polarization angle distributions in the violet, blue, and red bands under cloudy conditions are generally the same, but are slightly different from that of the clear sky. Most of the distributions lie in the blue and green region reflecting the fact that the polarization angles in cloudy weather are smaller than those in sunny weather.

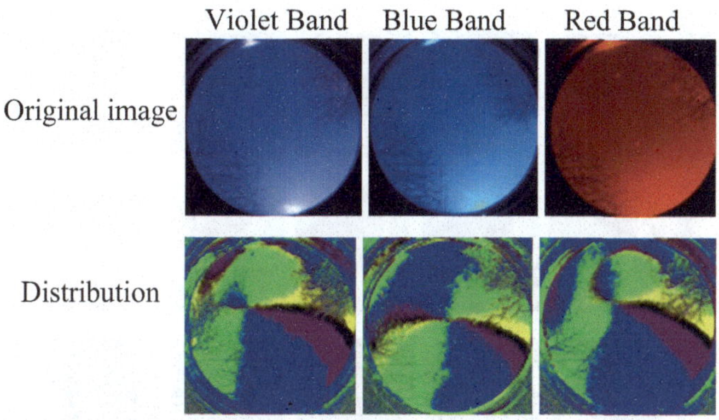

Fig. 7.29 Distribution of sky polarization angles in three bands under shading from trees

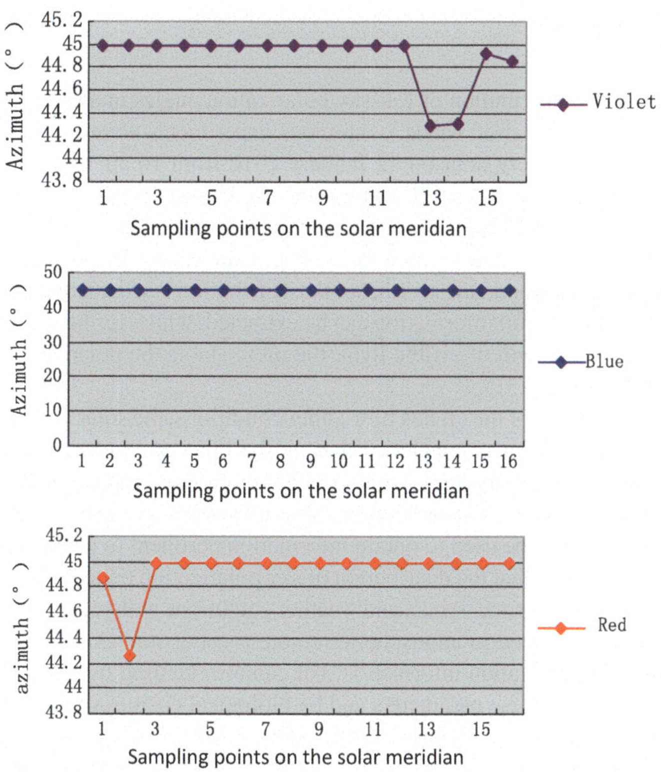

Fig. 7.30 Distribution of sky polarization angles in three bands (violet, blue, and red) under cloudy weather conditions (a gloomy, rainy sky): the three images in each row are captured at about the same time, with less than 1 min between them

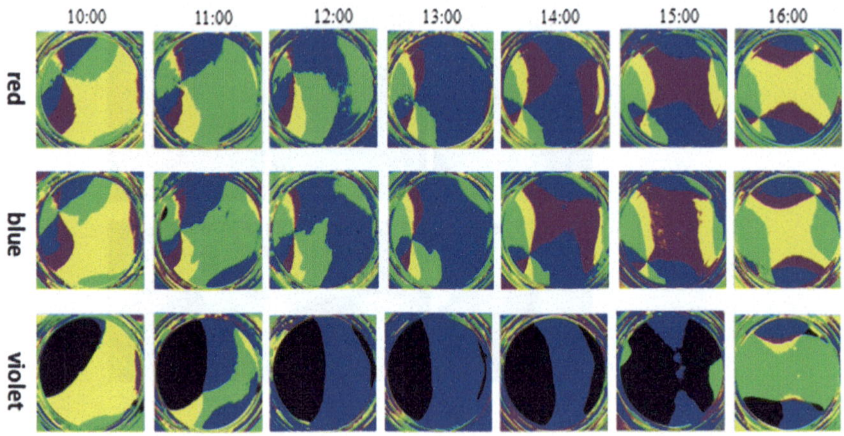

Fig. 7.31 Distribution of sky polarization angle in different bands under clear weather conditions

To illustrate the distribution of full-sky polarization angles in a certain period of time in different observation bands, continuous experiments were conducted in the Zhongguancun area every hour on 22 January 2010 from 10 am to 4 pm from the fifth floor of the Remote Sensing Building (Peking University). The shape and size of the distribution at different solar elevation angles in the blue and red bands are different. With the change in solar elevation angle, the entire sky polarization angle distribution revolves around the zenith, but the distribution is so irregular in the violet band that almost no useful information can be extracted, which further explains why, the longer the wavelength of visible light, the more stable the polarization angle is in clear weather (Fig. 7.31).

After consideration of the effects of weather conditions, the solar elevation angle, and observation bands, it is concluded that, in the blue band, characteristics such as the distribution of polarization angle and its rotation around the zenith and the symmetrical distribution about the solar meridian do not change under any conditions: the blue band can thus be used to obtain information pertinent to navigation.

These results prove that the distribution of sky polarization patterns is affected by the combined effects of time, space, and weather conditions with regularity in theory, and this can be employed in atmospheric remote sensing studies. Sky polarization not only provides navigation information, but can also be used by other navigation systems to calibrate course parameters and by biological creatures to find directions due to its regular and stable variations. Related information on how animals, such as insects, use sky polarization information to supplement their navigation and how sky polarization is meaningful to their survivals will be explained in detail in Chap. 10.

References

Collett M, Collett T, Bisch S et al (1998) Local and global vectors in desert ant navigation. Nature 394(6690):269–272

Fang J, Ning X (2006) The principle and application of astronomical navigation. Beijing University of Aeronautics and Astronautics Press, Beijing (in Chinese)

Guonan L (1985) Introduction to atmospheric radiation. Meteorological Press, Beijing (in Chinese)

Hartmann G, Wehner R (1995) The ant's path in integration system: a neural architecture. Biol Cybern 73(6):483–493

Ivanoff A, Waterman T (1958) Factors, mainly depth and wavelength, affecting underwater polarized light. J Mar Res 16:283–307

Ivanoff A, Waterman T (1985) Elliptical polarization of submarine illumination. J Mar Res 16:255–282

Jiang J, Wang W, Du H (2001) Space science and applications. Science Press, Beijing (in Chinese)

Peixuan S, Jietai M, Jianguo L et al (2003) Atmospheric physics. Peking University Press, Beijing (in Chinese)

Waterman T (1954) Polarization patterns in submarine illumination. Science 120:927–932

Waterman T (1955) Polarization of scattered sunlight in deep water. Deep Sea Res 3(Suppl):426–434

Waterman TH (2006) Reviving a neglected celestial underwater polarization compass for aquatic animals. Biol Rev 81:111–115

Yang J, Zhang S (1988) Navigation basis. Harbin Shipbuilding Technology Institute Press, Harbin (in Chinese)

Yao H (2006) Research on sky polarized light for biomimetic micro-nano navigation system. Master thesis, Dalian University of Technology (in Chinese)

Chapter 8
Atmospheric Remote Sensing 2: Neutral Point Areas of Atmospheric Polarization and Land-Atmosphere Parameter Separation

Then area with zero atmospheric polarization effect, and with the sun as its centre is discussed: this area is called the atmosphere polarization neutral point area. This area reflects a new atmospheric window, namely one in which polarization remote sensing reflectance light from land objects could not exist therein, and may be captured by a polarized remote sensor with only slight attenuation. This provides a possible method for atmospheric attenuation effect reduction, land object polarization information strengthening, and realisation of land-atmosphere separation. To be specific, the neutral point area theory of atmospheric polarization and its features provide the physical rationale behind a land-atmosphere separation method based on atmospheric polarization neutral points; basic land-atmosphere separation experiments based on neutral points provided real detection evidence; exploration of polarization-based land-atmosphere separation observation methods allowed consideration of sun-synchronous orbit satellite utilisation and other related methods.

8.1 Theory of Neutral Point Areas in Atmospheric Polarization

When solar radiation passes through the atmosphere and is scattered by atmospheric particles, it is polarized. Generally speaking, single scattering by atmosphere particles may lead to a positive value, while multiple scattering may lead spatial polarization have to negative values. The joint point of positive and negative polarization is a zero polarization point: this is the so-called atmosphere neutral point, where the degree of polarization (DOP) is zero. At beginning of the nineteenth century, three famous types of neutral points were observed: Arago Neutral Points (1809), Babinet Neutral Points (1840), and Brewster Neutral Points (1842). The discovery of atmospheric neutral points and studies of their features marks a milestone in atmosphere optics for that era. These studies mainly focus on reflecting atmospheric situations through location changes of atmospheric neutral points under different weather conditions.

L. Yan et al., *Polarization Remote Sensing Physics*, Springer Remote Sensing/Photogrammetry, https://doi.org/10.1007/978-981-15-2886-6_8

The author studied the space polarization distribution using a self-made "full-sky polarization measuring instrument" and found that atmospheric neutral points are not real points, but are areas centred on neutral points and the DOP of these areas is close to zero. So the author proposed atmospheric neutral point theory to resolve the obstruction posed by atmospheric polarization effects on land surface polarization (a problem for land-targeted polarization remote sensing). Atmospheric neutral point features of space-based observations, and those of ground-based observations, are the same under certain conditions. Since ground-based observation is more convenient than space-based observation, it is necessary to study atmospheric neutral point features from the perspective of ground-based observations so as to learn more of these atmospheric neutral point features for future space-based observations.

8.1.1 Introduction to Atmospheric Neutral Points

In 1809, Dominique Francois Jean Arago, a French astronomer, discovered space polarization. Shortly after that, he found a point in the sky whose DOP is zero. This point is known as the Arago Neutral Point to this day. About 30 years later, Jacques Babinet, a French meteorologist discovered the second neutral point. Several years later, Scottish physicist David Brewster discovered the third neutral point. These neutral points were studied by researchers through the nineteenth and twentieth centuries. The most important optical parameters, in sunny weather, can be well-described by Rayleigh scattering theory, but space light polarization differs from the ideal Rayleigh model. This difference is called the polarization defect, and is caused by multiple scattering through aerosol particles, differences in molecules in various directions, and particle distribution of aerosol, particle shape, and ground reflectance light. A significant feature of this defect is the neutral point. At this point, polarization of space scattering disappears (Coulson et al. 1960).

In fine weather, on the flat plane vertical to sun, three atmospheric neutral points appear in the sky (the Arago, Babinet, and Brewster points: see Fig. 8.1): observation of such points is more than that of other features of space polarization. Arago is

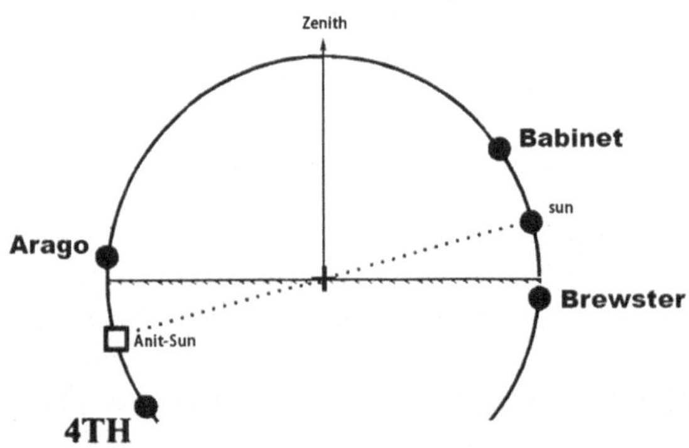

Fig. 8.1 Polarization neutral point in space (Coulson et al. 1960)

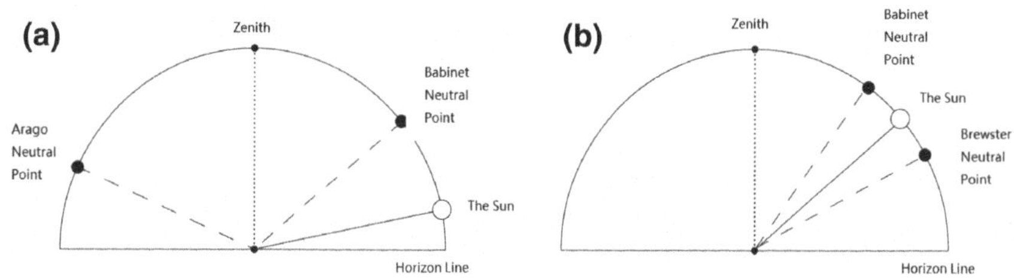

Fig. 8.2 Location of space neutral points on the principal plane at different solar altitudes

located at 20°–30° above the anti-sun point. Its location in space is easily observed and that is why the Arago point is much easier to be observed than other neutral points. Babinet is at about 20° above the sun. It is difficult to observe because it lies in the same direction as the sun. Arago is located at 20°–30° below the sun. The area between the sun and horizon is the brightest and smog near the horizon often causes low DOP values, so it is difficult to see the Brewster point. With normal weather, only two points can be observed at the same time, namely Babinet and Arago (Fig. 8.2a) or Babinet and Brewster (Fig. 8.2b): when the sun is high in the sky, Babinet and Brewster points are close to each other, and when the sun is at zenith, the two points are almost coincident. Given the symmetry of neutral points, theoretically, there should be a fourth neutral point just under the anti-sun point: this point has been discovered by space-balloon-based observations (Coulson 1988). Researchers also carried out model simulation of atmospheric neutral points in the sky to this effect (Horvath et al. 2002).

A reasonable explanation of atmospheric neutral points (Berry et al. 2004) is as follows: with single scattering, space light polarization is normally above zero, but multiple scattering by atmospheric particles causes negative polarization. Where positive and negative polarization regions coincide, the DOP is zero, and this forms an atmospheric neutral point. Generally, an atmospheric neutral point lies in the main plane of the sun and the zenith. The stronger the multiple-scattering, the more negative the polarization in the atmosphere, and the further the neutral points are from their ideal location.

8.1.2 Features of Atmospheric Neutral Points

Based on the above observation, the location of atmospheric neutral points is closely related to solar elevation, atmospheric conditions, observation band, etc. A theoretical explanation of the effects, and features, of atmospheric neutral points follows.

8.1.2.1 Relationship Between Atmospheric Neutral Points and Solar Elevation

Figure 8.3 shows the relationship between location of atmospheric neutral points and sun incidence angle at the principal plane when the aerosol optical depth is 0.10 (Horvath and Varju 2004). The abscissa is the sun zenith angle and the ordinate is the angle distance between the corresponding neutral point and zenith. This is the neutral point location theory as derived by Chandrasekhar based on atmosphere radiative transfer calculations.

The Babinet point may be 18° above the sun when the solar elevation is zero (namely when the sun starts to rise), it then rises as the sun rises. When the solar elevation is 90°, it will be in the same location as the sun, thus the Babinet point follows the sun from sunrise to sunset.

For a Brewster point, it appears at a solar elevation of 70°. When the sun moves towards its zenith, the Brewster point and the sun are almost collinear.

For an Arago point, it will appear in the sky only when sun zenith angle is between 70° and 90°. That is to say, it can only be observed in the early morning and early evening, and its angular distance from the sun is almost constant.

Figure 8.3 also shows that, under normal conditions, only two neutral points can be observed at the same time in clear sky. When the sun zenith is somewhere between 70° and 90°, Babinet and Arago points co-exist at the same time; when the sun zenith is below 70°, Babinet and Brewster points appear at the same time.

Figure 8.4 shows the DOP relationships between location of atmospheric neutral points and sun incidence angle at the principal plane when the aerosol optical depth is 0.20. The horizontal axis denotes the zenith angle, and the vertical axis shows DOP: in Fig. 8.4, 1, 2, 3, 4, and 5 denote zenith angles of 90°, 80.8°, 60.0°, 30.7°,

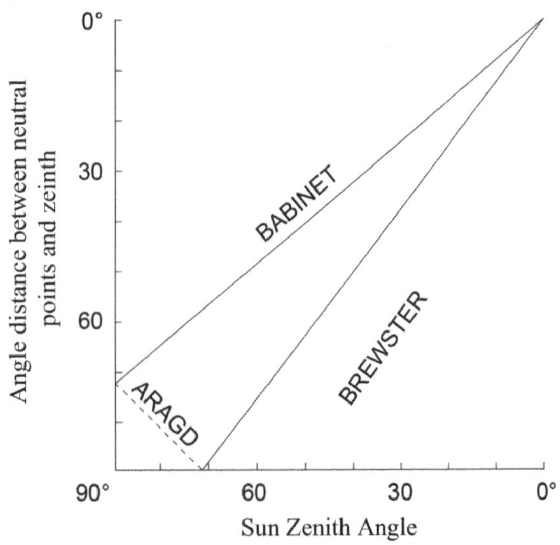

Fig. 8.3 Atmospheric neutral point location and solar altitude: aerosol optical depth of 0.10 (Horvath and Varju 2004)

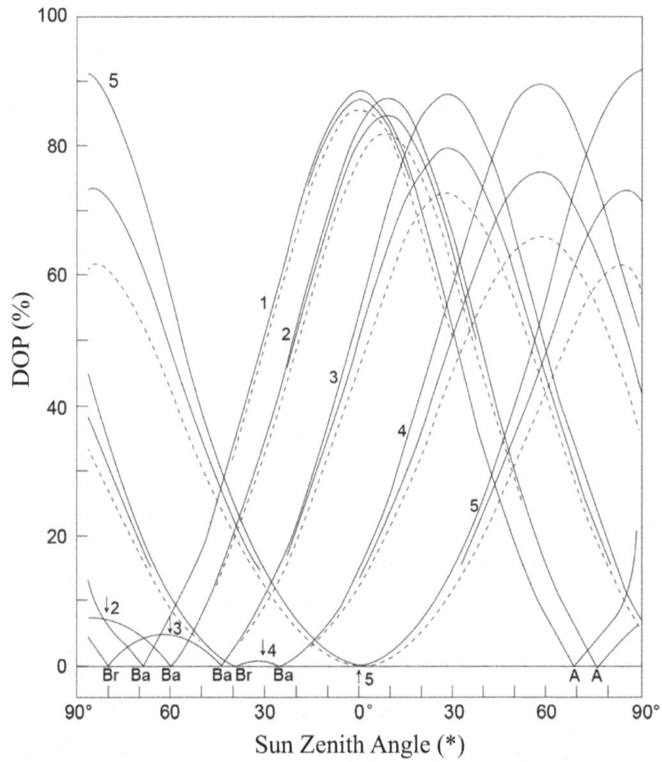

Fig. 8.4 DOP distribution with different incidence angles on the principal plane for an optical depth of 0.20 (Chandrasekhar and Elbert 1954)

and 0°, respectively. The thicker full line has not been calibrated by ground-truthed data. The dashed line plots the result when the ground refractive index is 0.20. The thinner full line represents the result when the ground refractive index is 0.10. The point, where the curve and horizontal axis cross, gives the location of a neutral point.

When the sun zenith angle is 90° or 80.8°, there are two neutral points in the sky, namely the Babinet and Arago points. When sun zenith angle is 60.0° or 30.7°, Babinet and Brewster points appear at the same time. When the sun zenith angle is 0°, namely when the sun incidence is vertical to the ground, the neutral point and the sun coincide, and all neutral points gather at the location of the sun.

8.1.2.2 The Relationship Between Atmospheric Neutral Points and Atmospheric Conditions

Figure 8.5 shows the DOP distribution on a plane vertical to the sun with three different optical depths, when the sun zenith angle is 53.1° and the ground reflectance is zero, The crossing points of the curve and line of zero DOP are Babinet and

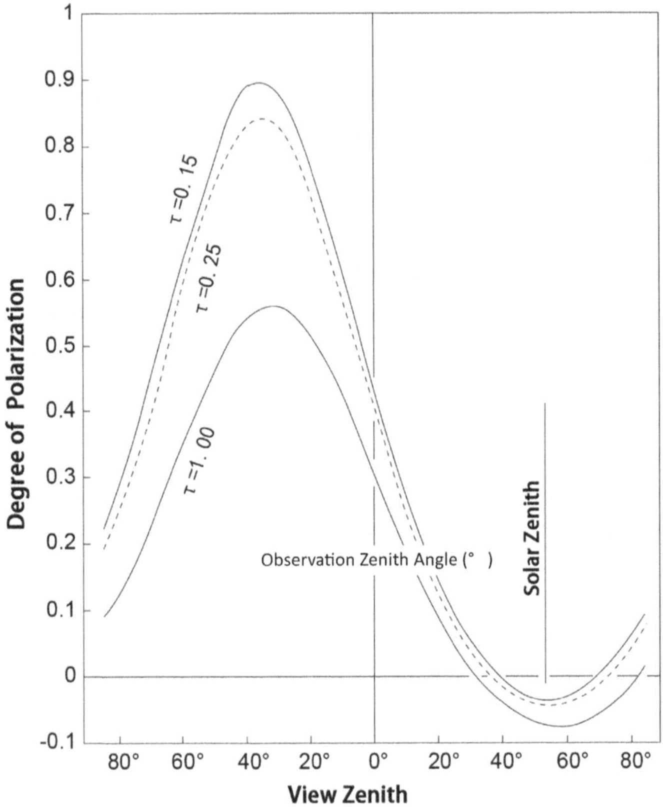

Fig. 8.5 DOP distribution curve in atmosphere on the principal plane at three optical depths (Chandrasekhar and Elbert 1954)

Brewster points. A small area with zero DOP in an area near the sun means that the component of electric vector that is horizontal to sun vertical plane is higher than the vertical component. With greater optical depth, the effect of multiple scattering is more prominent: its major effect is DOP reduction, and a secondary effect causes an increasing area wherein the DOP is negative (if we do not use absolute value in DOP calculation, the negative value of DOP have physical significance, telling the strong direction of polarization) and the location of the neutral points also change slightly.

Figure 8.6 shows the DOP distributions in atmosphere at different optical depths at a sun zenith angle of 79.15°, and a land surface reflectance of 0. The abscissa represents the cosine of the observation nadir angle, and the ordinate represents the DOP.

When the observation nadir angle is 79.15° ($\mu = 0.18817$), for neutral points with zero DOP derived from single scattering particles, the two neutral points will gradually appear on the horizontal line opposite the sun as the optical depth increases.

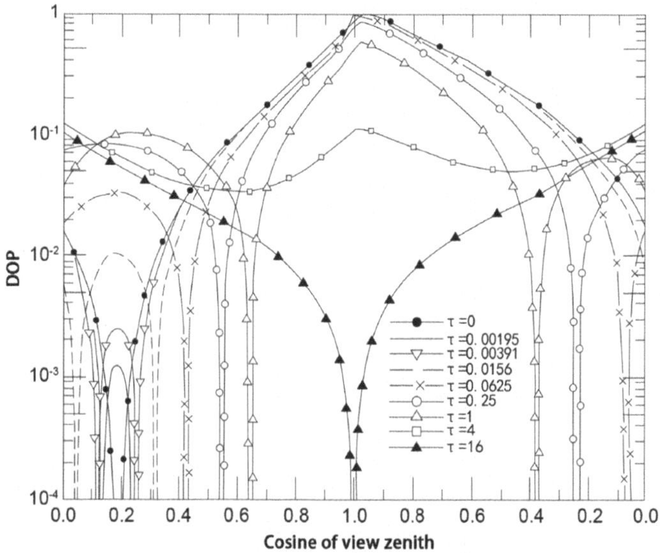

Fig. 8.6 DOP distribution in the atmosphere at different optical depths (Sekera 1956)

Assuming that $1 < \tau < 2.5$, four neutral points appear; when $\tau > 2.5$, no neutral point can be observed due to multiple scattering enhancement. This also shows that, when the aerosol optical depth increases to a certain extent, atmospheric conditions have a crucial effect on the location and nature of neutral points, and neutral points cannot be observed from the ground. Figure 8.6 shows that aerosol optical depth has a crucial impact on the maximum DOP of the sky.

8.1.2.3 Relationship Between Atmospheric Neutral Points and Observation Band

Figure 8.7 shows Arago neutral point DOP locations observed in different bands by Horvath (Kattawar et al. 1976). The black point in the figure denotes the location of the Arago neutral point.

Figure 8.7 shows Arago neutral point locations observed at different bands: under normal weather conditions, the shorter the light's wavelength, the smaller the spatial DOP. When wavelengths are relatively large ($\lambda > 500$ nm), certain wave spectrum characteristics appear, and for short wavelengths, DOP exhibits strong scattering (caused by multiple scattering in atmosphere). With a short wavelength, the strength of negative polarization is enhanced, which caused positive polarization zones to move towards neutral points. Negative polarization areas around the sun, or opposite the sun, are larger at shorter wavelengths and vice versa. In Fig. 8.7, the Arago point observed in the red band is the closest to the ground line, in the green band it is further from the ground line, and in the blue band, the furthest from the ground line.

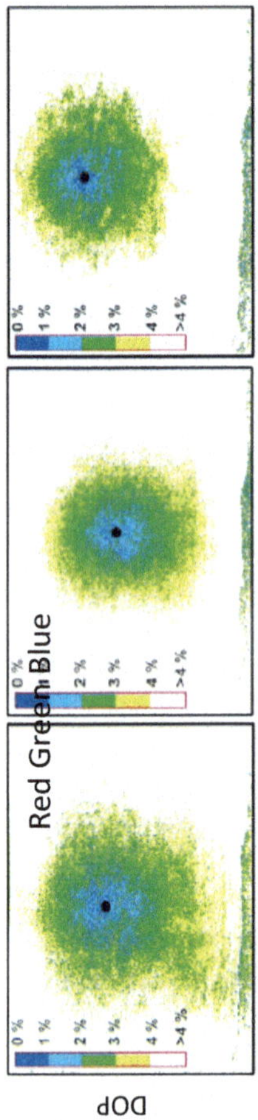

Fig. 8.7 Arago neutral point DOP location (Kattawar et al. 1976)

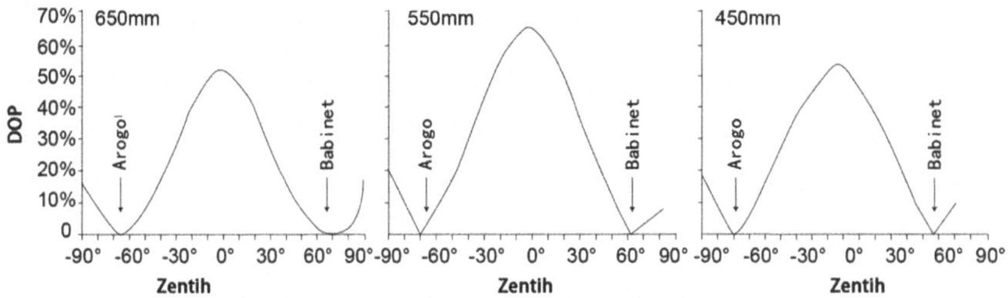

Fig. 8.8 Atmosphere DOP separate point distributions (Horvath et al. 2002)

Figure 8.8 shows separate point distributions of spatial DOP in three bands, namely red (650 nm), green (550 nm), and blue (450 nm) (Horvath et al. 2002) as captured in a desert environment in Tunisia. The abscissa denotes the observation zenith angle, and the ordinate denotes the DOP. These data were shot on 26 August 1999 at 06:00 (local time). From the figure, it can be seen that the largest space polarization occurs in the green band and the smallest in the red band. This is because the ground surface is a desert which is yellow or red, and red and yellow lights are the largest components in that light reflected by the ground. This leads to a decrease in spatial DOP in the red band. For the neutral point, there are certain differences in locations when using different observation bands (intercepts on the abscissa). This means that neutral points have certain wave spectral characteristics.

Figure 8.9 shows atmosphere polarization distributions observed in different observation bands. The crossing point of each curve and the line $y = 0$ is a neutral point. Different numbers in the figure denote different wavelengths in microns. From the Figure, we can see that neutral point locations observed in different bands are different, but in direction of the sun, the change in location of Babinet neutral points is smaller than that of the Brewster neutral point.

8.1.2.4 The Relationship Between Atmospheric Neutral Points and Land Surface Reflection Index

Figure 8.10 shows the spatial DOP distribution on the vertical plane of the sun in a Rayleigh atmosphere at a solar altitude of 53.1°, an optical depth of 0.25, and five different land surface albedo values (0, 0.10, 0.25, 0.50, and 0.80, respectively). Albedo is the proportion of total reflectance radiation flux to incidence radiation flux with regard to a surface. It is a physical parameter denoting the reflectance features of a land surface to solar short-wave radiation. From Fig. 8.10, crossing points of curve and points with zero DOP remain almost unchanged despite changes in land surface albedo: this means that changes in land surface albedo have little significant effect on neutral point locations.

From Fig. 8.10, land surface albedo has no effect on the prediction of neutral point location. It is not difficult to understand the independence of neutral point from land surface reflectance. Rayleigh scattering theory is concerned with the biggest

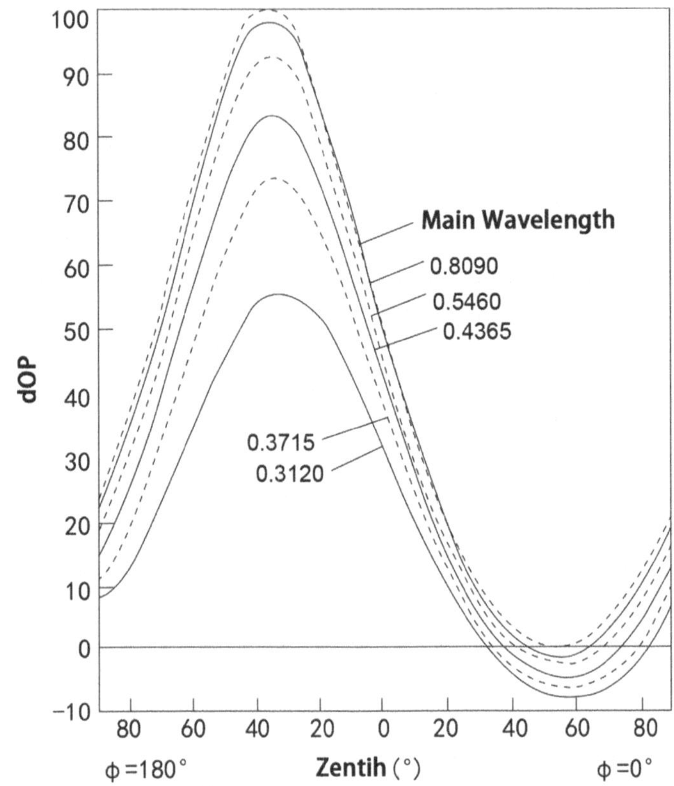

Fig. 8.9 Atmosphere DOP distribution at different wavelengths (Horvath et al. 2002)

possible polarization of scattering light and other theories pertain to smaller degrees of polarization. According to Lambert theory, polarization generated by land surface reflectance only matches Rayleigh scattering, so the effect of reflectance polarization of a land surface is just a part of a certain expressed natural light: land surface reflectance has certain effects on polarization distributions in the atmosphere and it significantly affects the maximum spatial DOP. With ground calibration, the maximum polarization change between 94 and 89% is dependent on atmosphere optical depth.

8.1.3 Neutral Line of Atmospheric Polarization

In the nineteenth century, whether neutral points exist and their appearance became a major topic in atmosphere optics. Many scholars put a lot of energy into the study of neutral points. Among them, Carl Dorno did much prominent work: he not only observed the neutral points of the principal plane, but also studied the continuous

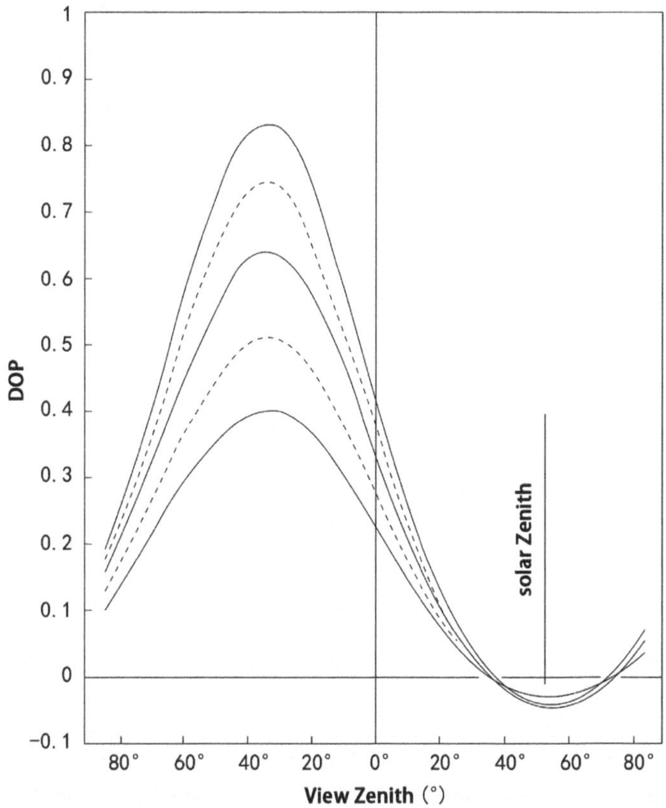

Fig. 8.10 Spatial DOP distribution on the vertical plane of the sun in a Rayleigh atmosphere over different land surface albedo values

changes in those neutral points on the hemisphere. He called these trajectories neutral lines: these separate negative and positive polarization regions. Neutral lines are closely related to the location of the sun. The left-hand part of Fig. 8.11 shows major observations made by Dorno as of 17 May 1917, when he studied polarization lines (the different lines in Fig. 8.11 represent different observation times): when the sun is close to the ground horizon, a neutral line connects Babinet and Arago points through a double-headed symmetric closed curve. This is also called the Busch double-headed line. When the sun rises, this double-headed line will be more symmetric than in other situations. When the incidence angle is close to 70°, the double headed line will open and part of the track will be below the sun along the ground horizon, passing through the rising Brewster point. The neutral line is made up of such two separation lines. Only when the incidence angle is almost 45°, do the two ends facing each other connect to form a new closed curve. For even smaller angles of incidence, neutral lines point toward the centre and they merge to one point at solar zenith. The right-hand part of Fig. 8.11 shows the value of a neutral

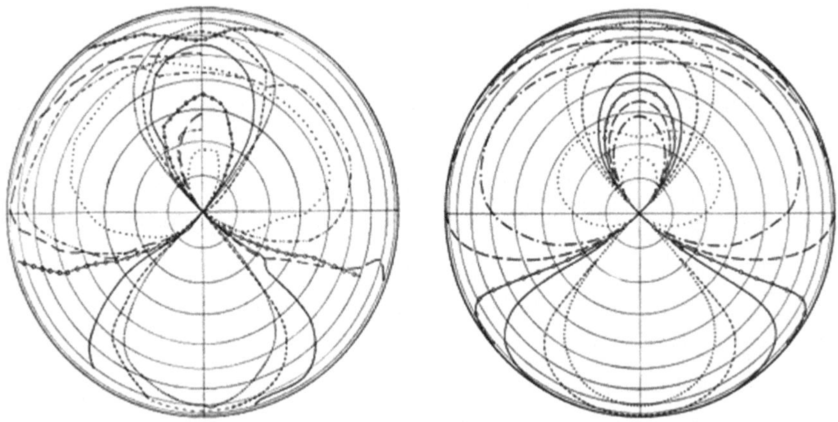

Fig. 8.11 Comparison of the observed value of the neutral line (left) and its value based on theoretical calculation (right)

line based on theoretical calculations. Different lines denote the distance between different zeniths and the sun (······· = 90°, ——— = 58.7°, – – – = 43.9°, - - - - - - = 76.1°, -o-o-o = 50.2°, —·—·· = 36.9°, and ······ = 19.9°). When comparing this to the left-hand part showing observations by Dorno, we can see the two figures match each other well.

Based on observation and theory in relation to polarization neutral lines, this section not only proved the existence of polarization points, but also showed the regular distribution of neutral points. Such a continuous distribution of polarization neutral line can be used for observing ground-targeted polarization remote sensing data.

8.2 Observation of Neutral Point Areas in the Atmosphere Based on Polarization Pattern

The DOP distribution and location of neutral points in the full sky are not fixed: DOP is distributed as a ring around polarization neutral points, and full-sky DOP distributions are formed by the overlapping of two rings. Polarization angles are distributed as strips around two angle gathering points. The full-sky DOP distribution has a certain symmetry. The location of an atmosphere neutral point is closely related to solar elevation, atmospheric conditions, observation band, etc. At the same time, the location of gathering points of polarization angles overlaps that of polarization neutral points. Studies of polarization angle show that polarization angle is less affected by atmospheric and other external factors and its distribution is more regular.

8.2.1 *Regular Distribution of DOP According to Polarization Neutral Points*

Taking the experimental data shown in Fig. 7.17 as an example, under different weather conditions, from the standpoint of a single DOP distribution image, the atmospheric DOP distribution is like a ring around polarization neutral points, and the DOP value increases from neutral points to edge of the ring. Around the black area, the DOP value will be at a maximum and then gradually decrease towards another neutral point. This means that, at larger solar elevation angles, we can only see two polarization neutral points and all DOPs are distributed around them. At smaller solar elevation angles, we can only see one polarization neutral point and all DOPs are distributed around one polarization neutral point. This is in line with polarization neutral point distribution theory as introduced in Chap. 3.

In different observation bands, from the DOP distribution with clear sky (Fig. 7.19), we can see the DOPs of purple, blue, and red wavebands all surrounding polarization neutral points (the red area in the figure) with a regular distribution. This is close to modelled results, but the neutral point locations of three wavebands are slightly different. Areas with maximum DOP (the black area in the figure) are also different to each other: this means that the location of the neutral point varies based on observation waveband. From spatial DOP distributions on cloudy and rainy days (Fig. 7.20), it can be seen that two neutral points appear and the DOP is distributed around them. Although DOP distributions are similar when using different bands, the location of polarization neutral points varies across each of the three bands. Moreover, the DOP value is relatively small and those DOPs are mostly distributed in yellow or green areas.

Figure 8.12 shows about sky polarization distributions at different bands in Zhong Guancun on 22 January 2010, with a clear sky. These are generated by shooting the full sky, at hourly intervals from 10 am to 4 pm from the Fifth Floor of the Remote

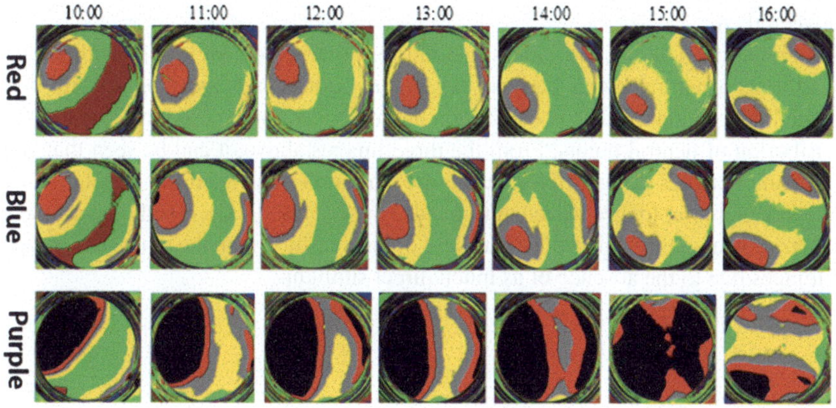

Fig. 8.12 DOP changes in different bands based on solar altitude in fine weather

Sensing Building of Peking University. When viewed from single DOP distribution image, the atmospheric DOP distribution is like a ring around polarization neutral points, and the DOP value increases from neutral points to edge of the ring. Then the DOP value will gradually decrease towards another neutral point. The full-sky DOP distribution is closely connected to solar elevation, but during the whole observation period, both polarization neutral points can be observed. This means that two ring-shaped distributions can be observed and that the full-sky DOP distribution is formed by the overlapping of two rings.

8.2.2 Regular Distribution of Polarization Angle According to Polarization Neutral Point Position

From the image taken under a clear sky, it can be seen that space light polarization angles are regularly and symmetrically distributed around polarization neutral points. From polarization neutral points to outer areas, and in a clockwise direction, the polarization angle gradually increases from 0° in the blue area to somewhere between 30° and 60° in the purple area, it then increases from −60° in the yellow area to −30° in the green area and returns to 0°. Polarization angles gather at one neutral point and go towards another polarization neutral point to gather again (due to the angle of the field of view, this neutral point is not shown in the figure). This is similar to modelled calculations. Images captured on cloudy days have a certain similarity to those from sunny sky: space polarization angles are distributed symmetrically around polarization neutral points, but the distribution is more irregular and angles do not change gradually but have overlapping areas of different sizes (with no clear edges, and yellow and purple areas also overlap).

Figures 8.13, 8.14 and 8.15 show distributions of sky polarization angles in the purple, blue, and red bands under clear sky, in rainy weather, and under shade: the three images in each figure were shot at almost same time, with no more than 2 min between them.

As mentioned above, the gathering point for polarization angles lies in the same location as a neutral point; because sky polarization angles are more evenly distributed than spatial DOP, so polarization angle images can be used to represent the distribution of neutral points. From the three images above, it can be seen that sky polarization angles using different bands are similar in trend but differ in detail: they are more stable at longer wavelength bands (the red band). On cloudy days, space polarization angles are distributed in a less orderly fashion than the other two groups of images due to the absence of as much direct sunlight

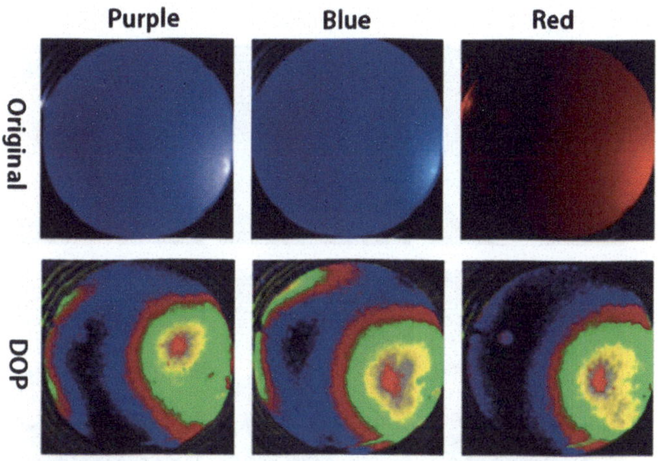

Fig. 8.13 Sky polarization angle distribution in different bands under a clear sky

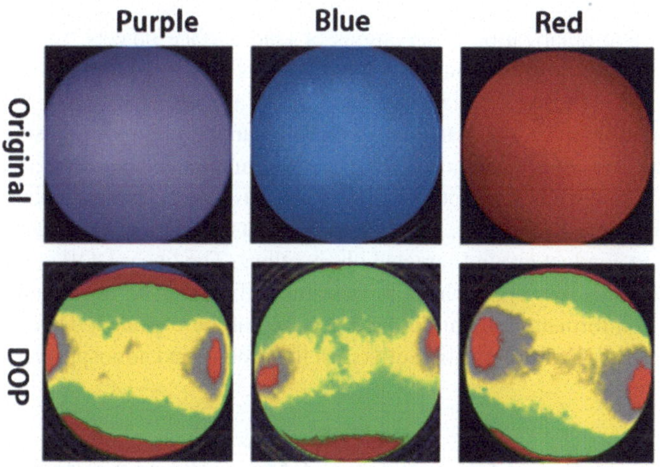

Fig. 8.14 Sky polarization angle distribution in different bands during rain

8.3 Separation of Object and Atmospheric Effects Based on Neutral Point Areas of Atmospheric Polarization

In earth observation, especially in the visible light band, atmospheric scattering light has strong polarization features, so satellite-borne polarization sensors adopted internationally are mainly used for detecting the nature of the atmosphere, aerosol optical depth and size of aerosol particles, etc. Objects on the land surface also have a strong polarising nature for example, water ponds, vegetation, buildings, etc.: such polarization can be detected by remote sensing, so the French POLDER sensor was designed

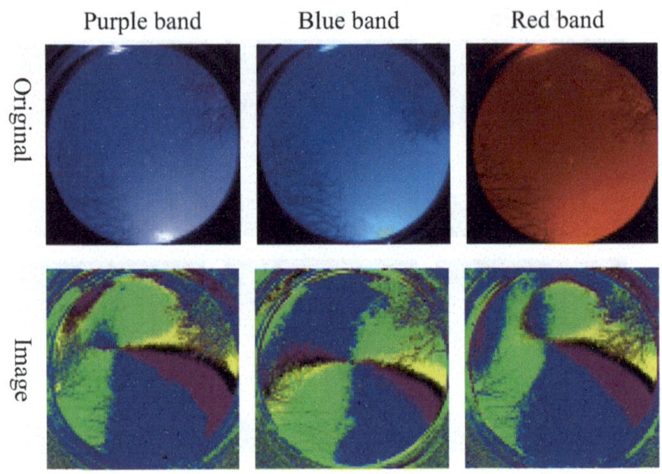

Fig. 8.15 Sky polarization angle distribution in different bands under shade

for multi-angle polarization observations of land surface objects. In practice, the strong atmosphere polarization always overwhelms polarization reflection information about land objects. Polarization effects in the atmosphere become a bottleneck issue preventing ground-targeted detection in polarization remote sensing. How to separate land object polarization reflectance information and atmospheric polarization effect in polarization remote sensing images has long been problematic. There is no easy solution and therefore, we placed sensors at the location of atmospheric neutral points to conduct ground-targeted observations. The linear DOP at these points is zero, so in this way atmospheric polarization effects can be best eliminated, and polarization information about the land surface can be obtained. This can help to achieve separation of polarization effects in the ground and atmosphere.

8.3.1 Space-Based Observations of Atmospheric Neutral Points

Remote sensing is a process from the upside to the downside and as such, runs contrary to ground-based observation. When ground-based polarization remote sensing is conducted by utilising polarization neutral points, the basic issue is whether, or not, atmospheric neutral points can be observed with satellite-based methods.

Figure 8.16 shows atmospheric neutral points as shot by Gabor Horvath with a full polarization space detection camera on the ground and at 3500 m above ground. Neutral points are marked in cyan. Local solar elevation angles were all 0° when images were captured. The black point is the sun and the red area is over-exposed. The shooting wavelengths are 450 nm, 550 nm, and 650 nm (based on red, green, and

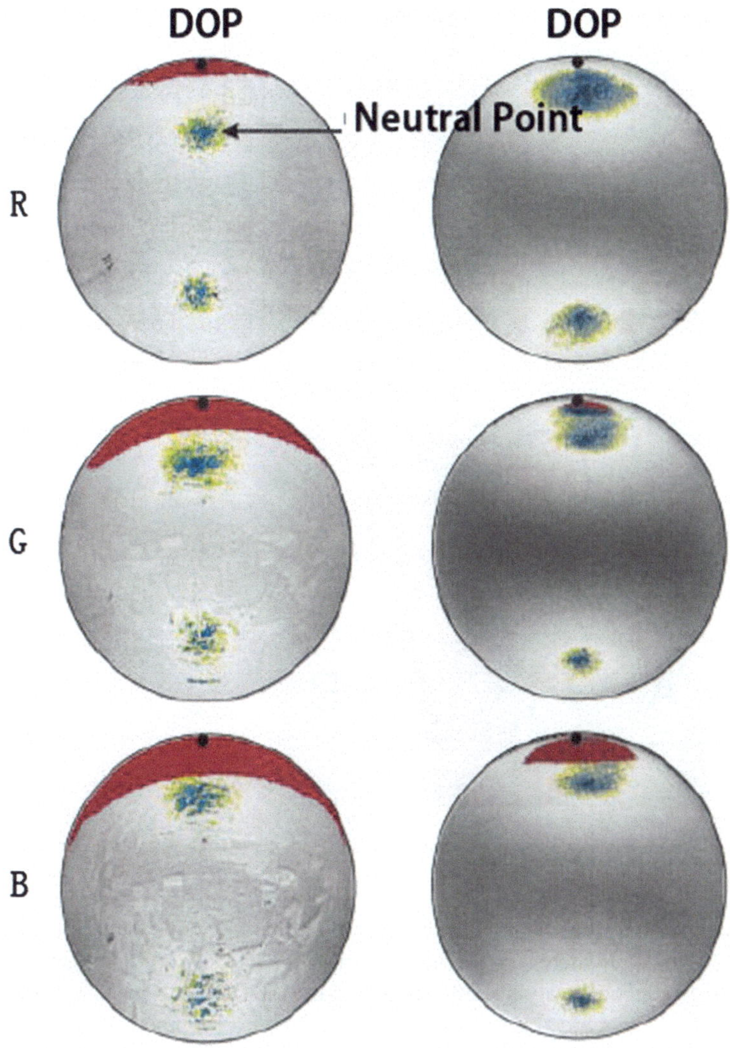

Fig. 8.16 Comparison of atmospheric neutral points observed from ground (left) and from 3500 m altitude (right)

blue bands, respectively). The three images on the left were shot from the ground, and the three on the right were shot from a balloon (albeit not simultaneously): neutral point locations differed but only slightly. The three images on the right proved that atmospheric neutral points can be observed from the top of the atmosphere.

Figure 8.17 shows a DOP distribution of remote sensing images from a polarization sensor. These images are from a series of images taken by POLDER on 5 October 2003 (Pass No. 034, Level-1 data). The wavelength of the band used is 443 mm, and it employed near-vertical detection. Areas A and B are land and ocean, respectively.

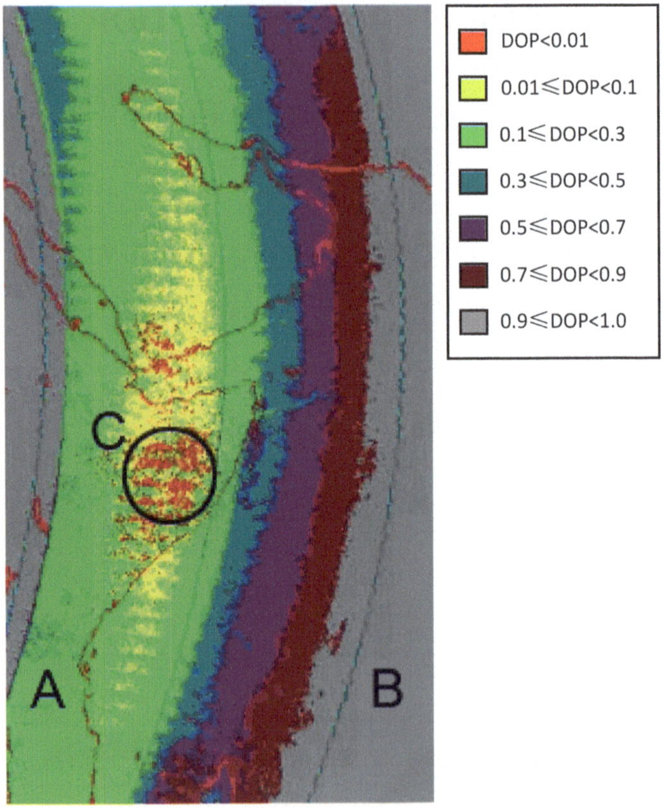

Fig. 8.17 DOP distribution of remote sensing images from the POLDER polarization sensor

As shown in the figure, the DOP is distributed regularly and the DOP in the middle is smaller. The DOP increases gradually towards the two sides of the images: this is similar to ground observation results. In the figure, the DOP of Zone C is equal to, or less than, 0.01, which is deemed to have been near-zero for practical purposes, and so this zone can be viewed as a polarization neutral point area.

Figures 8.16 and 8.17 both demonstrate that satellite-based polarization sensors can be used in atmospheric neutral point observation.

8.3.2 *Features of Atmospheric Neutral Points Based on Space-Based Observations*

8.3.2.1 Upward and Downward Radiation at Neutral Points

Studies have been conducted on light spectral intensity and polarization states of molecular atmospheric light. Most this work has been based on the Stokes Parameter Method as proposed by Chandrasekhar (2003) in 1950. Coulson (1959) presented detailed lists of values in 1960 by taking into account light intensity and DOP after multiple scattering. These include such parameters as optical depth (seven values from 0.02 to 1.00), sun zenith angle (taken from 0° to 180° in 30° increments), and ground reflectance (0, 0.25, and 0.80). Based on combinations of these parameters, Stokes parameter, DOP, and polarization azimuth angle are calculated and 142,688 combinations were provided for upward and downward radiation. The following concerns Coulson's calculation of upward and downward radiation:

I_e, I_r denote the intensity of vertical and horizontal directions through a polariser, then:

$$I_{e,r}(\tau;\mu,\varphi) = I_{e,r}^{(0)}(\tau;\mu,\mu_0) + I_{e,r}^{(1)}(\tau;\mu,\mu_0)\cos\varphi + I_{e,r}^{(2)}(\tau;\mu,\mu_0)\cos 2\varphi \tag{8.1}$$

Azimuth irrelevance (Formula (8.1)) implies the following for scattering radiation at the bottom of the atmosphere:

$$\begin{cases} I_e^{(0)}(\tau;-\mu,-\mu_0) = \dfrac{C}{\mu-\mu_0}[K\xi(\mu) + L\eta(\mu) - M\psi(\mu) - N\phi(\mu)] \\ I_r^{(0)}(\tau;-\mu,-\mu_0) = \dfrac{C}{\mu-\mu_0}[K\sigma(\mu) + L\theta(\mu) - M\chi(\mu) - N\zeta(\mu)] \end{cases} \tag{8.2}$$

For scattering radiation at the top of the atmosphere:

$$\begin{cases} I_e^{(0)}(0;\mu,-\mu_0) = \dfrac{C}{\mu+\mu_0}[K\psi(\mu) + L\phi(\mu) - M\xi(\mu) - N\eta(\mu)] \\ I_r^{(0)}(0;\mu,-\mu_0) = \dfrac{C}{\mu+\mu_0}[K\chi(\mu) + L\zeta(\mu) - M\sigma(\mu) - N\theta(\mu)] \end{cases} \tag{8.3}$$

Among which,

$$\begin{cases} K = \psi(\mu_0) + \chi(\mu_0), \quad L = 2[\phi(\mu_0) + \zeta(\mu_0)] \\ M = \xi(\mu_0) + \sigma(\mu_0), \quad N = 2[\theta(\mu_0) + \eta(\mu_0)] \end{cases} \tag{8.4}$$

$$C = \frac{2}{32}\mu_0 F_0 \tag{8.5}$$

$\psi, \phi, \chi, \zeta, \xi, \eta, \sigma, \theta$ are the eight functions of linear combinations of two of X_e, Y_e, X_r, Y_r, and they satisfy the following non-linear integral equation system:

$$\left.\begin{aligned} X_i(\mu) &= 1 + \mu \int_0^1 \psi_i(\mu')\left[X_i(\mu)X_i(\mu')\right] - \left[Y_i(\mu)Y_i(\mu')\right]\frac{d\mu'}{\mu+\mu'} \\ Y_i(\mu) &= e^{-\tau/\mu} + \mu \int_0^1 \psi_i(\mu')\left[Y_i(\mu)X_i(\mu')\right] - \left[X_i(\mu)Y_i(\mu')\right]\frac{d\mu'}{\mu-\mu'} \end{aligned}\right\}(i = e, r) \tag{8.6}$$

where $\psi_e(\mu) = \frac{3}{4}(1-\mu^2)$ and $\psi_r(\mu) = \frac{3}{8}(1-\mu^2)$.

The azimuth relevance (Formula (8.6)) gives the following for scattering radiation at the bottom of the atmosphere:

$$\begin{cases} I_e^{(1)}(\tau; -\mu, -\mu_0) = \dfrac{4C}{\mu-\mu_0}\mu\mu_0\left(1-\mu^2\right)^{\frac{1}{2}}\left(1-\mu_0^2\right)^{\frac{1}{2}}W^{(1)}(\mu,\mu_0) \\ I_r^{(1)}(\tau; -\mu, -\mu_0) = 0 \\ I_e^{(2)}(\tau; -\mu, -\mu_0) = \dfrac{-C}{\mu-\mu_0}\mu^2\left(1-\mu_0^2\right)W^{(2)}(\mu,\mu_0) \\ I_r^{(2)}(\tau; -\mu, -\mu_0) = \dfrac{C}{\mu-\mu_0}\left(1-\mu_0^2\right)W^{(2)}(\mu,\mu_0) \end{cases} . \tag{8.7}$$

For scattering radiation at the top of the atmosphere:

$$\begin{cases} I_e^{(1)}(0; \mu, -\mu_0) = -\dfrac{4C}{\mu+\mu_0}\mu\mu_0\left(1-\mu^2\right)^{\frac{1}{2}}\left(1-\mu_0^2\right)^{\frac{1}{2}}M^{(1)}(\mu,\mu_0) \\ I_r^{(1)}(0; \mu, -\mu_0) = 0 \\ I_e^{(2)}(0; \mu, -\mu_0) = -\dfrac{C}{\mu+\mu_0}\mu^2\left(1-\mu_0^2\right)M^{(2)}(\mu,\mu_0) \\ I_r^{(2)}(0; \mu, -\mu_0) = \dfrac{C}{\mu+\mu_0}\left(1-\mu_0^2\right)M^{(2)}(\mu,\mu_0) \end{cases} \tag{8.8}$$

where,

$$\left.\begin{aligned} W^{(j)}(\mu,\mu_0) &= X^{(j)}(\mu_0)Y^{(j)}(\mu) - Y^{(j)}(\mu_0)X^{(j)}(\mu) \\ M^{(j)}(\mu,\mu_0) &= X^{(j)}(\mu_0)X^{(j)}(\mu) - Y^{(j)}(\mu_0)Y^{(j)}(\mu) \end{aligned}\right\}(j = 1, 2) \tag{8.9}$$

And for $X^{(j)}, Y^{(j)}$ The following equation is satisfied:

$$\psi^{(1)}(\mu) = \tfrac{3}{8}(1-\mu^2)(1+2\mu^2),\ \psi^{(2)}(\mu) = \tfrac{3}{16}(1+\mu^2)^2 \tag{8.10}$$

The U component of the Stokes parameter can be expressed as:

$$U(\tau; -\mu, \varphi) = \frac{2C}{\mu - \mu_0}\begin{bmatrix} 2\mu_0(1-\mu^2)^{\frac{1}{2}}(1-\mu_0^2)^{\frac{1}{2}}W^{(1)}(\mu, \mu_0)\sin\varphi \\ -\mu(1-\mu_0^2)W^{(2)}(\mu, \mu_0)\sin 2\varphi \end{bmatrix} \quad (8.11)$$

$$U(0; \mu, \varphi) = \frac{2C}{\mu + \mu_0}\begin{bmatrix} 2\mu_0(1-\mu^2)^{\frac{1}{2}}(1-\mu_0^2)^{\frac{1}{2}}M^{(1)}(\mu, \mu_0)\sin\varphi \\ +\mu(1-\mu_0^2)M^{(2)}(\mu, \mu_0)\sin 2\varphi \end{bmatrix} \quad (8.12)$$

where I is the total intensity of incident light and the second Stokes component Q is:

$$I = I_e + I_r \quad (8.13)$$

$$Q = I_r - I_e \quad (8.14)$$

The following concerns Coulson's calculation of DOP and the polarization angle of upward and downward radiation.

8.3.2.2 Comparison of Different Solar Elevations

Figure 8.18 shows the polarization neutral point for atmospheric upward and downward radiation based on calculations with a plane-parallel atmosphere model as undertaken by Coulson (1959). The figure shows angle distance from the sun to

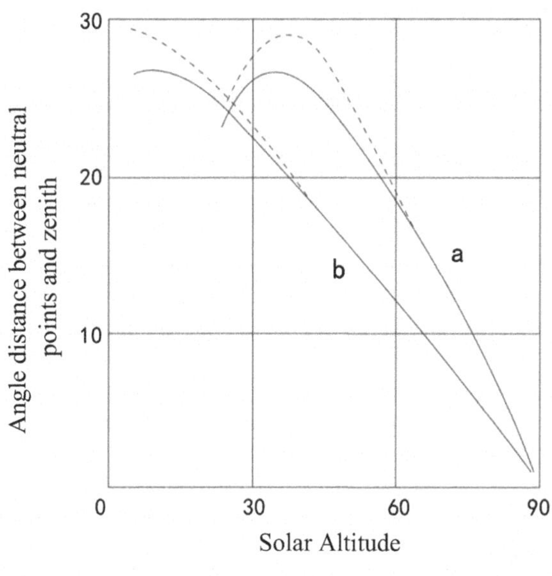

Fig. 8.18 Relationship between angle distance from the Sun to neutral points and solar altitude

neutral points and solar elevation and the abscissa represents solar elevation, and the ordinate represents the angle distance from a neutral point to the sun. The atmosphere optical depth is 1.0. The solid line denotes upward radiation and the dotted line, downward radiation.

Curve (a) shows the upward and downward radiation curve of a Brewster neutral point. Curve (b) shows the upward and downward radiation curve of a Babinet neutral point. For a Brewster neutral point, when the solar elevation is between 60° and 90°, the location of neutral points of upward and downward radiation is identical. When the solar altitude is between 25° and 60°, although the angle distance between a neutral point of upward and downward radiation follows a similar trend, they have already separated. When the solar elevation is below 25°, the location of the sun is too low and no Brewster neutral point appears in the sky. For a Babinet neutral point, when the solar elevation is between 32° and 90°, the location of neutral points for upward and downward radiation is identical. When the solar elevation is between 5° and 25°, the neutral point locations, for upward and downward radiation, start to differ. When the solar elevation is below 5°, there is no Babinet neutral point for either upward or downward radiation.

Figure 8.18 shows that a Brewster neutral point can be observed at the top of the atmosphere. When the solar elevation is between 60° and 90°, the location of neutral points of upward and downward radiation is identical. For a Babinet neutral point, as observed when the solar elevation is between 32° and 90°, neutral point locations of upward and downward radiation are similar. That is to say, atmospheric neutral points and atmospheric neutral points observed on the ground are the same when the solar altitude is relatively high. In general, when resource detection is being undertaken using a satellite in a ground-targeted manner, the solar elevation when the satellite flies by is relatively high: at this time, neutral point locations for both upward and downward radiation are similar.

8.3.2.3 Comparison Under Different Atmosphere Optical Depths

Figure 8.19 shows neutral point locations for downward radiation based on atmosphere depth under a Rayleigh atmosphere at a sun zenith angle of 31.43°. The abscissa denotes atmospheric optical depth, and the ordinate represents the cosine of the observation nadir angle. The observation nadir angle of Babinet and Brewster neutral points is positive ($\phi = 180°$) and that for the Arago neutral point is negative ($\phi = 0°$). Land surface albedo (A) in the figure is 0, 0.2, or 1, respectively.

Figure 8.20 shows neutral point locations for upward radiation based on atmosphere depth under a Rayleigh atmosphere at a sun zenith angle of 31.43°. The abscissa represents atmospheric optical depth, and the ordinate represents the cosine of the observation nadir angle. The observation nadir angle of Babinet and Brewster neutral points is positive ($\phi = 180°$) and that for the Arago neutral point is negative ($\phi = 0°$). The dotted and full lines denote land surface albedo (A) values of 0 or 1, respectively.

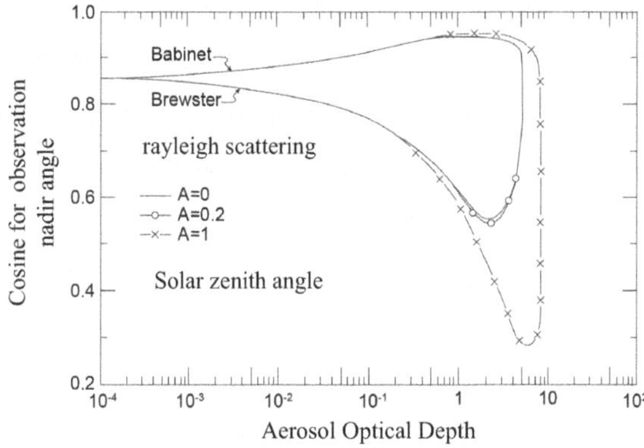

Fig. 8.19 Neutral point location for downward radiation based on optical depth under a Rayleigh atmosphere

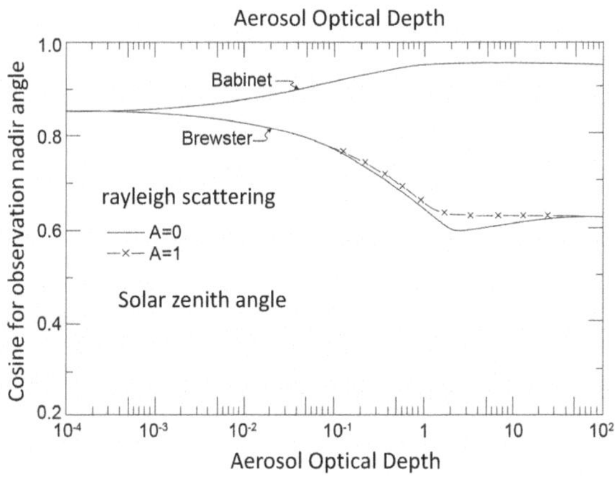

Fig. 8.20 Neutral point location for upward radiation based on optical depth under a Rayleigh atmosphere

Figure 8.20 shows the location of neutral points when the sun zenith angle is 31.43°. This is a typical condition when the sun is close to zenith. For upward radiation, the zenith angle has two neutral points at the most, namely the Babinet and Brewster neutral points. To compare the location of upward radiation and downward radiation neutral points, firstly the situation when land surface albedo is zero should be analysed.

For the Brewster neutral point, curves of the upward and downward radiation are similar when the optical depth is less than 0.1. As the optical depth increases, the shape of upward and downward radiation curves changes. When the optical depth is between 10^{-3} and 5, the curve of downward radiation changes faster than that of the upward radiation, and when the optical depth exceeds 5, the curve of upward radiation gradually becomes horizontal, and the Brewster neutral point of downward radiation gradually overlaps the Babinet neutral point. This also demonstrates that when the optical depth exceeds 6, the Brewster neutral point cannot be seen in downward radiation.

For the Babinet neutral point, in downward radiation, when the optical depth is about 1.3, it reaches its highest value and then gradually moves back. When the optical depth is 5.2, the Babinet neutral point overlaps the Brewster neutral point, thus forming a single neutral point. In downward radiation, the Brewster neutral point cannot be seen when the optical depth exceeds 6. In upward radiation, the Babinet neutral point moves towards zenith when the optical depth increases. When the optical depth is about 1, its location will not change when the optical depth increases further.

In the figure, it can be found that the Babinet neutral point changes slightly when the optical depth increases: under the same conditions, the Brewster neutral point changes significantly. When the optical depth is less than 5, the location of the Babinet neutral point for upward radiation is almost the same as its location in downward radiation. This also showed that the location of a Babinet neutral point is the same for ground-based, and space-based, observations.

8.3.2.4 Comparison of Different Ground Surface Reflectances

Figure 8.21 shows that the DOP distribution of diffuse reflection and transmission radiation on principal plane (Plass et al. 1973) is apparently dependent on ground reflectance and atmospheric turbidity. T1 and T2 in the figure demonstrate different turbid atmospheres and R denotes the data plot for a molecular atmosphere. With increasing ground reflectance, the DOP will decrease. As aerosol loading increases, the turbidity increases and the DOP of diffuse transmissivity radiation decreases, but that of diffuse reflection increases slightly: this is because, as the aerosol loading

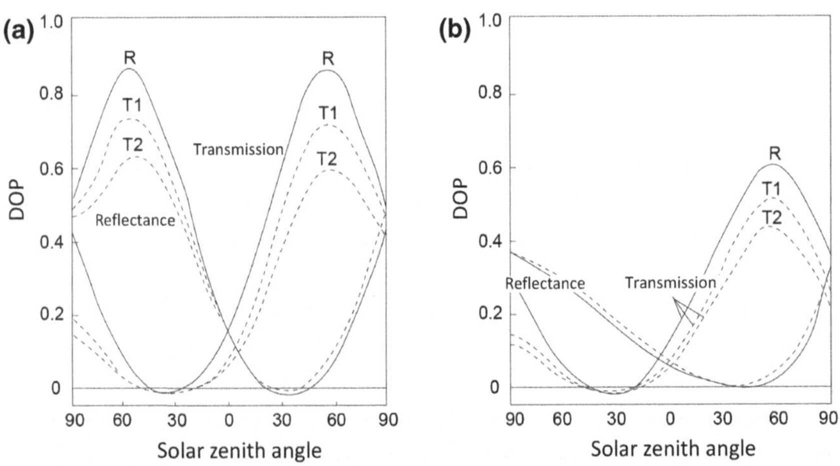

Fig. 8.21 DOP distribution of diffuse reflection and transmission radiation on principal planes of different ground reflectance

increases, the aerosol optical depth also increases, and atmospheric scattering radiation reflected to space will exceed the non-polarized radiation reflected directly from the ground to space. Figure 8.21 shows that the transmissivity radiation curve has a negative polarization zone around the sun, and neutral points appear therein. The ground reflectance (Fig. 8.21a) is zero and that in Fig. 8.21b is 0.25. In Fig. 8.21a, neutral points are almost located symmetrically. That is to say, neutral points for ground-based observation and space-based observations lie in almost the same area; and neutral points in Fig. 8.21b have different locations since upward radiation has one neutral point and downward radiation has two. The only neutral point of upward radiation and one of the two neutral points of downward radiation are located in almost the same place. This neutral point is probably a Babinet neutral point. Figure 8.21 shows that differences in ground reflectance have a certain effect on the spatial polarization distribution, and especially on the maximum polarization distribution of the sky. With increasing ground reflectance, the maximum polarization of the sky will decrease by a large margin because ground reflectance light increases multiple scattering in the atmosphere and generates a depolarising effect: however, the increase of ground reflectance has a lesser effect on the location of the neutral point. The neutral point location for transmission is almost unchanged and for reflectance radiation, the Babinet neutral point location is practically unchanged.

8.3.3 Selecting Applicable Atmospheric Neutral Points

The Arago, Brewster, and Babinet neutral points are three neutral points commonly found in fine weather. Since neutral points are severely affected by external conditions such as the location of the sun, atmospheric conditions, ground reflectance, and observation band, ideal neutral points for ground-based polarization remote sensing observation should: firstly, have a favourable location for remote sensing observation, secondly, be less severely affected by external interference or influences. The following text covers different factors affecting these three neutral points.

The location of the Arago neutral point is relatively low in the sky, with the angle between it and the ground horizontal line being less than 20°; because it is not on the same side of the sun and is easy to detect either by the naked eye or by use of a simple device, detection thereof is most common among the three neutral points. In the remote sensing of visible light and near-infrared band data, to ensure sufficient lighting, conditions with high solar elevation and good lighting under normal conditions are preferred. An Arago point can then only be observed in the early morning and late at night and its location is relatively low and thus not conducive to ground-based remote sensing polarization observation.

For a Brewster point, it appears when the solar elevation is 20°. When the sun moves toward zenith, the Brewster point and the sun are nearly collinear and consistent. When the sun reaches zenith, the Brewster and Babinet neutral points will be in the same location as the sun. The Brewster neutral point is only on one side of the sun, and its location is always lower than the sun. In low-latitude areas, the solar altitude is relatively high. The location of the Brewster neutral point will be relatively high as will the sun, but in high-latitude areas, the sun will not reach zenith, so the location of the Brewster neutral point will be much lower. This advantage of low location in the sky is similar to the advantage conferred when using the Arago neutral point. Secondly, according to Figs. 8.6 and 8.17, the Brewster neutral point is affected by atmospheric conditions and ground surface reflectance, so the Brewster neutral point is not conducive to good ground-targeted polarized remote sensing.

Relatively speaking, the Babinet neutral point is ideal for ground-based polarization remote sensing, because:

(1) The Babinet neutral point is on the same side of the sun, and it may be 18° above the sun when the solar elevation is zero (i.e., when the sun starts to rise). Then it will rise with the sun. When the solar elevation is 90°, it will be in the same location as the sun. So a Babinet neutral point always follows the sun from sunrise to sunset and always lies above the sun. A Babinet neutral point, which can be observed all day and has relatively large elevation angle, is conducive to good lighting in remote sensing practice and is convenient for remote sensing observations.
(2) The Babinet neutral point is only slightly affected by atmospheric conditions. Figure 8.6 shows that, in upward radiation, the Babinet neutral point moves towards zenith when the optical depth increases. When the optical depth is about 1, its location will not change as the optical depth increases. This is to say, the location of the Babinet neutral point changes slightly when the optical depth is less than 1 and its location is unaffected at optical depths exceeding 1: this means that the Babinet neutral point is not sensitive to atmospheric conditions and its location is stable when they change.
(3) The location of the Babinet neutral point is less affected by choice of observation band than the Brewster neutral point, as shown in Fig. 8.7. Observation at the Babinet neutral point is of great significance to multi-spectrum remote sensing.

8.3.4 Application of Atmospheric Neutral Points to Polarization Remote Sensing

8.3.4.1 Studies of Aerial Remote Sensing

Application of atmospheric neutral points to aerial remote sensing requires observation of the ground surface at a Babinet neutral point. Since a Babinet neutral point is mainly affected by solar elevation, the major issue is to decide the timing of any observation.

Solar elevation at a given location changes during a day. The solar elevation of a location can be calculated by the following solar altitude formula:

$$\sin h = \sin\phi \sin\delta + \cos\phi \cos\delta \tag{8.15}$$

where, ϕ is geographic (al) latitude, and by consulting an astronomical calendar, the declination angle δ of the sun can be found, and t is calculated as degree based on time. Based on the aforementioned method of calculation, the solar elevation at any point (longitude, latitude) on the earth at any time can be calculated.

For a Babinet neutral point, its location has a certain functional relationship with solar elevation. When the optical depth is about 0.10, the relationship is as shown in Fig. 8.16: for convenience, a Babinet neutral point curve can be viewed as two sections of a bi-linear plot, namely, fitting the section when solar altitudes are between 0° and 30° as one line segment, and that when the solar elevation is between 30° and 90° as another. That is:

$$\begin{cases} y = 0.9x + 18 \quad (0 < x < 30) \\ y = 0.75x + 22.5 \quad (30 \le x < 90) \end{cases} \tag{8.16}$$

where, x is solar altitude, and y is the elevation angle of the neutral point.

For the equatorial zone, solar elevation at noon is 90°. All neutral points are concentrated around sun points. If the ground surface is observed vertically at this time, the polarization sensor is in a neutral point area (it should be noted that this is a special case).

For non-equatorial zones, the solar altitude at noon will be less than 90°, and the local solar elevation can be calculated on the basis of local time. Then using the calculated solar elevation (Formula (8.15)), the location of the corresponding neutral point can be calculated and then proper remote sensing timings can be ascertained. One thing must be clarified, even if a remote sensor cannot be located in the centre of a neutral area of the sky, there are areas with low DOP around neutral points: such areas are good for ground-based polarization observation.

8.3.4.2 Studies of Aerial Remote Sensing Application

In applying atmospheric neutral points to aerial remote sensing, the major issue is how to design the requisite satellite orbit. Considering that the locations of the Babinet neutral point and location of the sun are relatively consistently correlated, a sun-synchronous orbit is the best choice for ground-based remote sensing relying on neutral points.

The direction of the sun shining orbital plane of a sun-synchronous orbit is almost the same throughout a given year, that is to say, the angle between the orbital plane normal and the shadow on the equatorial plane in the direction of the sun remains almost constant and the local time when the satellite passes over an equatorial node does not change. In sun-synchronous orbit, for locations at the same latitude, the satellite passes by the same location at the same time on each day, so the angle of incidence of the sunlight is consistent. This helps when using a passive remote sensor that takes the advantage of sun reflectance light, because observation conditions are stable. As the illumination angle of each of Landsats-1 to 5 is 37° 30′, and the satellite and sun are synchronous, so a satellite can pass by above the ground at the same local time each day. When Landsats-1 to 5 move in descending mode, it will pass by above the equator at 9:42 local time.

A sun-synchronous orbit has a quasi-stable sun incidence angle. This orbital feature enables a neutral point to be possible in ground-targeted detecting by remote sensing. Figure 8.22 shows sun-synchronous orbit θ. The angle is the illumination angle of the sun-synchronous orbit. This angle remains unchanged as the satellite moves. When this orbit was designed, firstly, the local time when the satellite passes by should be confirmed, then the solar altitude and location of a Babinet neutral point in the sky can be calculated. The angle between the sun and the Babinet neutral point can be calculated, and this is the angle distance between the two. If the orbit illumination angle is set to be the same angle between the sun and the Babinet neutral point in a standard atmosphere, the polarization sensor can observe the land surface at the location of this neutral point throughout the period when the satellite passes by,

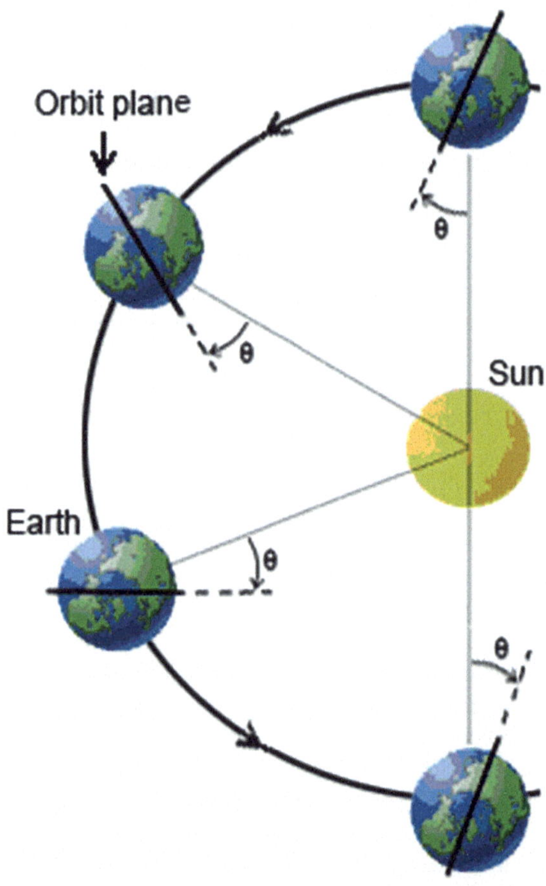

Fig. 8.22 Sun-synchronous orbit

so as to eliminate atmospheric polarization effects and to obtain polarization features of relevance to the land surface.

Of course, it is necessary to work together with researchers involved in satellite remote sensing to design orbits to match such neutral points. Here, we only provide a new method featuring a satellite-borne polarization sensor, in the hope of eliminating atmospheric polarization effects from ground-based polarization remote sensing observations.

This section theoretically analysed the existence of atmospheric polarized points and zones: its significance to remote sensing lies in that a new remote sensing observation window helps to make ground-targeted detection possible and provides new theoretical, and methodological, bases for obtaining land surface information.

8.4 Basic Experiment to Measure Separation of Object, and Atmospheric, Effects Based on a Neutral Point

This section covers the design of experiments to accomplish separation of object, and atmospheric, effects based on the theory of neutral point area observation in Sect. 8.8, so as to validate the feasibility of the method of their separation by exploitation of the advantages of a neutral point.

To validate the feasibility of the method of separating object, and atmospheric, effects by exploiting the advantages conferred by a neutral point, the author designed a ground-based validation experiment. This experiment is concerned with polarization imaging of the same area at neutral points and non-neutral points, and then comparing the polarization parameter results of the imaging. Figure 8.23 shows the observation geometrical diagram for ground experiment validation, and (a) is the observation at neutral points, while (b) is that at non-neutral points. The DOP of the line connecting the ground surface and neutral points is zero. The imaging device adopted in the experiment is a Nikon D200 single-lens reflex camera. In the experiment, an iodine polaroid is installed in front of camera, and the angle between the transmission axis and reference axis is changed to obtain data at 0°, 60°, and 120°.

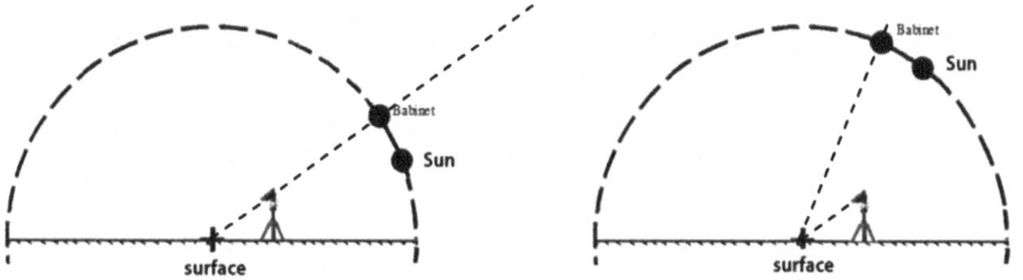

(a) Neutral point observation (morning) (b) Non-neutral point observation (noon)

Fig. 8.23 Observation geometrical diagram for ground experiment validation

Figure 8.23 shows a comparison of polarization parameters when imaging at neutral, and non-neutral, points. The observation times were 7:00 and 11:20 am on 29 April 2010. The observations were made under a clear sky on a calm day. The observations were captured from the platform on the fifth floor of the Remote Sensing Building, Peking University. The targeted area is the Summer Palace Resort. In Fig. 8.23, the left-hand image is aimed at non-neutral points, and the right-hand image is aimed at a neutral point location.

Figure 8.24a shows images shot without a polaroid, with higher atmosphere visibility. The images observed both at a neutral point zone and at a non-neutral zone are clear; Fig. 8.24b shows Q-component polarization images based on calculation. The Q-components of the two images are both less seriously affected by atmospheric effects and land object information can be easily identified. Artificial architecture and exposed soil on mountains are readily visible. Figure 8.24c shows U-component polarization images based on calculation. The images (on the right) observed at neutral points provide a better demonstration of land object information than those captured at non-neutral points.

Figure 8.24d shows the DOP of land objects: the separation of land-atmosphere effects is difficult. On that DOP image based on observation at neutral points (the image on the right), both the Fo Xiang Loft of the Summer Palace (about 3.1 km from the shooting location) and West Mountain (about 6–8 km from the shooting location) are easily identified. In particular, information pertaining to the soil on the mountain (or road) is consistent with that in non-polarized light images: however, for the DOP image based on observation at non-neutral points (the image on the left), information about near land objects is well-demonstrated (for example, the Fo Xiang Loft), but information about the mountains located further away was poorer. This means that polarization information of more distant objects cannot be obtained at non-neutral points. This also means that, with increasing distance between observer and target, atmospheric polarization effects will increase and the polarization information pertinent to land objects will be poorer.

Comparing the two images in Fig. 8.24d, information about land objects in the DOP image based on observation at neutral points (right) is far richer than that based on observation at non-neutral points, especially for distant land objects. This proves that imaging in a neutral point zone can help to eliminate atmospheric polarization effects and enhance land object polarization effects. So, the feasibility of the method of separating object, and atmospheric, effects by exploiting the advantage conferred by imaging at neutral points is validated.

Fig. 8.24 Comparison of polarization parameters when imaging at neutral, and non-neutral, points

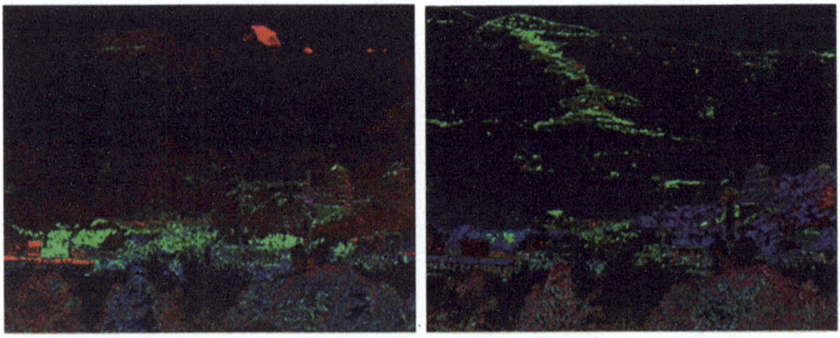

(d) DOP image

Fig. 8.24 (continued)

References

Berry M, Dennis M, Lee R (2004) Polarization singularities in the clear sky. New J Phys 6(162):1–14
Chandrasekhar S (2003) Radiative transfer. Dover Publications Inc, New York
Chandrasekhar S, Elbert D (1954) The illumination and polarization of the sunlit sky on Rayleigh scattering. Trans Amer Philos Soc 44:643
Coulson K (1959) Characteristics of the radiation emerging from the top of a Rayleigh atmosphere 1. intensity and polarization. Planetary Space Sci 1(4):265–276
Coulson K (1988) Polarization and intensity of light in the atmosphere. A. Deepak Publisher, Hampton
Coulson K, Dave J, Sekera Z (1960) Tables related to radiation emerging from a planetary atmosphere with Rayleigh scattering. University of California Press, Berkeley
Horvath G, Varju D (2004) Polarized light in animal vision: polarization patterns in nature. Springer, Berlin
Horvath G, Bernath B, Suhai B et al (2002) First observation of the fourth neutral polarization point in the atmosphere. J Opt Soc Amer Opt Image Sci Vis 19(10):2085–2099
Kattawar G, Plass G, Hitzfelder S (1976) Multiple scattered radiation emerging from Rayleigh and continental haze layers. 1: Radiance, polarization, and neutral points. Appl Opt 15(3):632–647
Plass G, Kattawar G, Catchings F (1973) Matrix operator theory of radiative transfer. 1: Rayleigh scattering. Appl Opt 12(2):314–329
Sekera Z (1956) Recent developments in the study of the polarization of sky light. Adv Geophys 3:43–104

Chapter 9
Atmospheric Remote Sensing 3: Atmospheric Polarization Characteristics and Multi-angular Three-Dimensional Chromatography

Quantitative studies of atmospheric aerosol particles are key to atmospheric polarization remote sensing physics, because aerosol particles are decisive factors affecting the remote sensing atmospheric window. Furthermore, atmospheric polarization characteristics can be described through space polarization diagram theory and atmospheric polarization neutral point zone theory, which is equally effective to space polarization field theory when describing atmospheric polarization characteristics. To be specific, polarization is the physical basis for atmospheric scattering given the importance of its polarization features and proves the single scattering characteristics of non-spherical aerosol and polarization observations, to quantify the atmosphere through polarization methods. It also underpins multiple scattering based on non-spherical models and multiple flat distributions of the full-sky polarization field, to achieve atmospheric three-dimensional chromatography at various observation angles, separation inversion of non-spherical aerosol, and land surface information based on space polarization field theory. The theories expounded in Chaps. 7 and 8 are combined to constitute polarization field vector features for use in remote sensing.

9.1 Physical Basis of Atmospheric Scattering

This section introduces the physical basis for aerosol inversion by utilising polarization remote sensing information, namely the scattering characteristics of atmospheric particles, including Rayleigh scattering and Mie scattering.

9.1.1 Single Scattering in the Atmosphere

To quantify atmospheric scattering, it is necessary to calculate all input and output Stokes parameters. When R, which is the distance between particles and observation

L. Yan et al., *Polarization Remote Sensing Physics*, Springer Remote Sensing/Photogrammetry, https://doi.org/10.1007/978-981-15-2886-6_9

locations, is too large, namely larger than the wavelength, a scattering matrix $P(\xi)$ can be utilised to demonstrate this process:

$$I(\xi) = k_e\omega\frac{dv}{4\pi R^2}P(\xi)I_0 \tag{9.1}$$

where I_0 is the incident Stokes vector, I is the scattering Strokes vector, ξ is the scattering angle between incident and outgoing light, and dv is small cubic unit including scattering elements. I_0 and I have the same units as the irradiance; k_e is an extinction coefficient, denoting the extinction role of scattering elements on light in the unit volume. The following relationship holds:

$$k_e = \frac{\sigma_e}{dv} = \frac{1}{dv\sum_i \sigma_{e,i}} \tag{9.2}$$

Terms in Formula (9.2) have units of σ_e as m^2, denoting the area of photons that each particle can capture. The total particle cross-sectional area per unit volume is found by summing those of all particles. Extinction caused by scattering and absorption can be embodied by single scattering albedo ω, denoting the ratio of scattering in total extinction:

$$\omega = \frac{\sigma_s}{\sigma_e} = \frac{\sigma_e - k_a}{\sigma_e} = \frac{k_s}{k_e} = \frac{k_e - k_a}{k_e} \tag{9.3}$$

where k_s is a scattering coefficient, k_a is an absorption coefficient, and σ_s is the scattering cross-sectional area. For situations where the single scattering albedo is 1, a scattering element is fully scattered and there is no absorption. For situations where the single scattering albedo is 0, a scattering element provides full absorption and there is no scattering.

For particles with the same features in all directions, k_s and ω do not change with scattering angle, but for both air particles or other macro-particles, their scattering of electromagnetic waves in all directions is different. So a matrix embodying the scattering phase function is needed to demonstrate the difference in scattering capacity of particles in all directions. The matrix of the scattering phase function demonstrates the radiation capacity in all different scattering angles. Namely, it acts as a rotating matrix transferred from incident I_0 to outgoing I. In P, the scattering phase function, component I is normalised.

$$\int_{4\pi} P_{1,1}(\xi)\frac{d\Omega}{4\pi} = 1 \tag{9.4}$$

where $d\Omega$ is a solid angle. Another physical factor used to measure the features of scattering related to direction is non-symmetrical factor g. In normal situations, scattering elements are believed to be spherical and their components vertical to the scattering cross-section are equal. Uncertain elements can be believed to be a scattering weight for front and back directions, and can be viewed as equivalent to a phase function, but for large-particle aerosol, it is not normally spherical. Then, a non-symmetrical element is no longer equivalent to the phase function and non-symmetrical factor can be expressed as:

$$g = \int_{4\pi} \cos P_{1,1}(\xi) \frac{d\Omega}{4\pi} \tag{9.5}$$

If $g = 1$, then we believe frontward scattering is dominant: if $g = -1$, then backward scattering is dominant (if $g = 0$, they are similar.)

$P(\xi)$ is a 4×4 scattering matrix:

$$P(\xi) = \begin{bmatrix} a_1(\xi) & b_1(\xi) & 0 & 0 \\ b_1(\xi) & a_2(\xi) & 0 & 0 \\ 0 & 0 & a_3(\xi) & b_2(\xi) \\ 0 & 0 & -b_2(\xi) & a_4(\xi) \end{bmatrix} \tag{9.6}$$

where, when the scattering matrix is simplified, it includes six independent variables and 16 dependent variables. To simplify the scattering matrix into the forma above, at least one of the following conditions should be met:

Scattering particles are distributed randomly. The scattering plane is symmetrical (e.g. spherical);

Scattering particles are distributed randomly and they have same amount of mirroring particles;

The size of scattering particles is much less than the incident light wavelength, so Rayleigh scattering is the decisive factor affecting the scattering matrix.

In the studies in this book, atmospheric scattering can be viewed as Rayleigh scattering or Mie scattering and both meet the aforementioned conditions so Mie scattering can be simplified to Formula (9.6) and Rayleigh scattering occurs when scattering particles are much smaller than the incident light wavelength, while Mie scattering occurs when the particle size is almost equivalent to the incident wavelength. Mie scattering can be used to describe particle scattering of spherical, or non-spherical, shapes. Aerosol is not spherical, especially large-particle aerosol. Now studies about non-spherical dust aerosol scattering have been started, for instance, to simulate scattering of aerosol through the *T*-matrix method.

9.1.2 Rayleigh Scattering

When electromagnetic waves interact with tiny particles so that scattering occurs, dipoles can be caused to vibrate and emit electromagnetic wave, namely, scattering radiation. This kind of scattering is called Rayleigh scattering. The condition where Rayleigh scattering occurs is when the size of the particles is much smaller than the wavelength of the incident electromagnetic wave. The size of particles is defined as x, and their relationship with incident electromagnetic wavelength is given by:

$$x = \frac{2\pi r}{\lambda} \tag{9.7}$$

where, λ is the wavelength of the incident electromagnetic wave, and r is the geometrical radius of the particles. For particles measuring between 0.002 and 0.02, the scattering process can be simulated by Rayleigh scattering, i.e., as a scattering process of visible light and near-infrared radiation in the atmosphere.

The Rayleigh scattering coefficient is given by:

$$k_s = \frac{8\pi^3 \left(m^2 - 1\right)^2}{3\lambda^4 n}\left[\frac{6 + 3\delta}{6 - 7\delta}\right] \tag{9.8}$$

where m is the refractive index, n is number of molecules per unit volume, and δ is used to correct the slight asymmetry of molecular scattering (atmospheric molecules are not ideal dipoles, and will not produce completely polarized light at 90°). For air, in the visible light band, δ is almost equivalent to 0.03, but m is almost equivalent to 1.000277. If the refraction index is plural, then the role of absorption must be taken into account in Rayleigh scattering models. In the near-infrared band of a research scanning polarimeter (RSP) and aerosol polarimetric sensor (APS), such a scenario occurs, for instance, the absorption of carbon dioxide in the 1.6 μm band, and the absorption of methane in the 2.2 μm band. For other bands of RSP, the single scattering albedo is about 1, namely, absorption by air can almost be neglected. The Rayleigh scattering coefficient is almost inversely proportional to the biquadrate of the incident electromagnetic wavelength. For RSP, this means that Rayleigh scattering in the blue light band is stronger than that in the near-infrared band.

Since dipole vibration is vertical to the spreading direction of electromagnetic waves, scattering is direction-related. This process can be described by the simple scattering matrix included in Formula (9.6), but parameters need to be modified for this direction-related scattering matrix. Among these,

$$a_1(\xi) = \Delta_r \frac{3}{4}\left(1 + \cos^2 \xi\right) + (1 - \Delta_r)$$

$$a_2(\xi) = \Delta_r \frac{3}{4}\left(1 + \cos^2 \xi\right)$$

$$a_3(\xi) = \Delta_r \frac{3}{2} \cos \xi$$

$$a_4(\xi) = \Delta_r \Delta_{r'} \frac{3}{2} \cos \xi$$

$$b_1(\xi) = -\Delta_r \frac{3}{4} \sin^2 \xi \tag{9.9}$$

Other parameters can all be set to 0 in Formula (9.6) and in the formula above, Δ_r and $\Delta_{r'}$ are related to the isomerism of all directions of molecules as demonstrated in the following formula:

$$\Delta_r = \frac{1-\delta}{1+\frac{\delta}{2}}$$
$$\Delta_{r'} = \frac{1-2\delta}{1-\delta} \tag{9.10}$$

For the RSP studied here, strength of Rayleigh scattering is in direct proportion to the total amount of atmospheric molecules in the track. The reason for this is that, if atmospheric molecular absorption can be neglected, scattering coefficient is constant. For an atmosphere in static equilibrium, the Rayleigh optical depth from the atmosphere's top layer to a certain level can be accurately calculated according to the air pressure at that elevation. Molecule and aerosol scattering can thus be separated in radiation transmission models, so the optical features of aerosol are important parameters requiring optimisation and accurate inversion in radiation transmission models.

9.1.3 *Mie Scattering*

If particle size and incidence wavelength are equivalent, namely when the particle size is 0.2–2000 (e.g., the size of aerosol particles and cloud droplets), it would be inaccurate to calculate atmospheric scattering as if it were Rayleigh scattering. In the beginning of the twentieth century, Ludvig Lorenz and Gustav Mie both independently deduced what became known as the Mie scattering formula by way of the Maxwell equation. Here, we calculate the scattering feature of spherical particles based on density, size, and complex refractive index of spherical particles in a unit volume by taking advantage of the most important part of Mie scattering theory.

The phase function of Mie scattering can be demonstrated by Formula (9.6), with parameters given by:

$$a_1(\xi, x, m) = a_2(\xi, x, m) = \frac{1}{2}\left(S_1 S_1^* + S_2 S_2^*\right)$$
$$a_3(\xi, x, m) = a_4(\xi, x, m) = \frac{1}{2}\left(S_1 S_2^* + S_2 S_1^*\right)$$
$$b_1(\xi, x, m) = \frac{1}{2}\left(S_1 S_1^* - S_2 S_2^*\right)$$
$$b_2(\xi, x, m) = \frac{1}{2}\left(S_1 S_2^* - S_2 S_1^*\right) \tag{9.11}$$

where S_1 and S_2 are scattering functions, as deduced by use of the Mie scattering equation. They form a complicated boundless array and S_1 and S_2 have values dependent on scattering angle, particle size, and refraction parameter. The * in the formula means complex conjugate. The scattering function is:

$$S_1(\xi, x, m) = \sum_{j=1}^{\infty} \frac{2j+1}{j(j+1)} \left[a_j \pi_j(\cos\xi) + b_j \tau_j(\cos\xi)\right]$$
$$S_2(\xi, x, m) = \sum_{j=1}^{\infty} \frac{2j+1}{j(j+1)} \left[b_j \pi_j(\cos\xi) + a_j \tau_j(\cos\xi)\right] \quad (9.12)$$

where, a_j and b_j are the sizes of scattering particles and are functions of the refraction parameters; π_j and τ_j are functions of scattering angle, whereby π_j and τ_j are related to the Legendre polynomial expansion:

$$\tau_j(\cos\xi) = \frac{d}{d\xi} P_j^1(\cos\xi)$$
$$\pi_j(\cos\xi) = \frac{d}{\sin\xi} P_j^1(\cos\xi) \quad (9.13)$$

where, P_j^1 is the Legendre polynomial expansion. Relatively complex calculation of a_j and b_j is best done as part of the Mie scattering value calculation. These scattering parameters are related to scattering particle sizes and the refraction index. Multiple iterative operations are necessary and the calculation process is complicated.

Mie scattering parameters can be used for calculating scattering and extinction coefficients of particles. Scattering coefficients and extinction coefficients are both composed of an infinite sequence related to scattering particle size and refraction parameters. The volume scattering coefficient is given by:

$$k_s = \frac{n\lambda^2}{2\pi} \sum_{j=1}^{\infty} (2j+1)\left(a_j a_j^* + b_j b_j^*\right) \quad (9.14)$$

The volume total extinction parameters are:

$$k_e = \frac{n\lambda^2}{2\pi} \sum_{j=1}^{\infty} (2j+1) R\left(a_j + b_j\right) \quad (9.15)$$

In Formulae (9.14) and (9.15), particle n is the amount (number) density of the unit volume, and R denotes the real number part of the complex refractive index. The imaginary part of the complex refractive index represents the absorption-related features of particles, and single scattering albedo is given by:

$$\omega = \frac{\sum_{j=1}^{\infty} (2j+1)\left(a_j a_j^* + b_j b_j^*\right)}{\sum_{j=1}^{\infty} (2j+1)R\left(a_j + b_j\right)} \tag{9.16}$$

In practice, to confirm the parameters above, we should confirm j, the total number of arrays: j is dependent on available computer power. Generally speaking, j is a little larger than x, the particle size parameter. So, calculation of Mie scattering may be increasingly difficult due to increasing particle size. If too much calculation is involved, error due to misconvergence may arise, so, when x, the particle size parameter, exceeds 2000, geometrical optical methods can be used to carry out radiation calculations. In the calculation of aerosol and molecule scattering, the particle size parameter is much less than 2000, so normally it is not necessary to use geometrical optical calculations.

When the particle size is too small, Mie scattering calculation results are similar to those assuming Rayleigh scattering. Namely, backward and forward scattering are almost the same ($g = 0$), and the scattering component normal to the spreading direction is the smallest. As the particle size increases, the forward scattering component increases, and the asymmetrical component also increases. When the particle size increases to that of cloud droplets, scattering side-lobes start to appear, and cloud bows will be generated. Cloud bows can be accurately described by polarization radiance and are sensitive to the size of cloud droplets thus distinguishing scattering generated by cloud from that by aerosol. Scattering generated by aerosol does not generate significant side-lobe effects.

This section introduces calculation methods used to analyse single scattering features of atmospheric particle sets and the main scattering feature of atmospheric particles, and Rayleigh scattering and Mie scattering, laying the theoretical basis for aerosol inversion through polarization information.

9.2 Single Scattering Characteristics of Non-spherical Aerosol and Polarization Observation Evidence

Why it is necessary to study the single scattering characteristics of aerosol? Interaction between particles in the atmosphere (molecules and aerosol) and light is a multiple-scattering process, which is constituted by countless single scattering events. Calculation of multiple scattering events should be based on known single scattering characteristics. That means, single scattering forms the basis for multiple scattering, and will serve as the input for calculating multiple scattering and radiation transmission.

Atmosphere aerosol and molecules are key scattering elements: for atmospheric molecules, their size is much smaller than the wavelength of the incident wave. Atmospheric molecules have a stable composition, and studies on their scattering feature have been thorough. The size range of aerosol particles expands over five scales. For a large proportion of aerosol particles, their size should be equivalent to the incident wavelength, and even larger than the radiant wavelength. When aerosol is assumed to be spherical, Mie scattering theories are generally adopted to obtain aerosol scattering characteristics. Generally speaking, three assumptions are necessary: (1) ideal spheres, (2) an even composition and random distribution of particles, and (3) a mirror surface on each particle. Although some aerosol particles in coarse mode may not strictly conform to Mie scattering theory regarding requirements on particle shape, Mie scattering theory remains a good predictor of real scenarios.

9.2.1 Polarization Scattering Characteristics of Non-spherical Particles

In some circumstances, there are significant effects arising from the presence of non-spherical dust aerosol particles (for example, when the predominant component of atmospheric aerosol is dust). Since dust is generated mostly through mechanical means, so the particles are relatively large (generally above 1 μm), and its shape normally cannot be described by a normative sphere. Internationally, inversion of dust-type aerosol is simulated by spherical particle ratio, to assume aerosol is composed of spherical and non-spherical particles, while non-spherical particles are based on an average combination of oval aerosol of different eccentricities. This method generally adopts a ground photometer from the Aeronet Robotic Network to carry out observation and the results are relatively accurate. In aerosol inversion, the inversion algorithm of MODIS ground aerosol also includes non-spherical dust aerosol (Levy et al. 2007). For non-spherical dust aerosol, *T*-matrix and geometrical optical methods are introduced to calculate single scattering characteristics thereof.

Actually, there are no accurate models available for describing dust-type aerosol. Generally, aerosol models which combine *T*-matrix calculations and geometrical optical algorithms in Aeronet observation simplified dust-type aerosol into 25 groups of oval aerosol particles based on their eccentricities (namely, 0.33, 0.37, 0.40, 0.44, 0.48, 0.53, 0.57, 0.63, 0.69, 0.76, 0.83, 0.91, 1.00, 1.09, 1.20, 1.31, 1.44, 1.58, 1.73, 1.89, 2.07, 2.49, 2.73, and 2.99), and relative proportions overall. The *T*-matrix is used to calculate scattering characteristics of symmetrical particles on a non-spherical surface by directly solving Maxwell's equation (Martonchik and Diner 1992). It is one of the most powerful tools available for calculating non-spherical particle behaviour. Generally, it can help to resolve the scattering characteristics of particle groups where the particle size parameter is within 50. It can also be used to calculate scattering characteristics of various non-spherical particles, but when the shape of particles becomes more non-spherical, the *T*-matrix method will become slow and can only be applied to fewer particle sizes. Furthermore, misconvergence may arise.

In a real process, when the incident wavelength is 550 nm, use of the *T*-matrix becomes difficult at an aerosol radius exceeding 5 μm, which is problematic given the fact that aerosol particles can reach 20 μm in diameter. When particle sizes increase, their scattering characteristics deviate from Mie scattering behaviours and tend to that predicted by geometrical optics. By utilising a method combining geometrical optics and the *T*-matrix, the scattering characteristics of particles when the size is small and that of non-spherical dust particles whose size is relatively large can be realised.

Single scattering characteristics of particles calculated based on non-spherical particles differ greatly from those based on spherical particles. The major difference is demonstrated in the backward direction which is the main direction for applying ground-targeted observation remote sensing. To take 860 nm as an example (at 550 nm a similar result ensues so this will not be explained further, analysis was conducted of the differences in F_{11}, F_{12}, and F_{33} (three main parameters affecting linear polarization scattering) and the effect on the aerosol scattering matrix for different particle shapes.

Figure 9.1 shows the difference in scattering phase function F_{11}, among all shapes of aerosol particles. The left-hand figure shows the scattering characteristic differences between spherical particles and aerosol particles (all 25 eccentricities). The right-hand figure shows the scattering characteristic differences between spherical particles and particles of eight different shapes. Among them, the parameters of aerosol are: logarithmic normal distribution; median radius 0.5 μm, and standard deviation 1.09. The calculation model is a combination of geometrical optical models and the *T*-matrix model (total scattering angles calculated: 181). The right-hand figure shows that there is a difference in backward section scattering characteristics of oval and spherical particles (these directions are the major observation directions for ground-targeted observation). By mixing particles of different shapes in certain proportions, the particle scattering characteristics obtained differ from those of purely spherical particles so, it is necessary to conduct inversion through a non-spherical particle model.

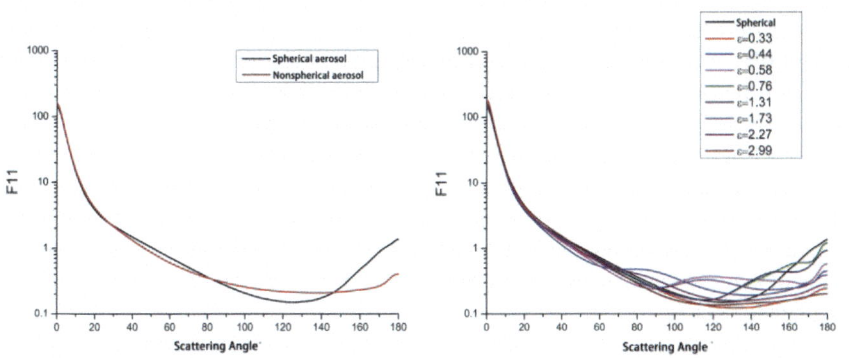

(a) Average difference between spherical and non-spherical models

(b) Difference between different non-spherical models

Fig. 9.1 F_{11}, the phase function of aerosol scattering with different particle shapes at the 860 nm band for a geometrical optical model and the *T*-matrix method

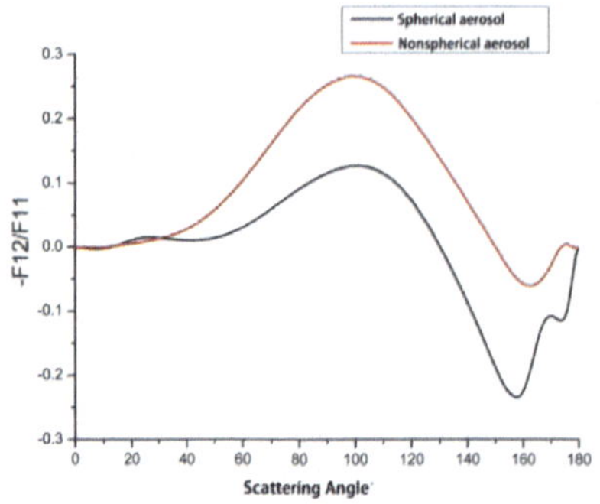

(a) Average difference between spherical and non-spherical models

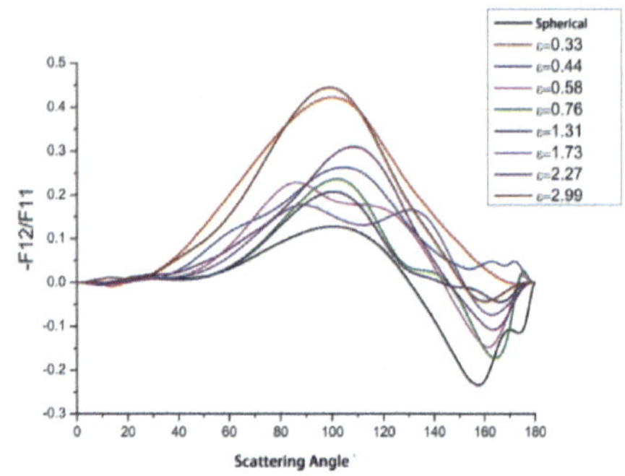

(b) Difference between different non-spherical models

Fig. 9.2 F_{12}, the phase function of aerosol scattering with different particle shapes at the 860 nm band for a geometrical optical model and the *T*-matrix method

Figure 9.2 shows the difference of scattering phase function F_{12}, among all shapes of aerosol particles. The left-hand figure shows the scattering characteristic differences between spherical particles and aerosol particles (all 25 eccentricities). The right-hand figure shows the scattering characteristic differences between spherical particles and particles of eight different shapes. Among them, the parameters of aerosol are: logarithmic normal distribution; complex refractive index 1.52–0.008i, median radius 0.5 μm, and standard deviation 1.09. The calculation model is a combination of geometrical optical models and the *T*-matrix model (total scattering angles calculated: 181). F_{12} is the major polarising factor, having a crucial impact on the polarization scattering characteristics of aerosol particles. This means that, regardless of the different shapes or mixtures of particles, their F_{12} scattering characteristics are all quite different from those of spherical particles. This also means it is necessary to adopt non-spherical particles in aerosol inversion when we utilise polarization for inversion of aerosol particles.

Figure 9.3 shows the difference of scattering phase function F_{33}, among all shapes of aerosol particles. The left-hand figure shows the scattering characteristic differences between spherical particles and aerosol particles (all 25 eccentricities). The right-hand figure shows the scattering characteristic differences between spherical particles and particles of eight different shapes. Among them, the parameters of aerosol are: logarithmic normal distribution; complex refractive index 1.52–0.008i, median radius 0.5 μm, and standard deviation 1.09. The calculation model is a combination of geometrical optical models and the *T*-matrix model (total scattering angles calculated: 181).

In F_{33}, there remains a relatively large difference between spherical and non-spherical scattering characteristics, so we need to use particle scattering characteristics calculated through non-spherical models to obtain reflectance and polarization reflectance signals at top of atmosphere (TOA).

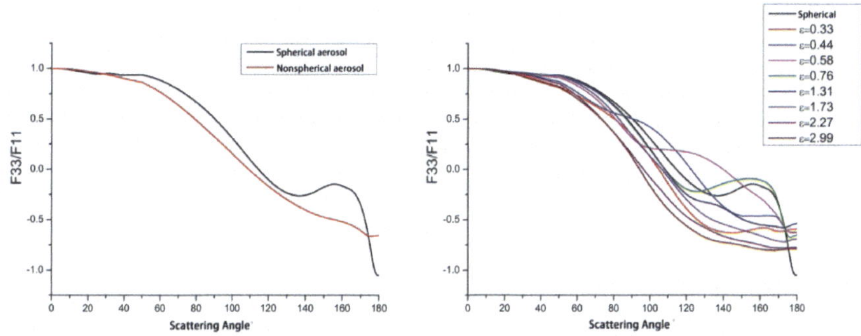

(a) Average difference between spherical and non-spherical models

(b) Difference between different non-spherical models

Fig. 9.3 F_{33}, the phase function of aerosol scattering with different particle shapes at the 860 nm band for a geometrical optical model and the *T*-matrix method

9.2.2 Polarization Scattering Characteristics of Coarse and Fine Aerosol Particles

The analysis above concerns coarse aerosol. Is non-spherical modelling necessary, or not, for fine aerosol? What kinds of aerosol particles should be calculated based on non-spherical models and what kinds need simpler simulation through spherical models? Mod4 and Mod9 (MODIS aerosol type) were selected to calculate the single scattering characteristics of fine and coarse aerosol, respectively. To take 860 nm as an example, calculation of the difference of F_{11} and F_{12} parameters was conducted in spherical and non-spherical scenarios, respectively. Figure 9.4 shows the difference

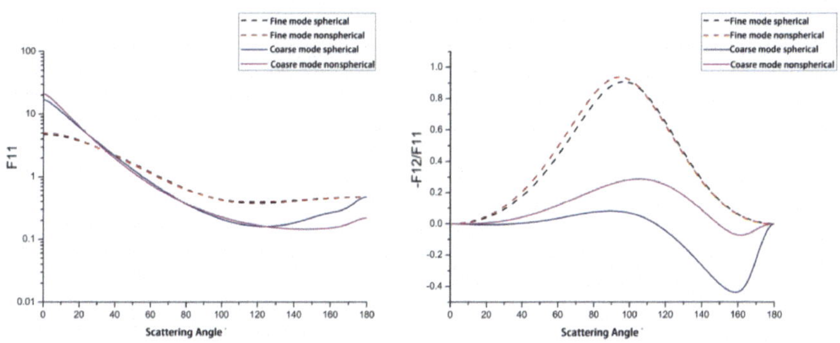

(a) Difference in F_{11}, phase function, between spherical and non-spherical models

(b) Difference in F_{12}, phase function, between spherical and non-spherical models

Fig. 9.4 Phase function of aerosol scattering of coarse and fine modes at the 860 nm band for a geometrical optical model and the *T*-matrix method in F_{11} and F_{12}, the aerosol scattering phase function, of fine and coarse modes under spherical and non-spherical scenarios. This figure shows that, for fine mode, the difference between spherical and non-spherical particles is small, so for fine particles, the calculated phase function data in spherical simulation and non-spherical calculation are consistent; however, in coarse mode, the difference between spherical and non-spherical particles is huge.

9.2.3 The Effect of Non-spherical Particle Models on Aerosol Inversion

Is there any difference for signals received by a sensor at TOA? Is the difference significant for signals received at TOA through non-spherical and spherical aerosol models? This section will adopt a logarithmic normal distribution mode to describe spectral distributions of aerosol, to analyse difference between reflectance index, and polarization reflectance, signals received at TOA by particles of different shapes in coarse or fine modes, respectively.

Figure 9.5 shows the inversion difference of AOD by spherical and non-spherical aerosol. Figure 9.5b shows the difference between the two when AOD is 0–0.6. So the difference is obvious when the AOD is below 1 and the difference between non-spherical and spherical aerosol models can reach 30%. Figure 9.5a shows the difference when AOD is 0–4. The results of the two scenarios are similar when the AOD is relatively large. To summarise, the results of the two models differ but their trend is similar, so we need to examine how crucial is this difference to climate measurement? According to IPCC2007, radiation caused by dusty-type aerosol generated from human activities is about -0.3 ± 0.1 W m^{-2}. The gaps between radiative forcing deviation of short-wave and long-wave signals obtained by calculation through

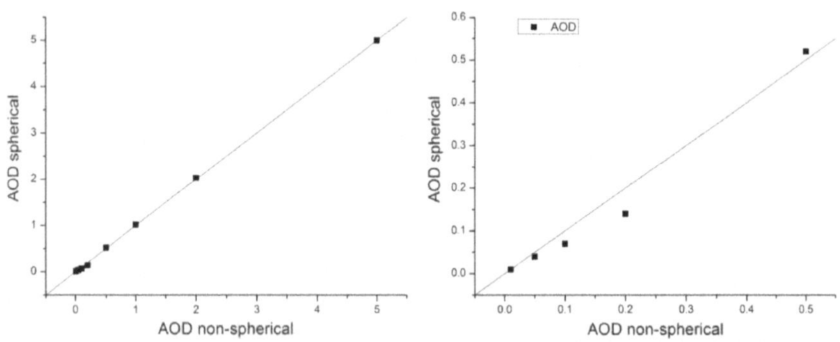

(a) When AOD is 0-4 (large range) (b) When AOD is 0-0.6 (small range)

Fig. 9.5 Inversion of AOD difference by spherical and non-spherical particles (abscissa is non-spherical, and ordinate is spherical) non-spherical dust-type aerosol and spherical dust aerosol are 13% and 6% respectively. The difference is significant, so it is necessary to adopt non-spherical dust aerosol models when simulating coarse mode aerosol.

The single scattering characteristics of aerosol particle group are the basis of aerosol inversion. So a proper aerosol model, as a priori knowledge of local aerosol type, is important. The error in selecting aerosol models can lead to significant inversion deviation or error. Factors that affect aerosol single scattering characteristics include aerosol size distribution, aerosol absorption characteristics, fine and coarse modes of aerosol, and its shape. In traditional inversion, aerosol particles are generally viewed as spherical to calculate the single scattering characteristics, but the aerosol is not spherical, or the majority thereof is non-spherical. The difference between results based on spherical models and TOA reflectance index data calculated based on non-spherical particles may be as much as 5%. The TOA DOP difference can be as high as 30%. This emphasises the need for adoption of non-spherical models.

This section introduced the theory governing aerosol inversion through polarization remote sensing, namely single scattering characteristics of non-spherical dust aerosol particles, which lays the foundation for calculating multiple scattering characteristics of atmospheric particles.

9.3 Multiple Scattering Based on Non-spherical Models and a Plane Distribution of the Full-Sky Multi-angular Polarization Field

This section mainly introduces the calculation of planar distribution characteristics of the full-sky polarization field based on single scattering model calculations (Sect. 9.2). This helps to provide visible models for inversion of aerosol characteristics based on polarization information.

9.3.1 TOA Multiple Scattering Calculation Based on Non-spherical Aerosol Models

Single scattering forms both the input, and basis, for multiple scattering calculations. The first factor that should be taken into account is whether, or not, single scattering characteristics of non-spherical dust aerosol calculated based on micro-level parameters includes polarization characteristics still having the corresponding polarization distribution after multiple scattering events. Will such a distribution be related to aerosol information? In addition, does the TOA signal based on non-spherical calculation bear any relation to spherical calculation results?

Actually, the scattering characteristics of spherical and non-spherical fine mode aerosol are similar, so here we only distinguish the TOA difference between spherical

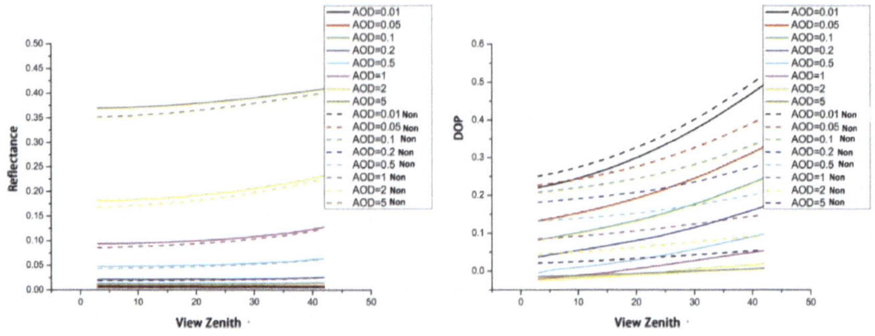

(a) Polarization reflectance index (b) DOP

Fig. 9.6 Calculation of polarization parameter difference between spherical and non-spherical coarse aerosol in the 860 nm band

and non-spherical coarse particles. We assume that, in the atmosphere, there is only coarse mode aerosol present and, by calculation at different AOD levels, we can obtain the difference of reflectance and DOP (860 nm) of TOA in the two situations shown in Fig. 9.6.

It could be seen that TOA reflectance in these two situations differs but the difference is generally within 5%: however, the polarization reflectance of TOA reaches 30%, so if the spherical assumption is adopted to calculate single scattering characteristics of aerosol, there will be a certain difference between the calculated result and reality. If spherical particles are adopted to establish coarse mode models, there will not only be differences in TOA reflectance calculations, but also relatively large deviations in TOA polarized reflectance calculations.

9.3.2 Dual-Distribution of Sky Polarization Patterns Based on Non-spherical Aerosol Models

The TOA polarization signal has a certain relationship with aerosol information. So, what is the distribution of polarization signals across the full-sky (full observation azimuth)? Remote sensors fall into two categories: space-based sensors (those carried on aircraft or satellite platforms), or ground-based sensors (e.g., ground photometry measurement instruments). Space-based sensors utilise the reflectance signal of either the atmosphere or land surface to carry out inversion of atmospheric, or land surface, parameters: ground-based sensors generally directly utilise direct or scattered sunlight to conduct atmospheric signal inversion.

In early research, Wu Xiatai and other researchers from Peking University conducted research on sky polarization patterns through ground-based polarization observation instruments. They found that relatively stable polarization pattern distributions do exist in the sky, and they also separated the effects between objects and atmosphere based on atmosphere polarization neutral point zones that have a relatively steady location in the sky polarization pattern. For space polarization patterns observed through ground-based observation, Guan Guixia, Yan Lei, and other researchers took advantage of bionic theories and carried out navigation studies based on steady polarization patterns in the sky. So, is there any rule for polarization signals detected by satellite-based sensor? Is this applicable to (1) ground-targeted observation research at atmospheric neutral points, and (2) inversion of aerosol information through use of polarization patterns? Here, we elaborate on these two issues. For the first issue, is the sky polarization pattern from ground-based observation also observed from space-based observation platforms? Can similar phenomena be observed thereon? That is to say, for satellite-based sensor observation, is there a relatively steady polarization pattern for the land surface (herein even land surface, and non-even ground surface issues are not covered), too? If yes, then do atmospheric neutral point areas observed during ground-based observation also exist in this situation?

To answer this question, we adopt a vector radiative transfer model to simulate the distribution characteristics of ground-based and satellite-based full-sky polarization signals observed in the sky over the ocean. The aerosol model adopted in this section is the standard radiative atmosphere (SRA) aerosol model. The model is a cloudless atmosphere aerosol model proposed by IAMAP (Radiation Commission of International Association of Meteorology and Atmospheric Physics) in 1983. In the troposphere, aerosol is the combination of four basic elements that all have a logarithmic normal distribution. The four elements are dust type, water-soluble type, marine type, and coat smoke type. The particle shape of the aerosol follows a non-spherical model.

The ocean surface can be viewed as an even land surface. Its reflectance characteristics constitute three parts: scattering by water volume, reflectance by water surface foam, and sun glint reflectance. In actual studies, water volume scattering (wastewater leaving reflectance) is viewed as non-polarized, and it is generated based on pigment concentration, salinity, and other conditions of the ocean (Dubovik et al. 2011). Water surface foam reflectance and solar flare reflectance are the two major sources of polarization, and are mostly caused by sea waves. In this section, polarization effects of the ocean are viewed as being caused by mirror reflection from broken units generated by water surface waves. In simulation, aerosol is assumed to be a scattering-type marine aerosol. By simulating the full-sky polarization pattern of space-based observations at TOA and full-sky polarization pattern of ground-based sky observations, the following contrasting relationships can be plotted (Fig. 9.7) where the polar radius means observation zenith and polar angle for observation azimuth angle. The sun incident zenith is 30° and the sun incidence azimuth angle is 0°.

According to Fig. 9.7, in the ocean surface scenario, satellite-based and ground-based full-sky polarization has certain a level of interworking (except in the mirror direction, where, due to significant effects arising from the mirror reflection effect, the existing radiative transfer model and land surface BRDF model cannot describe the reflectance signal and polarization reflectance signal in this direction). This section

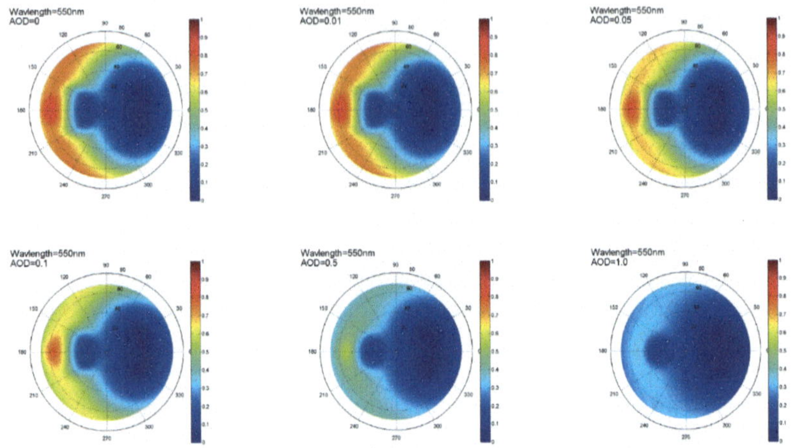

(a) Full-sky polarization distribution satellite base in ocean underlying surface situation (wavelength, 550 nm, AOD = 0, 0.01, 0.05, 0.1, 0.5, and 1.0)

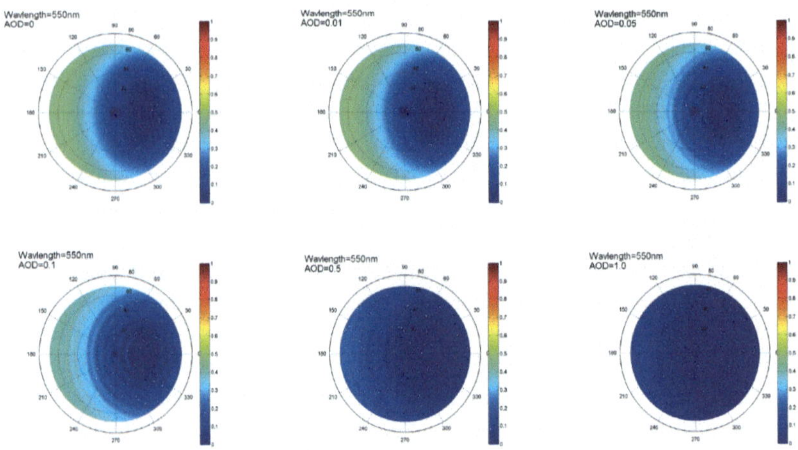

(b) Full-sky polarization distribution ground base in ocean underlying surface situation (wavelength, 550 nm, AOD = 0, 0.01, 0.05, 0.1, 0.5, and 1.0)

Fig. 9.7 Interwork-distribution of full-sky polarization distribution in ocean underlying surface situation (**a** satellite base; **b** ground base)

adopts SOSRT (successive order of scattering radiative transfer) model of aerosol inversion of AERONET (aerosol robotic network) of NASA (National Aeronautics and Space Administration), which is a radiative transfer model for AERONET realising global aerosol inversion, and the radiative transfer model for inversion of PARASOL/POLDER (polarization and directionality of the earth's reflectances). This shows that sky-based sky polarization patterns and ground-based sky polarization patterns have a certain interworked nature. Firstly, locations with lowest DOPs are all somewhere near the sun incidence direction, and that is to say, atmospheric polarization neutral points or neutral point zones exists near the sun incidence direction and this zone works both for sky-based and ground-based observations. Secondly, DOP is always centred around the sun and increases radially outwards. The zone with the highest DOP is in a large observation zenith zone in a direction opposite to that of the sun. In this zone, the atmospheric pathway is relatively long, atmospheric scattering is even larger, and light polarization effects increase. Ground-based and sky-based polarization distribution is on the same magnitude. This means upward and downward DOPs are equivalent. This is because the reflectance of the ocean at 550 nm is around 0.5% and oceanic contributions are negligible. In addition, with increased aerosol optical depth, ground-based full-sky polarization and sky-based full-sky DOP both decrease significantly. When the aerosol optical depth reaches 0.5, full-sky DOPs are below 30% and there is no polarization signal near the sun incidence direction. This means that an aerosol increase may reduce sky DOP.

Though sky-based and ground-based polarization patterns are related, they differ slightly. In ground-based observation, the section with a DOP of less than 0.1 is relatively widely distributed and the DOP is generally weak in the sun direction because this direction is severely affected by direct sunshine and any polarization effect is relatively small. In a pure Rayleigh sky, ground-based observation has a larger DOP than sky-based DOP. With increasing aerosol, DOP that can be observed by ground-based detection will drop sharply, and the rate of change in sky-based detection is smaller than for ground-based observation.

Analysis of polarization signal distribution characteristics over the ocean surface shows that full-sky polarization distribution patterns observed by ground-based instrument platforms can also obtain a similar signal distribution in sky-based observations. The common characteristics of the two are that both has the sun or neutral points around the sun as the centre of the circles, and that DOP will increase with increasing polar diameter from the centre of the circle. That is to say, there is certain relationship between ground-based and sky-based polarization signals. In other words, the atmosphere neutral point zone that can be observed on ground-based instrument platforms can also be detected on sky-based platforms. The only thing is its total area is smaller than ground-based platform detection limits and it is more concentrated.

Here we can see, not only that a sky-based platform can effectively detect atmospheric neutral points which enabled separation of the effects between objects and atmosphere through sky-based platform with atmospheric neutral points as its atmospheric window, but polarization signals detected by sky-based observation platforms also have certain distribution models. What is the relationship and correlation between such a mode and the model of aerosol optical depth and aerosol? Could this be used for aerosol inversion? The following section will focus on analysing the relationship between full-sky polarization signal distribution, aerosol model, and aerosol optical depth. During analysis, the land surface will be assumed homogeneous (heterogeneous land surfaces will not be taken into account).

9.3.3 Distribution of Sky Polarization Patterns Based on Non-spherical Aerosol Models

We previously took advantage of the SOS RT radiative transfer model to calculate an intensity signal and polarization signal received by sensors across the whole sky over ocean when the solar zenith is 30° and the observation zenith is between 0° and 70°. Even when there is no aerosol particles in the sky, air particles have strong polarization effects, so we need take into account sky intensity and polarization change at different AOD levels. This section considers the full-sky reflectance, polarization reflectance, and DOP distribution pattern at six AOD levels, namely 0.0, 0.01, 0.05, 0.1, 0.5, and 1.0. Figures 9.8 and 9.9 show full-sky DOP distributions at the 550 and 860 nm bands, respectively (the polar radius is view zenith and the polar angle is the view azimuth angle).

Figure 9.8 shows that, when the aerosol optical depth is small, a zone with a DOP greater than 0.8 will appear in the sky (these are for zones with relatively large zenith and generally will not be analysed). When the view zenith is 30°–50°, sky DOP often reaches 30%. This means that, in general, sky DOP will be relatively high. In the backwards direction, the DOP is very low, and even there is a zone in which the DOP approaches zero. This zone is an atmospheric polarization neutral point zone. As the aerosol optical depth increases, the DOP in the sky gradually decreases and starts to be even when seen over the full-sky's scope.

For absorptive soot aerosol, is there a difference in full-sky polarization pattern distribution?

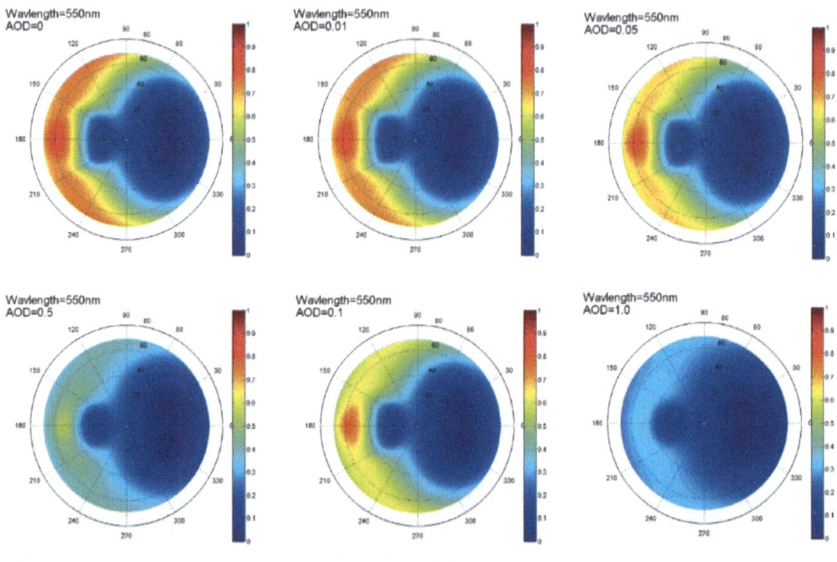

Fig. 9.8 Marine non-spherical dust aerosol full-sky DOP distribution at the 550 nm band

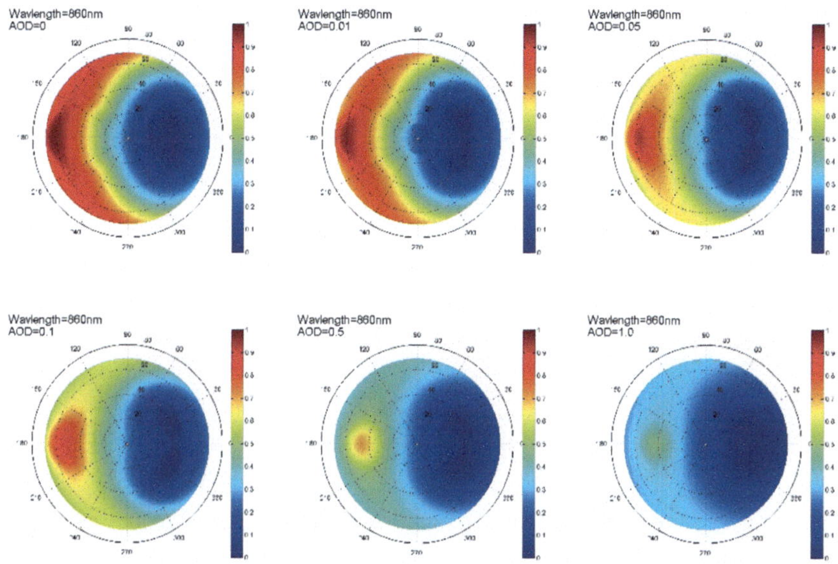

Fig. 9.9 Marine non-spherical dust aerosol full-sky DOP distribution at the 860 nm band

Figure 9.10 shows the full-sky reflectance index distribution and DOP distribution of soot aerosol at 550 nm. Compared to marine aerosol, soot aerosol has a more obvious weakening effect on intensity. It can be found that, when the AOD is either high

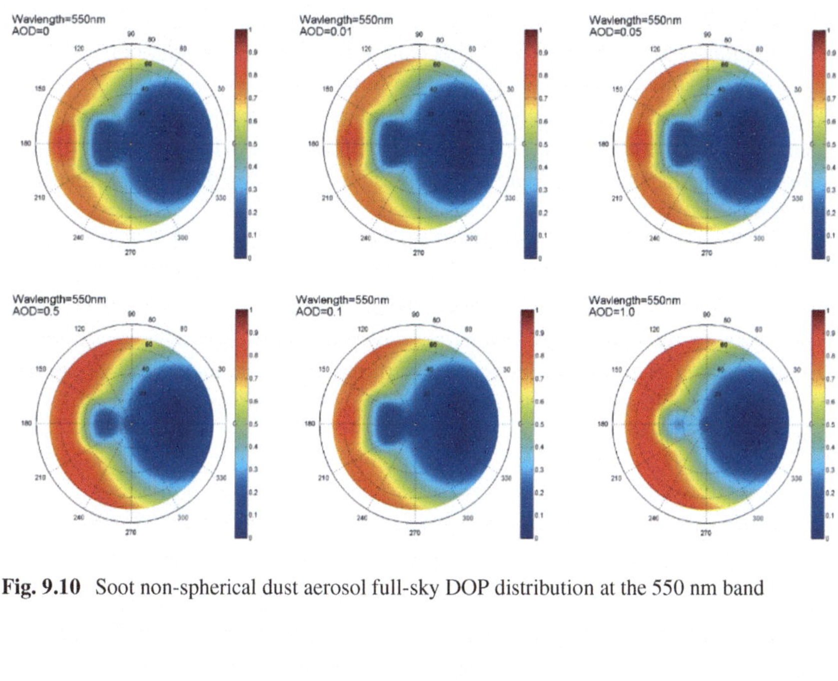

Fig. 9.10 Soot non-spherical dust aerosol full-sky DOP distribution at the 550 nm band

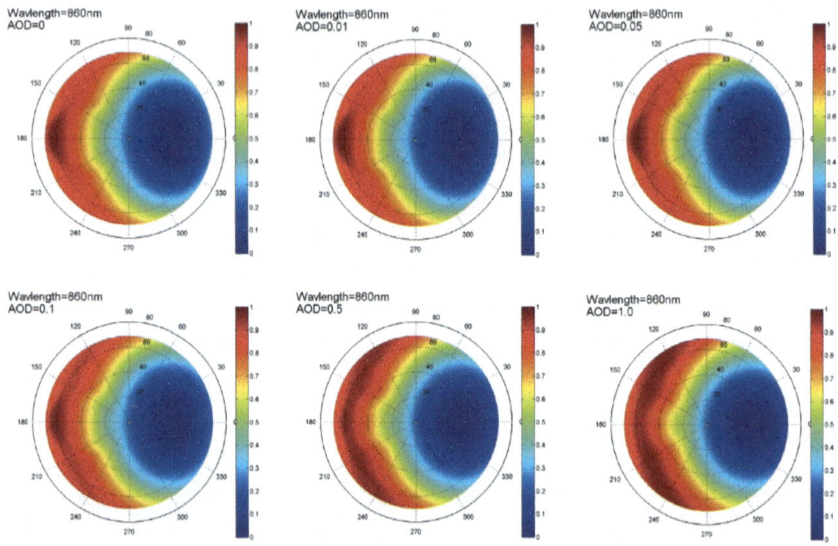

Fig. 9.11 Soot non-spherical dust aerosol full-sky DOP distribution at the 860 nm band

or low, except for the low DOP of atmospheric neutral point zones in a backwards direction, the DOP in other zones exceeds 30%. Especially for some large viewing zenith angles, the DOP of the sky will increase with DOD. This is in line with the discussion in Chap. 3 about aerosol with strong absorption and its obvious polarization characteristics. This also means that this polarization method can effectively eliminate aerosol with inversion absorption. That is to say, it can invert complex refractive indices to a certain extent. This cannot be done by aerosol optical feature inversion based only on intensity information in the traditional way.

Figure 9.11 shows full-sky DOP distributions of soot aerosol at 860 nm: reflectance at the 860 nm band is similar to that at 550 nm. From the view of DOP full-sky distribution, the DOP at 860 nm will increase with AOD, and in many directions, the DOP is close to 100%.

Different polarization effects from different angles showed the difference of aerosol in different directions. This can be viewed as a cubic section or by cubic multiple-angle chromatography and this laid the foundation for cubic block building of aerosol characteristics and quantitative analysis from different angles.

In this chapter, distribution characteristics of non-spherical model full-sky polarization fields are based on detection and observation of marine and soot type atmosphere polarization fields.

9.4 Separation Inversion of Non-spherical Aerosol and Land Surface Information Based on Sky Polarization Field Theory

Studies on aerosol inversion domestically and abroad have more than 30 years of history: because aerosol reflectance information is relatively weak, satellite remote sensing of aerosol information is about extracting atmospheric information from weak information. Generally speaking, several conditions are necessary for accurate aerosol inversion: (1) an accurate radiative transfer model, (2) a reasonable underlying surface model, and (3) a proper atmosphere aerosol model. A radiative transfer model is crucial to inversion accuracy, for instance MODIS (moderate resolution imaging spectroradiometer) selected a RT3 vector radiative transfer model to conduct marine aerosol inversion, which can effectively express air molecules in atmosphere and the polarization effect of aerosol. Since the signal obtained through backward scattering signals is weak in itself, the description of underlying surface features is also a crucial factor limiting the accuracy of remote sensing of aerosol. Currently aerosol inversion is limited in areas with low land surface reflectance, for instance, vegetation or the sky over ocean. For desert with high land surface reflectance and in urban areas, land surface reflectance signals constitute the main part of the signal received by a satellite, so aerosol contributions cannot be separated. Even when using multi-angle remote sensing, there is a certain limit on BRDF model establishment accuracy, which makes inversion more difficult. Aerosol model assumptions are also an important part of aerosol inversion. The main difference between satellite remote sensing aerosol and ground based photometry is that ground-based remote sensing is directly targeted at the sun and is less impacted by scattering light, while satellite remote sensing signals are weak and everything in the field of view is land surface objects which means it is more affected by environmental scattering light. Ground based photometer inversion does not need the assumption of an aerosol model, while satellite remote sensing is directly affected by assumptions about aerosol modelling. After 30 years of aerosol inversion modelling, there are many studies on aerosol models for regions with different environments. MODIS, MISR (multi-angle imaging spectroradiometer) and POLDER all have a set of aerosol models for inversion of global aerosol.

Aerosol separation includes both separation and inversion of particles, as well as the connection with the concept of separation of the effects between objects and atmosphere which serves for separation of the effects between objects and atmosphere, because the latter demonstrates the land surface inversion as the clear goal of remote sensing. At the same time, the sky polarization pattern includes a neutral point zone. Although the neutral point zone in sky polarization patterns is small, there is a neutral point zone therein, so it can be described as a sky polarization field. This sky polarization field includes the sky polarization pattern and an atmospheric polarization neutral point zone. That is about the relation between "plane" and "dot" and the two concepts are equally important. To be more accurate, it is the two factors that support atmospheric analysis and effect elimination and separation, as well as inversion of the effects between objects and atmosphere.

9.4.1 Studies on the Method of Separation of the Effects Between Objects and Atmosphere

This study is about two different land surface reflectance models, namely ocean and land, and this study explores separation methods of polarization reflectance signals for land surface and atmosphere based on polarization light information.

The ocean surface is more even and marine bodies are almost black body sources in the near-infrared band. In oceanic situations, the intensity inversion of aerosol also has high accuracy (for example MODIS). There are mature models for model building on clean sea surfaces with wave interruptions, and such models are accurate. Bidirectional reflectance distribution function effects of marine surfaces need take into account white caps, sunglint, and ocean chlorophyll concentration information. In solar reflectance bands, although the ocean has a low reflectance in the visible light and near-infrared bands, it has a certain polarization effect in the visible light zone. Polarization effects on marine surfaces are mainly because of mirror reflection effects on the ocean surface. Taking into account that scattering effects of hydrosol and water molecules on incoming sunlight may have certain polarising effects, we make use of RSP data to simulate aviation data from non-mirror directions. Ocean signals include two parts: a diffuse reflection of the ocean itself, the so-called water leaving radiation, demonstrating the scattering effect of the ocean on incoming light. The second is the mirror reflection effect. Here, water-leaving oceanic radiance is believed to have no polarization effect in models, only affecting the total intensity and having no effect on polarization light. Oceanic polarization effects are mainly caused by overall effects of mirror reflection from small mirror units on the ocean surface, so we adopt the Fresnel equation, which gives consideration to wave distribution, to build a BPDF effect model for the marine surface. For the ocean surface, it is not suitable to treat it as a Lambert plane for edge conditions, so we need to consider the distribution of water surface elements thereon.

A land surface is much more complex than the ocean surface, because the heterogeneity of land surfaces is greater, and BRDF effects on land surfaces are complicated. Currently there are many models for BRFD effects of a land surface, for instance, kernel-driven models are broadly applied to inversion of MODIS land surface reflectance, and the RPV model is widely applied to MISR data processing. Generally speaking, the RPV model can be adopted to improve multi-angle observations. The RPV model can be used for land surface model establishment. As for BPDF model establishment for a land surface, Maignan and other researchers in France proposed a land surface BPDF model which only has one parameter based on the models of Nadal and others in 2009: its accuracy is almost consistent with the two-parameter model proposed by Nadal (Maignan et al. 2009). The mechanism is such that the polarization reflectance of a land surface decreases with increasing vegetative cover, therefore, the normalised difference vegetation index is related to the polarization reflectance index of the land surface. In general remote sensing situations, NDVI is easy to measure. The polarization reflectance model of visible light and near-infrared radiation from a land surface can be shown as (Chen 2018):

$$R_p = \frac{C\exp(-\tan(\alpha))\exp(-v)F_P(\alpha, n)}{4(\mu_s + \mu_v)} \tag{9.17}$$

C is the only parameter with a degree of freedom, α is the angle of incidence of some reflectance element, v is the NDVI value measured through intensity data, n is the reflectance coefficient of the land surface, and μ_s and μ_v are the incidence, and outgoing, zeniths. So we have the model for land surface BPDF with a single parameter. The values needing to be confirmed are C, the experience value, and the NDVI as calculated in the near-infrared band. In general, the deviation of land surface polarization reflectance index based on this model is around 10%, and mostly within 20%. For different land surfaces, C, the experience value may take different values.

9.4.2 Research Scanning Polarimeter for Polarization Remote Sensing with Aviation Data

We adopted the research scanning polarimeter (RSP), an aircraft-borne multi-angle, multi-spectral polarization sensor developed by the APS project team at the Goddard Institute for Space Studies, NASA. The RSP is the aircraft-carried sample aircraft used for APS (aerosol polarimetry sensor) carriage and is carried by the USA's Glory satellite (launched into space on 4 March 2011). Unfortunately, the Glory satellite failed to be launched to its pre-established orbit, but the APS-2 Project has been proposed and will carry a satellite to space once again. RSP has accumulated a lot of aviation data in the past a few years. This book is based on aviation data from the RSP and we combined information from ground-based observation locations to conduct aerosol inversion and verification.

RSP is a prototype aircraft-carried polarization sensor developed by NASA. It instant angle of field of view is 14 mrad, and it can scan over 152 angles (in total) to within $\pm 60°$ in the orbital direction. RSP has nine bands, which fall into two groups: visible light near-infrared and short wave infrared. Its bands include: 410, 470, 550, 670, 860, 960, 1590, 1880, and 2250 nm. There is certain difference with APS settings. The RSP spectrum response function is as shown below: here, 410 and 470 nm will not be covered because they are severely affected by the ocean. We only adopt 550, 860, and 2250 nm for inversion of marine aerosol. For land aerosol, 550, 670, and 860 nm will be used. RSP can ensure high accuracy for polarization measurement, and its absolute radiation uncertainty is 3.5%, and polarization measurement uncertainty is 0.2%. RSP is composed of six sight-glass groups with accurate aim. RSP takes a polarization compensation scanning mirror instrument (namely a non-polarising instrument) to scan six sight-glass groups at a scanning scope of ±60°. Each sight-glass group is made of a pair of sight-glasses and each group measure three bands. One sight-glass from each group measures the linear polarization intensity at 0° and 90° to the instrument meridian plane. The other sight glass in the same group measures linear polarization intensity at 45° and 135° to the instrument meridian plane. This can ensure that each time of measurement will have polarization measurements of the same object. At each time of measuring, intensity, DOP and polarization azimuth angle were calculated in nine bands. General optical systems themselves have polarising effects. RSP adopts a double mirror system, and one reflectance mirror will have polarization compensation for the other: in this way, the scanning system of the RSP is neutral in polarization terms, which helps to avoid error from instruments themselves in polarization measurement.

On 28 September 2009, RSP conducted an aviation experiment over the Atlantic Ocean on the east of Chesapeake, USA. On the ground, there is a marine AERONET aerosol station, serving as verification for ground observation control locations. Meanwhile, the experiment also used MODIS satellite observation data. RSP data obtained are shown in Fig. 9.12. It should be noted that RSP itself is a non-imaging scanner. Figure 9.12 shows an image composed by combining 152 scanned data sets. This differs from the image generated by a real imaging sensor. In Fig. 9.12, scanning starts from the lower-right corner. Each column means 152 scanning corners at the same location. The black area means there is no observation at that angle. Different lines come from different observation locations. In this experiment, the RSP aviation elevation is about 9 km, and the corresponding ground resolution of nadir observations is about 140 m. The blue area in the figure denotes signals received from non-solar flare observation direction. While bright white flares show these observation angles are in the observation directions (in these directions, mirror reflection occurs). In flight, solar flares appear and disappear due to changes in flight direction. The area without solar flares in the middle is the area best used for inversion. At time of flight, the weather was fine around all ground photometry stations without disturbance from cloud and the wind speed on the ocean surface was about 2 m/s. Based on

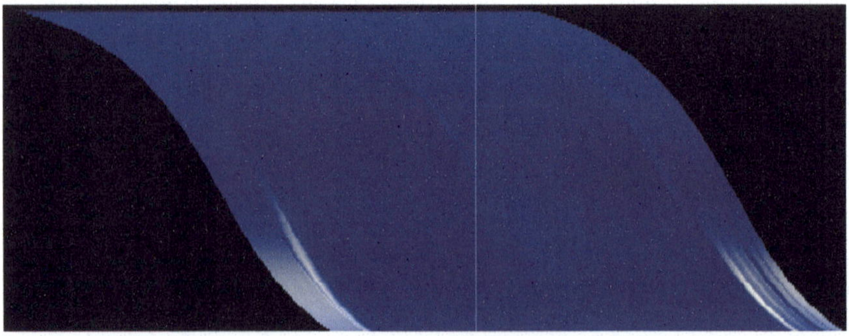

Fig. 9.12 Marine RSP aviation result (pseudo-colour combination: red (2250 nm), green (860 nm), and blue 550 nm) (white line is an example of multiple observation on one object)

Fig. 9.13 Land RSP aviation result (pseudo-colour combination: red (2250 nm), green (860 nm), and blue 550 nm) (white line is an example of multiple observation on one object)

land aerosol inversion, we conducted inversion based on RSP aviation experimental data collected at Southern Great Plains (SGP) in Oklahoma, USA (September 2005). This experiment took five days and obtained effective data on 16, 17, 19, and 20 September, and these data can be used to conduct inversion. The aviation elevation was about 6 km. During the flight, there was verification provided through Cart_Site data from ground AERONET aerosol stations. For example, RSP data obtained on 20 September are shown in Fig. 9.13. The black area in the figure denotes a part with no value after data distribution, while a straight line denotes data from the same location albeit at different observation zenith angles. Green areas denote vegetation, and yellow zones represent soil or mixed units of soil and vegetation. Locations with the same name have different results to some extent under different scanning angles.

9.5 Separation Inversion and Experimental Validation of Non-spherical Aerosol Over Ocean and Land Surfaces: Information Based on Sky Polarization Pattern Theory

Aerosol inversion through multi-angle and multi-spectral method must take into account scattering and polarization characteristics of different aerosol components at different bands. Based on the discussion above, coarse mode aerosol has similar scattering characteristics at different bands, while fine mode aerosol has little effect at 860 and 2250 nm where the wavelength is long. Such characteristics can be obtained through use of the extinction coefficient at different wavelengths in different aerosol modes. So, at 2250 nm, fine mode aerosol produces a weak result, similar to the air molecular effect, and generally it is difficult to distinguish the two effects in inversion. Coarse mode aerosol has a certain effect at this band, especially for dust-type aerosol, a large effect is seen in this band. For 860 and 550 nm, both coarse and fine aerosol play important roles. Atmospheric signals received by a sensor are a mix of coarse and fine aerosol signals. For polarization signals received by a sensor, atmospheric molecules play a weak role at 2250 nm: the polarization effect in atmospheric scenarios is small, while coarse mode aerosol has less effect on atmospheric polarization fields. So, the polarization effect generated by coarse and fine modes at 2250 nm can be neglected. So, atmosphere reflectance intensity signals at 2250 nm are only affected by coarse mode aerosol. Inversion of coarse mode aerosol characteristics can be done through intensity signal analysis at 2250 nm. Polarization information and coarse mode information work together for inversion of characteristics of fine mode aerosol. By establishing look-up tables, inversion of aerosol can be undertaken. The RSP aviation flight experiment can arrive at an inversion result as shown in Fig. 9.14.

The black line in the figure shows inversion values observed through ground AERONET, and can be viewed as aerosol real values at that time. The pink line shows RSP inversion values calculated using non-spherical dust aerosol models. The blue line shows RSP inversion values calculated using a spherical aerosol model. The red line shows inversion values from MODIS flying by at the same time. The results from multi-polarization and multi-spectral sensors agree with AERONET ground observation values and the deviation is around ±0.02 (550 nm). While at the same point, MODIS inversion has a difference of about ±0.12 (550 nm). Inversion through non-spherical dust aerosol models provides better accuracy than spherical aerosol models. It can be seen that aerosol inversion through multi-angle and multi-spectral methods offers more accurate aerosol inversion. Of course, differences between MODIS and RSP inversion may be caused by resolution. The ground resolution of the MODIS aerosol product is 10 km and that of the data used for RSP inversion is 140 m.

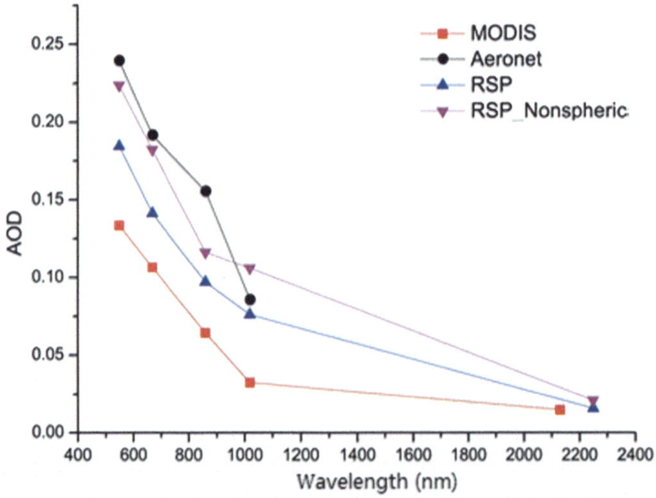

Fig. 9.14 Inversion result comparison between RSP and AERONET and MODIS

9.5.1 Separation Inversion and Experimental Validation of Non-spherical Aerosol Over Land Based on Sky Polarization Patterns

For aerosol inversion over the land surface, there are a series of methods that have been developed: the mostly common used is the dark target method adopted by MODIS. The basic mechanism therein is to utilise a prior linear relationship between visible light band data and short wave infrared band data in areas with dense vegetation to obtain land surface reflectance. Then through use of the look-up table, optimised calculation of aerosol optical depth at two visible light bands is conducted. Lastly, by interpolation method, aerosol optical depth was obtained at 550 nm: this method, however, is not applicable to land surfaces with bright targets thereon. MISR and other multi-angle observations followed by statistical optimisation through corresponding land surface BRDF (bidirectional reflectance distribution function) models are thus necessary. PARASOL POLDER assumed that air polarization information over land is the contribution of fine mode aerosol and conducted inversion of fine mode aerosol alone. Actually, there is still no good method for inversion of aerosol in the sky over land. This is because of the heterogeneity of the land surface and the BRDF effect, which make inversion difficult. We firstly conducted theoretical inversion analysis of the two kinds of aerosol, namely typical dust-type and soot aerosol, then conducted inversion of aerosol information using RSP land aviation data.

For aerosol over land, the traditional way of using intensity information cannot separate land surface information and atmosphere information easily (except in the sky over dark vegetation). BRDF effect of land surface restricted inversion of land aerosol through intensity information, especially aerosol over bright land surface. When land surface intensity reflectance information cannot be obtained effectively, polarization information has a special advantage in land aerosol inversion. The POLDER sensor has a pre-assumed condition that polarization signals received by sensors are contributed to by fine mode aerosol when it conducts land aerosol inversion. Actually, this assumption is not accurate. Coarse mode aerosol is non-negligible in its contribution to spatial polarization, especially when coarse mode aerosol information pertinent to non-spherical factors is taken into account.

So, when land surface intensity reflectance information cannot be obtained effectively, can aerosol inversion be conducted only with spatial polarization distribution data? The answer is yes: it is known that (1) polarization reflectance information of land surface is weak, and (2) polarization signals from the land surface can be calculated through land surface type and NDVI. That means that, if basic attributes (vegetation, soil, etc.) of underlying surfaces are known, land surface polarization reflectance signals can be obtained by using NDVI values. Actually, the effect of NDVI calculation deviation on the accuracy of inversion results is within acceptable limits, so the next issue is how to obtain NDVI values. NDVI is obtained after atmospheric calibration and inversion of aerosol data is undertaken before atmospheric calibration.

Aerosol over land and ocean are different from each other: the four aerosol models of the standard radiation atmosphere were chosen as basic aerosol types for establishing a set of look-up tables. In the inversion process, we assume that dust type and marine type are coarse mode aerosol, while water-soluble type and soot type are fine mode aerosol. Inversion was conducted by using look-up tables. The whole-day aerosol change in the four days selected on the date of experiment at the Cart Site observation station is shown in Fig. 9.15 (a, b, c, and d representing changes on 16, 17, 19, and 20 September 2005). Broadly speaking, the optical depth of aerosol over the four-day flight was low and the air was clean; however, there were certain aerosol fluctuations on 17 and 20 September, perhaps due to spatio-temporal changes in space weather, etc.

Inversion results are shown in Fig. 9.16 (a, b, c, and d show inversion results on 16, 17, 19, and 20 September 2005). The value of AREONET data, at 550 nm, comes from interpolation of AERONET 500 and 675 nm data. Inversion data from AERONET showed that, in the four days of this experiment, the aerosol optical depth of 550 nm is between 0.05 and 0.2. RSP inversion results adopted here are similar to inversion results of AERONET, but there is certain systemic deviation of about 0.03 therein. The inversion difference on 20 September is 0.06, the highest

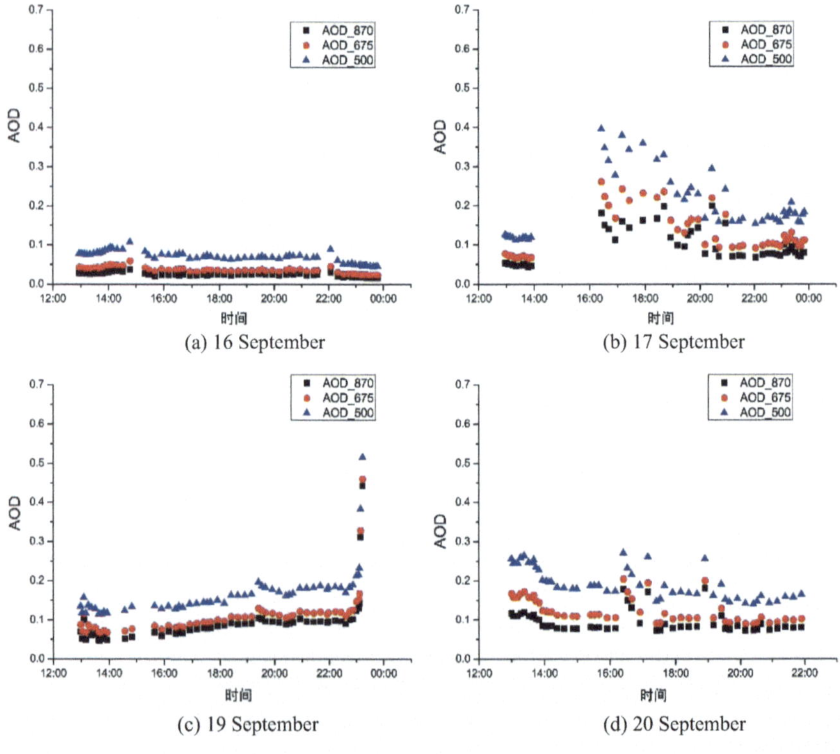

Fig. 9.15 Inversion value of RSP aviation flight AERONET

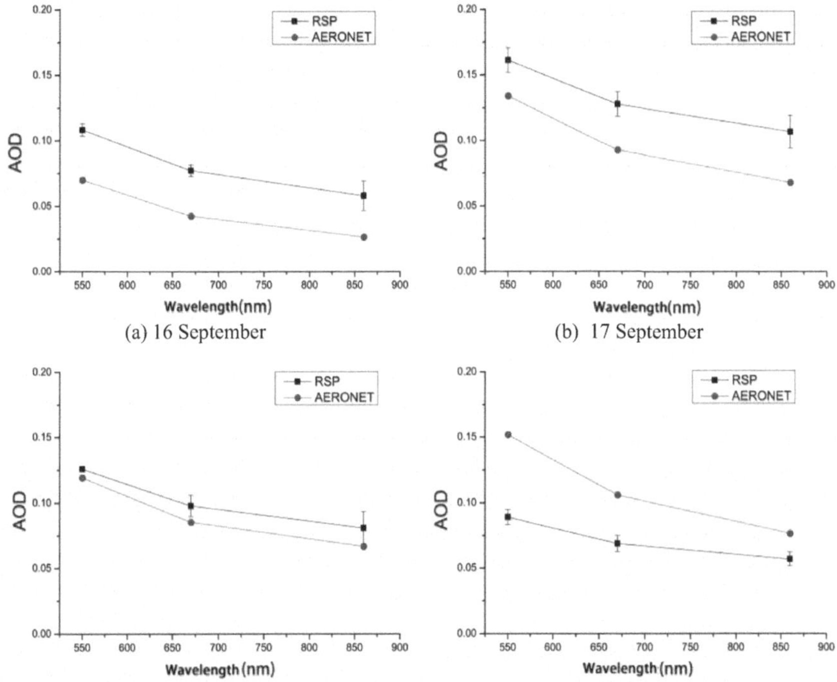

(a) 16 September (b) 17 September (c) 19 September (d) 20 September

Fig. 9.16 Land aerosol inversion results

value found. Preliminary RSP aviation flight results, and inversion results, show that, the inversion method based on polarization can be used for effective land aerosol inversion, however, it is noticeable that experiments conducted on this flight were limited, more verification experiments remain necessary. Such verification should also be conducted through global stations after APS-2 is launched.

References

Chen W (2018) Polarimetric characteristics of nonspherical aerosol and retrieval methods on its optical properties. PhD Dissertation, Peking University (in Chinese)

Dubovik O, Herman M, Holdak A et al (2011) Statistically optimized inversion algorithm for enhanced retrieval of aerosol properties from spectral multi-angle polarimetric satellite observations. Atmos Meas Tech 4:975–1018

Levy R, Remer L, Mattoo S (2007) A second-generation algorithm for retrieving aerosol properties over land from MODIS spectral reflectance. J Geophys Res 112(D13211).https://doi.org/10.1029/2006jd007811

Maignan F, Breon F, Fedele E et al (2009) Polarized reflectances of natural surfaces: spaceborne measurements and analytical modeling. Remote Sens Environ 113(12):2642–2650

Martonchik J, Diner D (1992) Retrieval of aerosol and land surface optical properties from multi-angle satellite imagery. IEEE Trans Geosci Remote Sens 30:223–230

Part IV
New Applications of Polarization Remote Sensing Physics

Bionic polarization automatic navigation based on earth's polarization vector field is provided in Chap. 10. Polarization remote sensing methods for advanced space detection and a global change study are provided in Chap. 11. These prove the objectiveness, uniqueness, stability, and repeatability of polarization remote sensing physics. This can help China to honour her responsibility to lead the theory, algorithms, and application of polarization remote sensing internationally. Based thereon, we explored the establishment of a set of new systematic polarization remote sensing methods, to provide solutions for land-atmosphere separation, and provide an innovative systematic physical analysis method, for advanced space detection, extremely weak signal attainment, and the third global natural field, namely sky polarization field navigation (the first and second being the gravitational, and geomagnetic, fields).

Chapter 10
New Areas in Polarization 1: Bionic Polarization for Automatic Navigation Using the Earth's Polarization Vector Field

Some insects (such as ants) use the angle between vector lines of sky polarization field and their body axis line to navigate. This is the most effective evidence of the stability and repeatability of remote sensing applications of polarization. This section includes: polarization pattern navigation mechanisms based on path integration behaviour of ants, to provide bionic characteristics of biologically sensitive sky polarization vector fields; bionic navigation theoretical basis and model analysis of remote sensing to analyse the physical nature of bionic polarization navigation; angle measurement design of bionic polarization navigation and function attainment, to build a hardware system for this purpose; bionic polarization navigation model and accuracy measurement based on cloud computing, to realise multi-object, multi-temporal, multi-angle polarized navigation mechanism that is both stable and repeatable; and experimental verification and accuracy analysis of bionic polarized navigation, to analyse its capacity for application in navigation.

10.1 Polarization Pattern Navigation Mechanism Based on Path Integration Behaviour of Ants

Polarized light navigation is a natural navigation method: many animals, such as ants, bees, crickets, and migratory birds, all have a polarization vision system. They can navigate by using the polarization characteristics of sunlight scattering in air. Polarized navigation is passive and is well-hidden. It is not disturbed by electromagnetism. No matter whether it is sunny or cloudy, or even before and after sunrise; polarized navigation can be used to set a direction. It is suitable for high-latitude areas where a magnetic compass cannot be used. Also, it is fast, at near to real-time navigation, and works when mobile. Polarized light navigation is accurate: according to reports from *Science* on 11 August 2006, migratory birds calibrate the cumulative difference of their magnetic compass through polarization light figure of sunrise and

L. Yan et al., *Polarization Remote Sensing Physics*, Springer Remote Sensing/Photogrammetry, https://doi.org/10.1007/978-981-15-2886-6_10

sunset, which enables them to find their nests and destinations accurately on each seasonal migration which may be a journey of thousands of kilometres.

This chapter explores a new navigation method based on polarization light that will not be easily affected by electromagnetism. The experiment proposed in this book can provide new navigation methods for aircraft and moving carriers, and enrich navigation theory and techniques. Meanwhile, it can also provide experimental and model bases for polarized, and bionic, navigation devices. In addition, polarized navigation technology can also provide new ideas for underwater object navigation. The blue spectrum of visible light can penetrate more than 200 m in water: by measuring polarization patterns of the underwater blue band, we can develop underwater polarized light navigation devices and achieve navigation of underwater objects. Combined with other navigation methods, accuracy of existing navigation methods can be improved by a large margin, GPS signal capture times of underwater objects such as submarines can be reduced, which is helpful to achieve continuous, seamless navigation of underwater objects. Bionic weaponry is a frontier field: automatic navigation and cybernetics is an important component, but vision capacity is another important guarantee of accurate control. Polarized navigation is an effective combination of the two factors. Navigation through polarization is in effect a presentation of artificial vision. Furthermore, polarized navigation can also be combined with other navigation methods (inertial navigation or satellite navigation), which can help to increase both navigation accuracy and the reliability of project performance.

10.1.1 Ants' Polarized Navigation Mechanism

Ants have the capacity to deduce path, memory, and calculate a route when they hunt for food. For ants that go foraging, the routes are strange and random, and they go back to their habitats directly after they find food rather than following their original paths. This is the role that path integration plays. Ants have kept in mind the vector relationship between their current location and home, and then they can return by taking a shortcut (Fig. 10.1). Despite the important role that path integration plays, it has its limitation: with on-going motion, vector relationships must be updated continuously. Such updating is not as accurate and rigid as when using mechanical devices, so error accumulates. Only after correction, can it better instruct space navigation. We think such correction information comes from polarization information in the sky. This is about making use of polarization light information to locate a source of light in sky (sunlight or artificial light). Experiments, in their early stages, have proved this speculation.

The study (Wehner and Menzel 1969; Rossel and Wehner 1986; Fent and Wehner 1985) of Wehner, a Swiss biological neurologist, shown that the ant retina is composed of hundreds of optic nerve rhabdomes that face in different directions. Each rhabdome is only sensitive to polarized light in its own direction. Each rhabdome is made of eight single-direction light sensitive parts that are perpendicular to each other (Fig. 10.2). And all polarization-sensitive nerve rhabdomes simulate

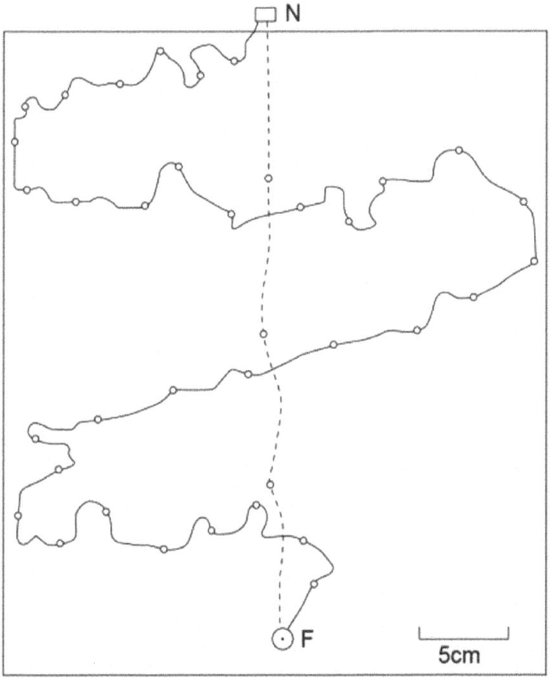

Fig. 10.1 Ants' shortcut to habitat after hunting for food the full line denotes the path that the ant followed when going out hunting food from Point N, its home, when food is found at F, the ant returned home almost following a straight line

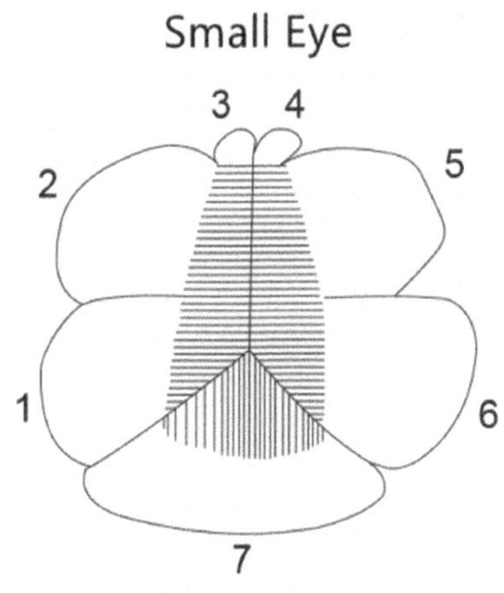

Fig. 10.2 Polarization sensitive nerve rhabdome

distribution of polarization light in sky to form a polarization sensitive array (Fig. 10.3). Outer short stems cater for polarization light distribution while inner short stems serve the nerve rhabdome array. Such a structure makes the incitement response of each rhabdome for polarization light in a certain direction a sine curve, which is a rigid direction analysis tool (Fig. 10.4).

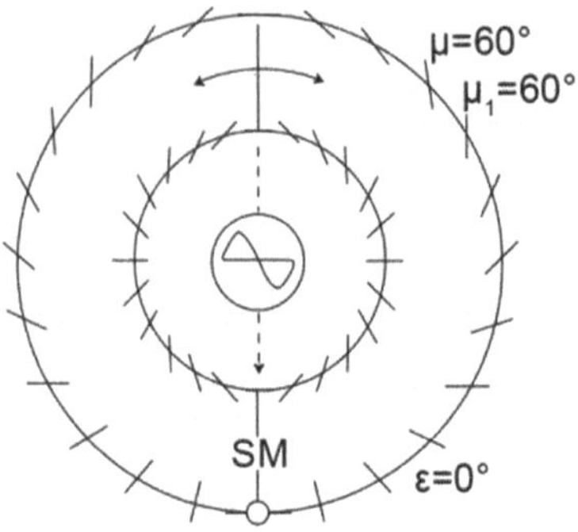

Fig. 10.3 Retina rhabdome array diagram

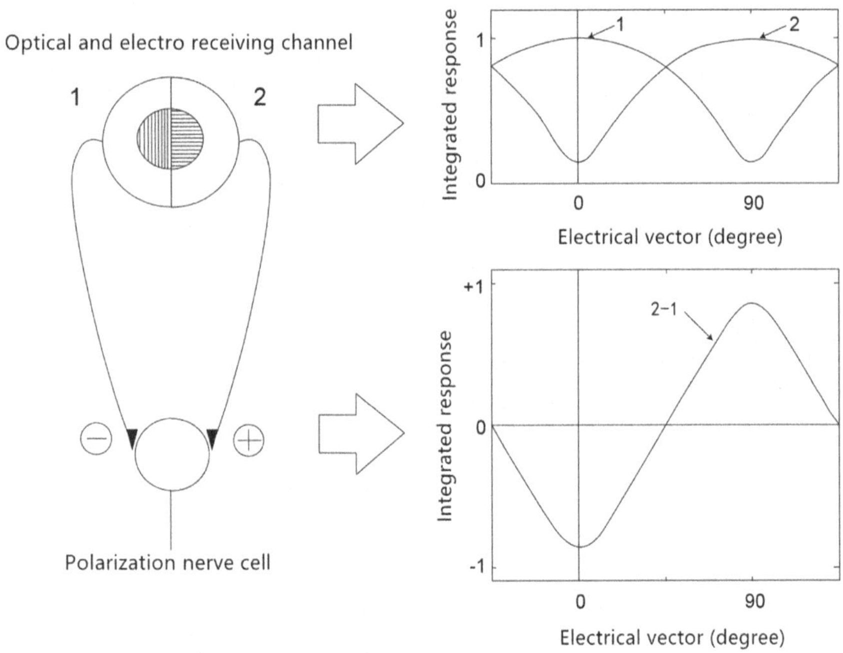

Fig. 10.4 The effect of a nerve rhabdome on the E-vector

In this way, when ants turn around, their nerves will response to systematic change. When body axis of the ant is in line with the sun meridian, the response of the retinal nerve rhabdome reaches its peak value.

The response of the retina polarization-sensitive nerve rhabdome serves as input into leaf segment of the optic nerve. Internal uncial neurons of the leaf segment of the optic nerve have three types, as polarized orthogonality nerve cells, and can accept input from optical nerve rhabdomes that are perpendicular to each other in most

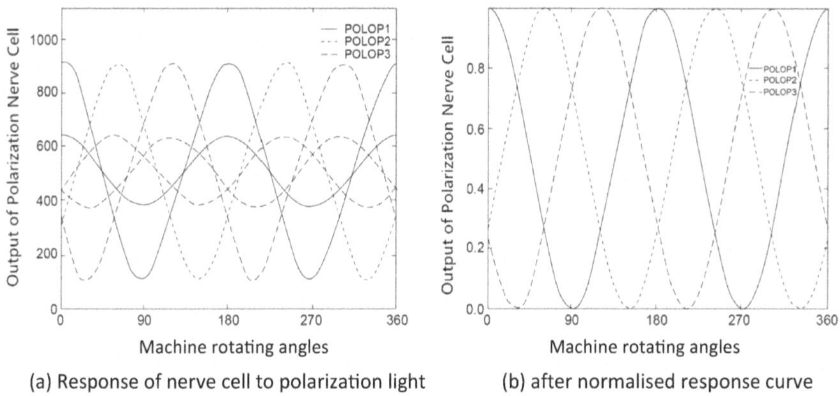

Fig. 10.5 Response curve of nerve rhabdome to polarized light

sensitive directions of the retina. The response of the three internal uncial neurons is also sinusoidal and the peak response is achieved when their angles with the sun meridian are 0°, 60°, and 120° respectively (Fig. 10.5). The response of internal uncial neurons will be integrated into compass nerve cells in the ant brain as input. After calculation and interpretation, the angle between the body axis and sun meridian can be obtained, which enabled navigation.

Studies by Marcus Coene of Zurich University proved that ants can not only identify directions through road marks, they also possess a back-up system called a "path integration device". This system enables ants to measure distances elapsed and relocate themselves through an internal compass from time to time. This enables ants to find a straight route to return to habitat even if the route leaving their habitat is labyrinthine, so as to minimise the return distance. When ants return home, if the ant's legs are stilted or stumped, then ants with stilts go beyond their nest while ants with stumps stop short; however, after several trials, all ants in the experiment can stop accurately (Wenher Wehner 2003; Wittlinger et al. Wittlinger et al. 2006). This means ants' calculation of distance is related to their step length (Fig. 10.6). When a navigation system can provide orientation, positioning, and self-calibration, it is regarded as a complete physical system and can be applied to human activities.

10.1.2 Experiment About Polarization Navigation and Path Integration

Creatures' activities are generally affected by multiple factors and strategies. To study the intrinsic biological mechanism thereof, it is necessary to separate each factor, strategically consider independent factors, abandon disturbing components, and discuss one item seriatim. We emphasise the effect and importance of polarized light navigation and path integration navigation, so it is necessary to separate and

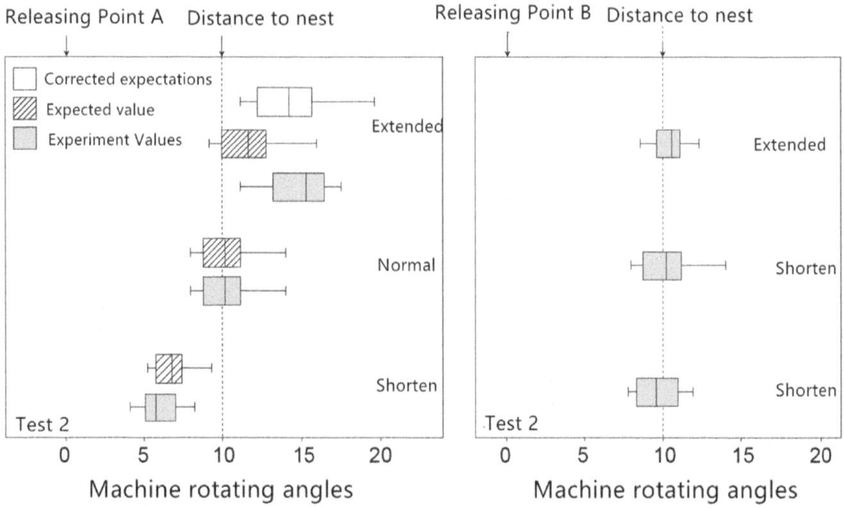

Fig. 10.6 Relationship between ants' shortcut to habitat and their step length

discuss the two methods independently, then summarise the relationship between the two methods. To realise this purpose, fine experimental settings are necessary.

The first issue to quantify routes to habitats after finding food in conditions such as direct sunshine, shelter, moonlight, artificial illumination, and darkness. By experimental methods such as polarization cover, passive replacement, light source variation, and movement recording and track analysis of ants, ant navigation calculations were conducted to deduce factors affecting their navigational use of polarized light, and to summarise the relationships between the two navigation methods and establish models thereof.

1. The polarization cover experiment: we place a large polaroid in the sky over the experimental area shown as Fig. 10.7, change the azimuth of the polaroid

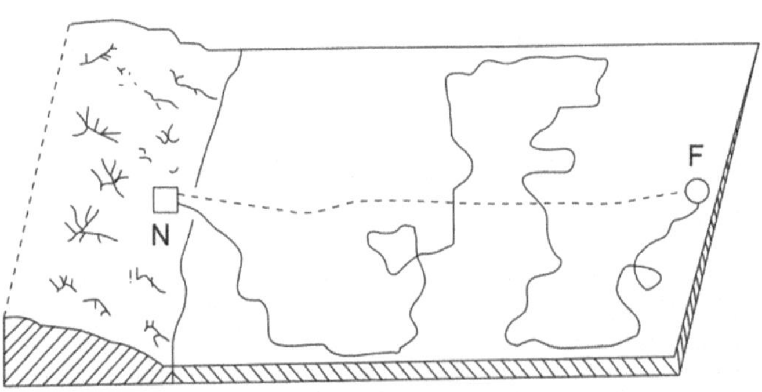

Fig. 10.7 Polarization cover experiment

and observe the activities of ants as they return to their habitats. If ants adopt a polarized light navigation strategy, their behaviour will vary according to polarization azimuth angle. Then the correlation between the two inputs can be found and analysis undertaken as to why. Based on this, analysis of polarization light patterns in the sky will be conducted. Figure 10.8 represents a polarization pattern in sky when the solar altitude is about 60°. The left-hand part of Fig. 10.8 shows a cubic figure, while the right shows a plane graph. The direction and width of the black bar denote the polarization direction and DOP, separately. This experiment was conducted under different lighting conditions including direct sunshine, sheltered sunlight, an artificial light source, and moonlight.

2. Moving source of light experiment: as shown in Fig. 10.9, an artificial source of light (15 W power) was placed in the eight locations marked L_0–L_7 successively and behaviour responses of the ants were observed. From L_0 to L_2, the azimuth

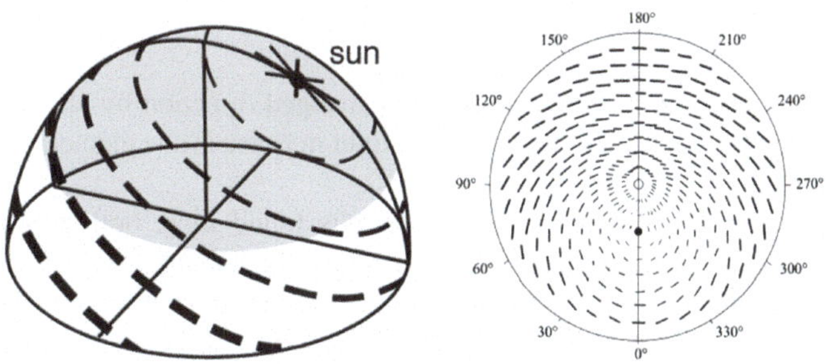

Fig. 10.8 Sky polarization pattern (solar altitude, 60°)

Fig. 10.9 Moving source of light

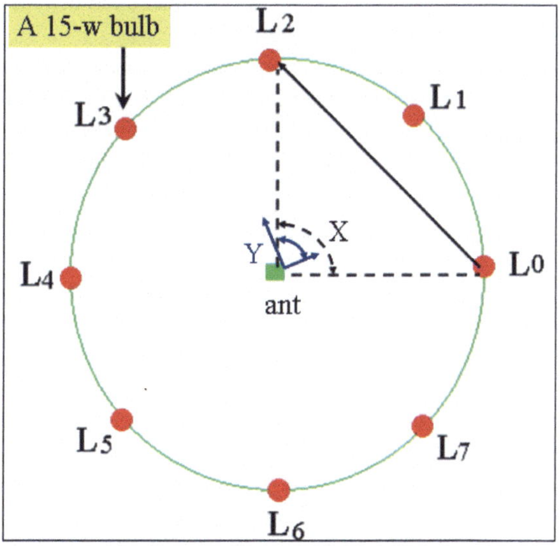

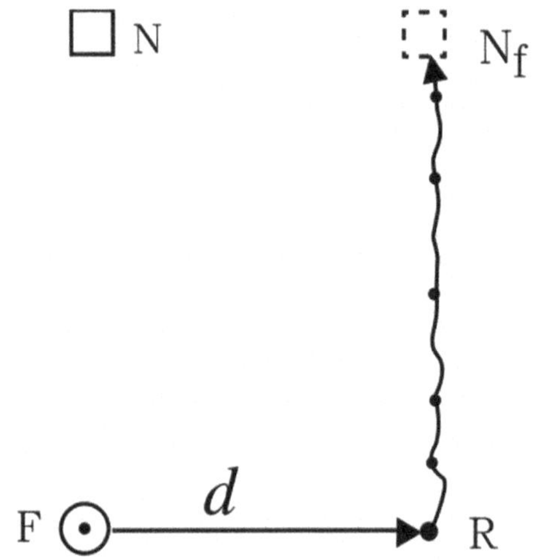

Fig. 10.10 Passive displacement experiment

angle was changed by 90° and ants also changed direction by about 90°, suggesting that the location of the source of light in space is one element supporting ants' navigation.

3. Passive displacement experiment: when ants found food, both ants and food were displaced by translation, and their return the habitat observed. Normally, ants would calculate direction and distance to return to habitat immediately they find food: after displacement, ants will proceed in the direction that is parallel to their calculated direction and move toward the location in their memory. They will stop after covering the original distance between the food location and their homes. Or they just stop at the imagined "habitat". As shown in Fig. 10.10 with direct sunshine: *d* stands for displacement distance, and the line between R and Nr denotes the actual path taken.
4. Sheltered sunlight experiment: comparing situations without direct sunshine including under shade and cloudy weather with the polarization cover experiment and passive displacement experiment with direct sunshine, to measure the difference and similarities in ant navigation models under different illumination intensities. In addition, an artificial source of light experiment can be undertaken, including indoor and outdoor artificial source of light experiments. Outdoor experiments were conducted mainly at night time, with an artificial source of light and without disturbing natural light. These experiments also combined the polarization cover experiment, passive displacement experiment, and moving source of light experiment. Indoor experiments are about repeating outdoor experiments by utilising artificial source of light in a dark room. Indoor experiments are undertaken as shown in Fig. 10.11.
5. Step length calculation experiment: when ants return home, if an ant's legs are stilted or stumped, then ants with stilts go further than their nest when they come back while ants with stumps stop short; however, after several times, all ants no matter whether with stilts or with stumps can stop at their habitat accurately

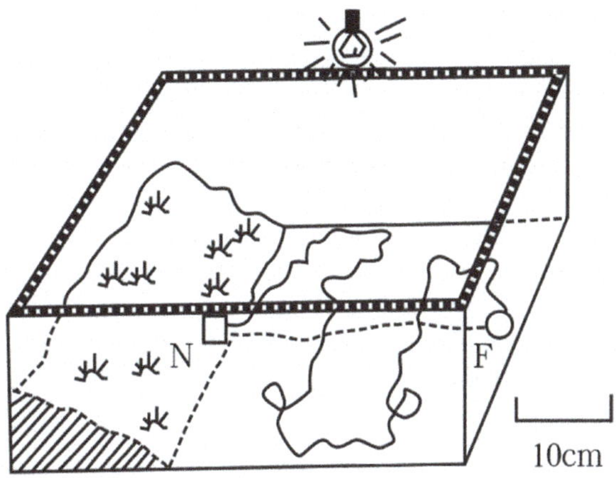

Fig. 10.11 Moving source of light experiment

(Wenher Wehner 2003; Wittlinger et al. Wittlinger et al. 2006). This means that ants' calculation of distance is related to their step length, as shown in Fig. 10.6. Based on this experiment, we planned to discuss how ants calculate step length then navigate along their path.

Figure 10.12 shows that the angular change in an ant's path has a linear correlation with the angle change in the polarization of the source of light. When the lighting azimuth angle is changed, ants undertake azimuth angle adjustment accordingly. Namely, the change in azimuth angle of their activity when returning home is related to change in light azimuth angle.

In addition, biological experiments show that the three major strategies that ants use are: path integration, vision navigation, and systematic searching. Path integration based on information about polarization pattern direction in the sky is the major

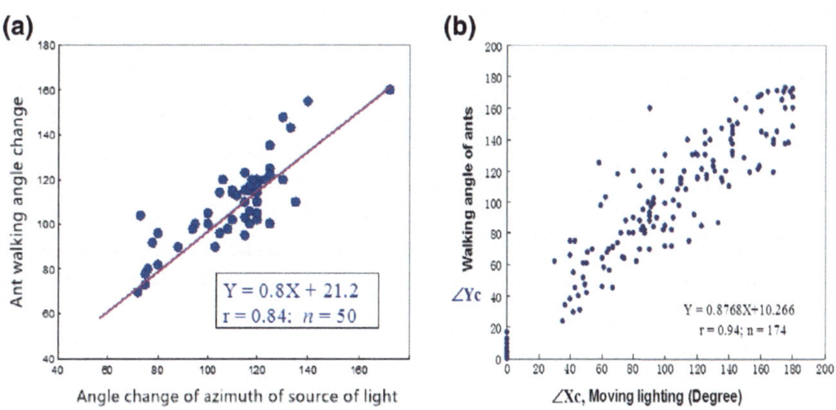

Fig. 10.12 Linear correlation between the angular change in ant paths and azimuth of the source of light

navigation strategy of ants. To choose the most proper navigation strategy at a given time, and how to switch between different navigation methods, are key to polarized navigation mechanism analysis. Based on the biological activities of ants, analysis of the factors affecting polarized navigation, navigation elements, and the mutual effect of different factors is necessary. Then the switching mechanism between different navigation patterns will be analysed to calculate polarization light distribution patterns in the sky at different times and locations. We then derive their DOPs and polarization azimuth angles, then analyse factors affecting polarization patterns (for instance, the wavelength of light, shelter from indoor situations, surroundings, etc.) and then we may be able to navigate and locate by taking advantage of polarized light pattern data.

10.2 Theoretical Basis for, and Model Analysis of, Bionic Polarization Navigation

The basic principle of polarized light navigation is: when sunlight or moonlight penetrate the atmosphere and get scattered by atmospheric molecules, a polarized pattern will be formed in the sky which is symmetrical to the line between zenith and the sun. Animals use their polarization vision to get direction information from this to navigate. What is the polarization state of sunlight after scattering by atmospheric molecules? What effect will the atmospheric environment have on sunlight scattering? How to obtain state information of polarization light scattered by atmospheric molecules? Analysis of the physical nature of bionic polarized navigation is the major theme of this section.

10.2.1 Scanning Mode

For scanning mode, the primary principle is to find the sun meridian by using the sensitivity of polarization orthogonality to polarization patterns in the sky, and secondly, we consider this direction as a zero reference direction for the whole system. In basic scanning mode, only one polarization sensitivity orthogonality unit is needed to identify the direction of a target. When rotated around their own long axis to scan the sky, polarization orthogonality units will generate the utmost response when the axis is parallel with the symmetrical line in the sky polarization pattern (the sun meridian line and anti-sun meridian line).

$$p(\varphi_{\max}) \geq p(\varphi), \forall\varphi \in [0^\circ, 360^\circ], \varphi_{\max} \in \{\varphi_{sol}, \varphi_{asol}\} \quad (10.1)$$

where, φ denotes the output angle of a polarization orthogonality unit relative to sun meridian. This function has two local maxima, which corresponds to sun meridian and the anti-sun meridian respectively. The sun meridian and anti-sun meridian can be identified through an intensity sensor installed on the head. This model has a simple structure, and is easy to implement, but the response curve of the local maximum is too flat, making direction-finding inaccurate. The solution of Lambrinos and other researchers is to use the signal extracted from the other two polarization orthogonality units to fine-tune areas with local maxima. The response directions of maximum E-vector of three basis polarization orthogonality units are 0°, 60°, and 120° respectively. Response curves of the scanning process can be represented by:

$$f(\varphi) = p_1(\varphi) - \|p_2(\varphi) - p_3(\varphi)\| \tag{10.2}$$

Basic models (response curves of POLOP1, POLOP2, and POLOP3) and improved models by Lambrinos and other researchers (response curve shown in dotted lines) are shown in Fig. 10.13.

Obviously, improved models have sharper curves at 0º and 180º when meeting local maxima, and $f(\varphi)$ is a quasi-linear function. Despite this, there are many angle values that are close to sectional maxima, so it is not easy to determine one maximum. A research team led by Professor Chu Jinkui found that the cross-points of response curves POLOP2 and POLOP3 are 90°, 180°, 270°, and 360°. We consider whether, or not, this situation is a coincidence or can we take advantage of these characteristics to discover the location of the maximum more accurately? So: we had $p_2(\varphi) = p_3(\varphi)$, then:

$$\lg\left[\frac{1 + d\cos(2\varphi - {}^{2\pi}/_{3})}{1 - d\cos(2\varphi - {}^{2\pi}/_{3})}\right] = \lg\left[\frac{1 + d\cos(2\varphi - {}^{4\pi}/_{3})}{1 - d\cos(2\varphi - {}^{4\pi}/_{3})}\right] \tag{10.3}$$

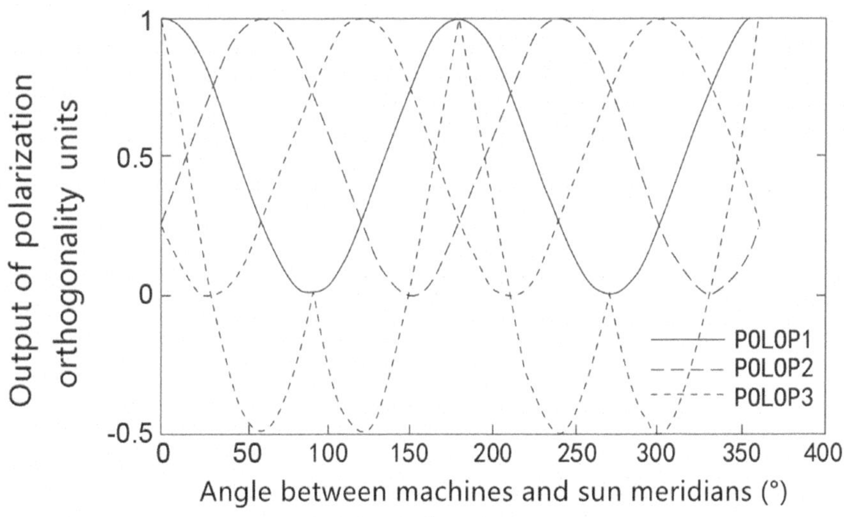

Fig. 10.13 Scanning modes with improvement by Lambrinos and other researchers

After reorganisation:

$$\cos(2\varphi - \frac{4\pi}{3}) = \cos(2\varphi - \frac{2\pi}{3}) \tag{10.4}$$

With $\varphi \in [0,\ 2\pi]$, we get $\varphi = 90°,\ \varphi = 180°,\ \varphi = 270°,\ \varphi = 360°$. Now, further improvement of scanning models can be undertaken: when the three polarization orthogonality units are rotated around the body axis by 360°, if output responses of $p_2(\varphi)$ and $p_3(\varphi)$ are the same, then the angles between axis and sun meridian are $\varphi = 90°,\ \varphi = 180°,\ \varphi = 270°,\ \varphi = 360°$. By comparing output responses of polarization orthogonality units, the reference direction can be found.

10.2.2 Concurrent Mode

For concurrent modes, the axis between the long axis of creatures to be navigated and the sun meridian is found by comparison of output results from three polarization sensors and a look-up table generated at the onset of movement. The two premium values found can be used to calculate the corresponding angle through judgement using an adjacent intensity sensor. Matching look-up table data has both increased the calculation burden and decreased the accuracy. Here, we improved all original concurrent modes, and a reference direction can be determined with one further step while neglecting the look-up table. The output of polarization unit can be described thus:

$$s(\varphi) = KI[1 + d\cos(2\varphi - 2\varphi_{\max})] \tag{10.5}$$

where, I is the total intensity, d is the DOP, φ is the current direction in which to navigate compared to the sun meridian, $\varphi_{\max}$ is the direction maximising $s(\varphi)$, and K is a constant. The output equation of polarization orthogonality units is:

$$p_1(\varphi) = \lg\left[\frac{1 + d\cos(2\varphi)}{1 - d\cos(2\varphi)}\right] \tag{10.6}$$

$$p_2(\varphi) = \lg\left[\frac{1 + d\cos(2\varphi - {}^{2\pi}/_{3})}{1 - d\cos(2\varphi - {}^{2\pi}/_{3})}\right] \tag{10.7}$$

$$p_3(\varphi) = \lg\left[\frac{1 + d\cos(2\varphi - {}^{4\pi}/_{3})}{1 - d\cos(2\varphi - {}^{4\pi}/_{3})}\right] \tag{10.8}$$

where, $p_1(\varphi),\ p_2(\varphi),\ p_3(\varphi)$ are the output values from polarization orthogonality units when the polarization axis is adjusted to 0°, 60°, and 120°.

After logarithmic proportional calculation, the output of polarization orthogonality units is not related to intensity but is relevant to DOP. DOP varies with sun location and weather conditions. Such changes directly affect the output of polarization orthogonality units. The solution is to eliminate the effect of DOP by normalising the output of polarization orthogonality units: we used a sigmoid function for anti-logarithmic transformation, then

$$\frac{1}{10^{p(\varphi)}+1}=\bar{p}(\varphi) \tag{10.9}$$

The three output equations became:

$$1-2\bar{p}_1(\varphi)=d\cos(2\varphi) \tag{10.10}$$

$$1-2\bar{p}_2(\varphi)=d\cos(2\varphi-{}^{2\pi}/_{3}) \tag{10.11}$$

$$1-2\bar{p}_3(\varphi)=d\cos(2\varphi-{}^{4\pi}/_{3}) \tag{10.12}$$

where, d and φ are unknown. With two unknowns in three equations, if the first two equations are selected, then:

$$d=\frac{1-2\bar{p}_1(\varphi)}{\cos(2\varphi)} \tag{10.13}$$

Combining (10.13) and (10.13), we get:

$$\tan(2\varphi)=\frac{3-2\bar{p}_1(\varphi)-4\bar{p}_2(\varphi)}{\sqrt{3}(1-2\bar{p}_1(\varphi))}$$

Furthermore:

$$\varphi=\frac{1}{2}\arctan\left\{\frac{\bar{p}_1(\varphi)+2\bar{p}_2(\varphi)-{}^{3}/_{2}}{\sqrt{3}\left[\bar{p}_1(\varphi)-{}^{1}/_{2}\right]}\right\} \tag{10.14}$$

$p_1(\varphi)$, $p_2(\varphi)$, $p_3(\varphi)$ are known (from the sensor), and $\bar{p}_1(\varphi)$, $\bar{p}_2(\varphi)$, $\bar{p}_3(\varphi)$ are calculated based on $p_1(\varphi)$, $p_2(\varphi)$, $p_3(\varphi)$, so values of direction angle φ can be obtained through sensor data. The φ value calculated is not related to DOP, and only two equations of the available three were used to calculate φ. So what is the role of the third equation? The following function transformation is applied to outputs $p_1(\varphi)$, $p_2(\varphi)$, and $p_3(\varphi)$ of the three polarization orthogonality units, namely 0°, 60°, and 120°:

$$1-\frac{2}{10^{p(\varphi)}+1}=\tilde{p}(\varphi) \tag{10.15}$$

In this way,

$$\tilde{p}_1(\varphi) = d\cos(2\varphi) \tag{10.16}$$

$$\tilde{p}_2(\varphi) = d\cos(2\varphi - {}^{2\pi}/_3) \tag{10.17}$$

$$\tilde{p}_3(\varphi) = d\cos(2\varphi - {}^{4\pi}/_3) \tag{10.18}$$

With random $d \in (0,\ 1]$,

$$\tilde{p}_1(\varphi)/\tilde{p}_2(\varphi) = \cos(2\varphi)/\cos(2\varphi - \frac{2\pi}{3}) \tag{10.19}$$

$$\tilde{p}_1(\varphi)/\tilde{p}_3(\varphi) = \cos(2\varphi)/\cos(2\varphi - \frac{4\pi}{3}) \tag{10.20}$$

$$\tilde{p}_2(\varphi)/\tilde{p}_3(\varphi) = \cos(2\varphi - \frac{2\pi}{3})/\cos(2\varphi - \frac{4\pi}{3}) \tag{10.21}$$

From the deduction above, $\tilde{p}_1(\varphi)$, $\tilde{p}_2(\varphi)$, $\tilde{p}_3(\varphi)$ can be obtained directly through outputs $p_1(\varphi)$, $p_2(\varphi)$, $p_3(\varphi)$ from the three polarization orthogonality units. Then only one of (10.21), (10.20), and (10.19) need be used to derive φ. This aligns with the conclusions drawn elsewhere by Wehner et al. (2002) that polarization angle φ is not related to DOP d. Then, through simultaneous transformation and trigonometric manipulation of (10.21), (10.20), and (10.19), the relationships between $\tilde{p}_1(\varphi)$, $\tilde{p}_2(\varphi)$, $\tilde{p}_3(\varphi)$ can be obtained:

$$\tilde{p}_1(\varphi) + \tilde{p}_2(\varphi) + \tilde{p}_3(\varphi) = 0 \tag{10.22}$$

This conclusion can be used for accuracy calibration.

It is noticeable that output from a sensor is a function with phase difference π, and the two φ values correspond to the angle between the carrier movement direction and the sun meridian and the angle between the carrier movement direction and the anti-sun meridian, respectively. To find the angle between the carrier movement direction and the sun meridian, we need an environmental light sensor: through output from environmental light sensors, estimation of relationships between the locations of carrier movement direction and the sun can be done roughly, to calculate the angle between carrier movement directions and the sun meridian.

This section introduced the rationale of bionic polarized navigation, namely relatively stable polarization patterns are generated in the sky due to atmospheric scattering. This section also analysed the physical nature of bionic polarization navigation and initially proved the rationale for realising this at hardware level.

10.3 Measurement Device Design and Function Attainment for Bionic Polarized Navigation

Biologists have shown that the reason why ants possess navigation and location capacity is because the structure of the optic nerves of ants, namely a compound eye, makes them quite sensitive to polarization of light in the sky. With increasing demand for navigation and location services, it is of significance to develop bionic polarized navigation angle measurement sensors based on the navigation and locating rationale of ants. To develop a bionic polarized navigation measurement sensor, firstly polarization light tests simulating ants need to be conducted. Based on polarized navigation angle measurement models simulating ants introduced in the previous section, we designed and built a simple experimental platform for bionic polarized navigation angle measurement based on polarized light in the sky, which laid the foundation for developing angle measurement experiments and the whole bionic polarized navigation system. Through analysis of ant navigation methods, the proposal is as shown in Fig. 10.14.

10.3.1 Overall Design of Angle Measurement for Bionic Polarized Navigation

The platform design for polarized navigation angle measurement experiments is shown in Fig. 10.15. The system is composed of opto-electrical receiving modules, signal processing modules, data collection modules, and angle calculation modules.

Optical and electrical receiving modules simulate the compound eyes structure of ants which are sensitive to polarization, to sense the direction and intensity of polarized light in the sky and transfer optical signals into an electrical signal. Signal processing modules convert the current signal into a voltage signal, to facilitate subsequent data collection. Data collection modules convert simulated signals into digital signals. Angle calculation modules are for adopting related navigation models to calculate the direction of movement of carriers.

10.3.2 Simulation and Realisation of Polarization Orthogonality Units

1. Optical and electrical stimulation of polarization orthogonality units

Based on the polarized navigation rationale of ants, we have tried to develop a polarization-sensitive compass model using inorganic optical and electrical materials. A polarization light sensitive compass is made up of polarized navigation units,

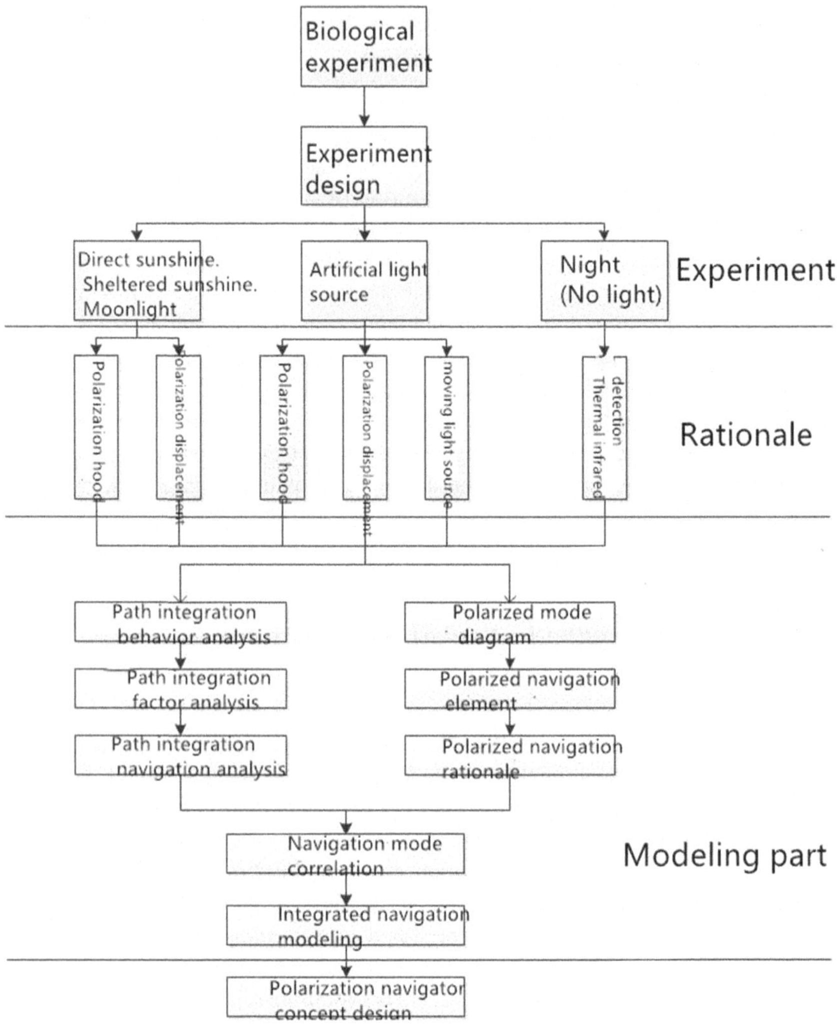

Fig. 10.14 Flowchart: experimental proposal

which have similar functions to the polarization nerve cells of insect nervous systems. Each polarization orthogonality unit is composed of a pair of polarization light sensors with a logarithmic amplifier (Fig. 10.16). The polarization light sensor is composed of an opto-electrical diode equipped with linear analysers and blue optical filters. In each polarization orthogonality unit, the polarization axes of two polarization sensors are perpendicular to each other, for simulating insects with perpendicular rhabdomes in their compound eyes. The output of each polarization light sensor will be amplified by logarithmic magnitude and then exported: the output of

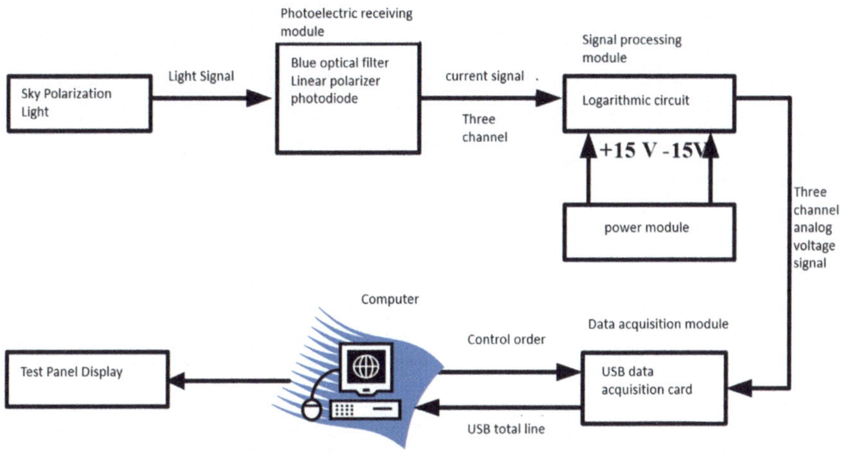

Fig. 10.15 The structure of a polarization-based navigation device

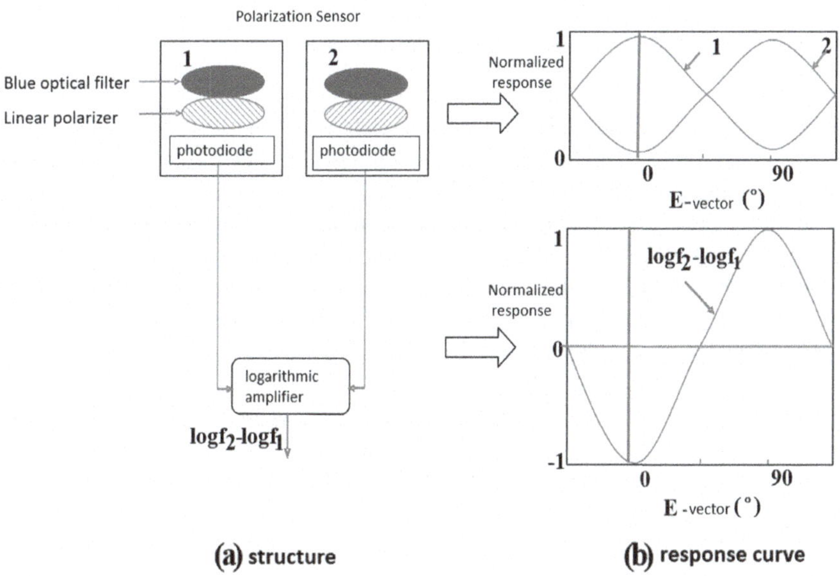

Fig. 10.16 Single-channel polarization orthogonality unit model

each polarization light sensor complies with the $\cos^2$ function, which is similar to the function of a rhabdome in the ant optical nerve.

The model simulating three kinds of ants-simulating polarization orthogonality nerve cells is composed of polarization orthogonality sensor units of three channels. We define the direction of polarization orthogonality sensor units of one channel at

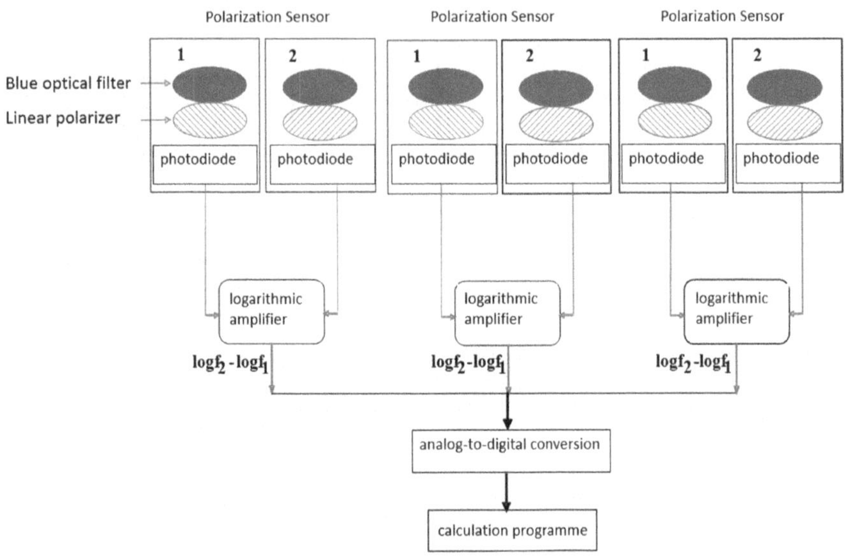

Fig. 10.17 Three-channel polarization orthogonality unit model

0°, and the main directions of the other two models as 60° and 120°. In this way, simulating models of three kinds of polarization orthogonality nerve cells compose three polarization direction analysers (0°, 60°, and 120°). The analyser in each direction is based on ant optic nerve rhabdome simulation in one direction. Simulating models of three kinds of polarization orthogonality nerve cells are shown in Fig. 10.17. If the analyser at 0° is defined as the main direction in the model, then the main direction of this analyser is the long axis direction of the object to be navigated.

The output from one polarization sensor is described as:

$$s(\varphi) = KI[1 + d\cos(2\varphi - 2\varphi_{\max})] \tag{10.23}$$

where, I is the total intensity, d is the DOP, φ is the current direction of the long axis to be navigated compared to the sun meridian, $\varphi_{\max}$ is the direction maximising $s(\varphi)$, and K is a constant. The output from polarization orthogonality units is:

$$p_1(\varphi) = \lg\left[\frac{1 + d\cos(2\varphi)}{1 - d\cos(2\varphi)}\right] \tag{10.24}$$

$$p_2(\varphi) = \lg\left[\frac{1 + d\cos(2\varphi - {}^{2\pi}/_{3})}{1 - d\cos(2\varphi - {}^{2\pi}/_{3})}\right] \tag{10.25}$$

$$p_3(\varphi) = \lg\left[\frac{1 + d\cos(2\varphi - {}^{4\pi}/_{3})}{1 - d\cos(2\varphi - {}^{4\pi}/_{3})}\right] \tag{10.26}$$

where, $p_1(\varphi)$, $p_2(\varphi)$, and $p_3(\varphi)$ are the output values of polarization orthogonality units when the polarization axis is adjusted to 0°, 60°, and 120° respectively. Since three equations in the formula contain two unknown numbers, d and φ can be obtained by inverse-logarithmic transformation and trigonometric calculation.

Based on the idea above, an angle measurement model for polarized navigation was designed as shown in Fig. 10.18. The polarization light detection platform is made up of six linear polarising films with different penetration directions. The angle between two adjacent polarising films is 60° and the penetration direction is also 60°. Transmission of polarization directions of two polaroids located opposite each other is mutually perpendicular.

A photoelectric converter is placed under each polaroid. The electrical signals output by the photoelectric converter were amplified by logarithmic magnitude and then calculated to obtain angle between the carrier and the sun meridian.

2. Realisation of polarization orthogonality units

Most polarization light in nature is linearly polarized light, which can be quantitatively described by linear polaroid. By rotating a polaroid, intensity changes with direction. Such intensity change was generated by polarized light, and in this way, polarized light in different directions can be measured, but can polarization intensities in different directions be obtained synchronously without rotating the polaroid? The answer is yes; Fig. 10.19 shows the method using a triangular Polaroid to achieve this effect (the arrow in the figure shows the direction of penetration of the polaroid). A rectangular polaroid is cut as shown in Fig. 10.19 and reassembled to the shape shown in Fig. 10.20, which can help in obtaining four polarization directions and the angle between two neighbouring polarization directions is 45°. Many creatures use this method for polarization of visual perception.

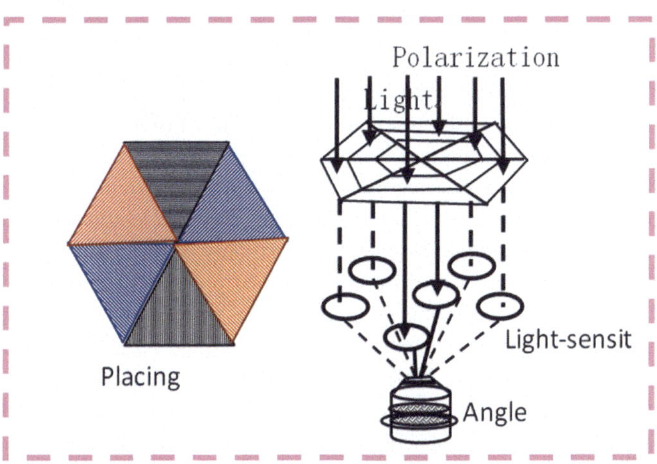

Fig. 10.18 Angle measurement experimental model of polarized light (shown below)

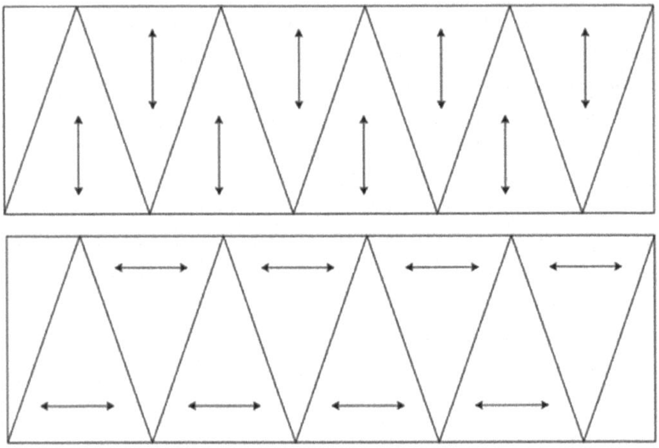

Fig. 10.19 Polaroid cutting pattern

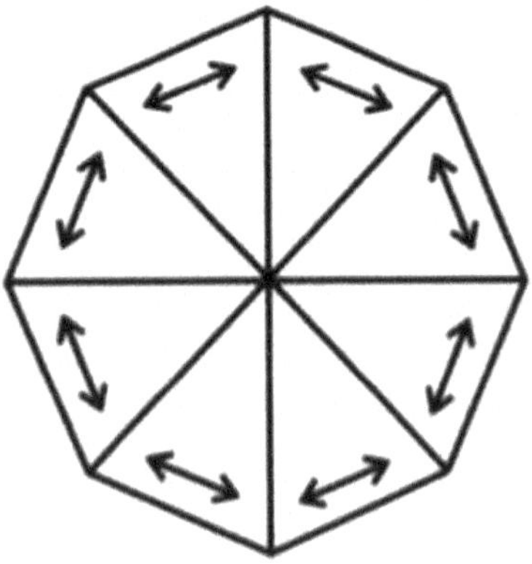

Fig. 10.20 Method with eight pieces

Of course, the method shown in Fig. 10.21 can also help to obtain six polarization directions simultaneously, with polarization directions of two counter-polaroid perpendiculars and an included angle of neighbouring polaroids of 60°. Such a structure was designed to meet the requirements of a three-channel system.

To simulate ant polarization light measurement, a photoelectric converter is necessary to transfer optical signal into electrical signals. Biologists have shown that different insects are sensitive to different bands of polarized light in the sky. Ant compound eyes are sensitive to blue and purple light, whose wavelength is about 370 to 520 nm, thus a photodiode sensitive to this band must be selected. We adopted light-sensitive component UDT-020UV manufactured by the UDT Sensor Company as the photoelectric converter here (Fig. 10.22). Such photoelectric detection platforms are very sensitive to light with wavelengths between 350 and 1100 nm: the wavelength peak of this sensitive range is 950 nm. Such platforms have small volume,

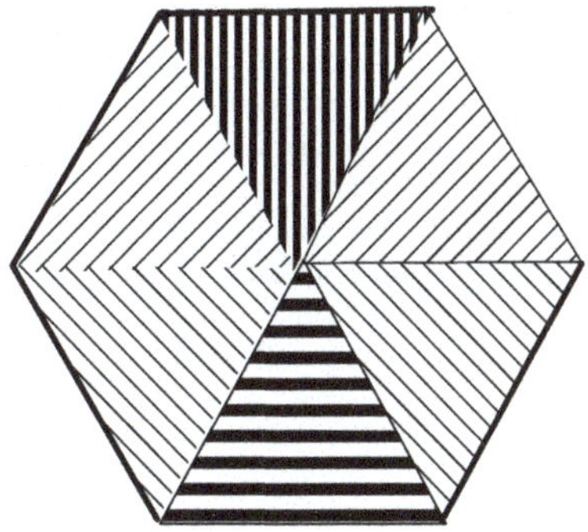

Fig. 10.21 Method with six pieces

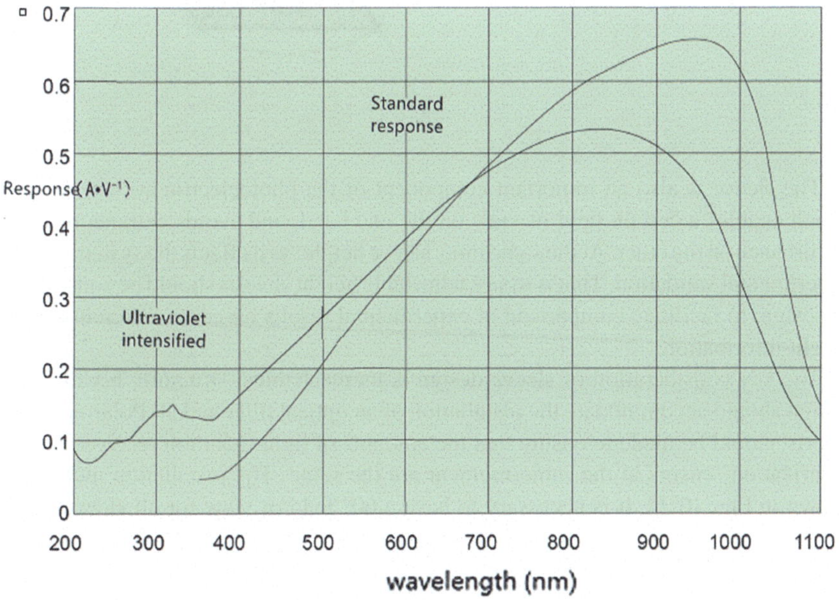

Fig. 10.22 UDT-020 UV sensitive wavelength range

high sensitivity, and high transformation accuracy. Except for the characteristics of its light spectrum, light-sensitivity area also should be considered when we select light-sensitive components, to ensure a certain signal intensity. The chosen UDT-020UV sensor has an internal amplifier, and its working voltage is 15 V, with an area of 16 mm^2, which can satisfy system testing requirements.

The polarization light detection platform is made from a mosaic of six triangular polaroids with a photosensitive diode placed under each polaroid. Since the system requires the volume to be as small as possible, each photosensitive diode is placed in the centre of each polaroid (Fig. 10.23).

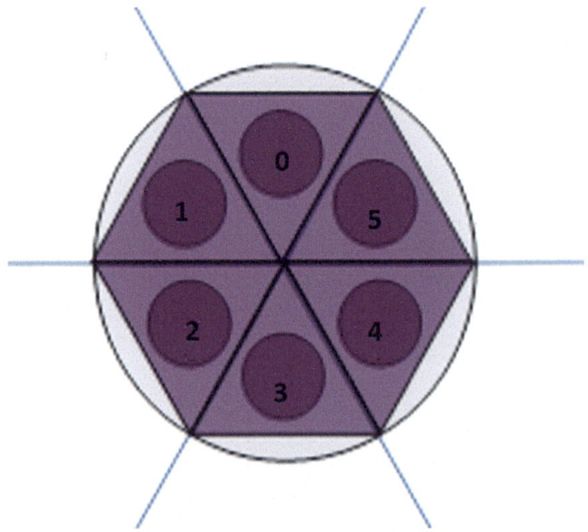

Fig. 10.23 Location of light-sensitive components (The photosensitive diode is composed of six polarization films, each being an equilateral triangle with each edge measuring 35 mm. The standard size is 77 mm including the outer metal ring.)

The sleeve is also an important component of the photoelectric system design, which enables a certain field of view on the one hand, and avoids extraneous light disturbance on the other. At the same time, sleeve height, as it affects the system, needs experimental validation. That is to say, adjustable height sleeves should be considered in design, to facilitate comparison of experimental results based on different sleeve height information.

The key consideration of sleeve design is its mechanical structure, because the design should accommodate the installation of an optical filter and all Polaroid. Best efforts should be made to ensure that the amounts of light radiation received by six polarization sensors at the same moment are the same. The installation method is shown in Fig. 10.24. It is necessary to have a 60° field of view for all sleeves, and

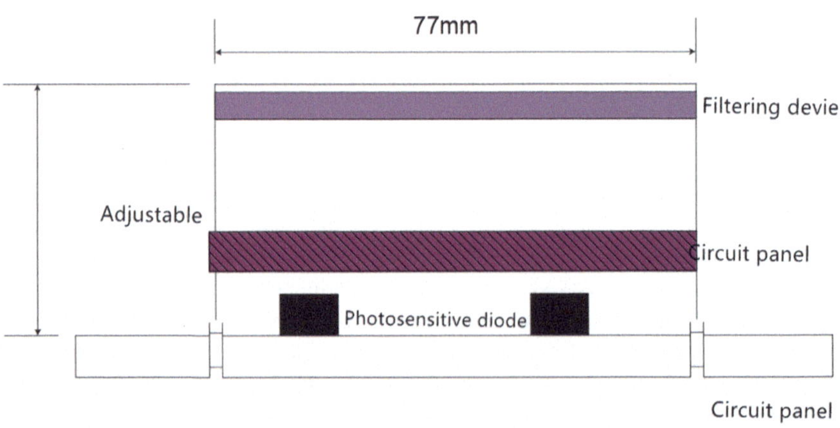

Fig. 10.24 Structural design of sleeves

diffuse reflection at its lowest possible level. A blue optical filter and a Polaroid facing in all six directions should be installed.

The signal output by photoelectrical receiving modules is a current signal: the value of this current signal is too small at between $1\mu A \sim 1mA$, and the range reaches 103, therefore direct analog-digital conversion is impossible. At the same time, general data collection modules are mostly used for collecting voltage signals. So, to facilitate data collection in future designs, this system adopted LOG102, an accurate logarithm calculation amplifier manufactured by TI Company in the USA. In each LOG 102, the logarithm calculation amplifier will use both routes of the light-sensitive current signals and then transfer them into voltage signals as output. In this way, a signal is calculated by logarithm ratio and the purpose of strengthening polarization comparison of incident polarized light was achieved. The logarithm ratio calculation circuit is shown in Fig. 10.25.

The scaling operation of LOG102 can eliminate effects of external intensity changes on calculation angle and, at the same time, logarithm calculation can greatly reduce the range of useful signals to facilitate subsequent data collection. Apart from this, LOG102, the logarithm calculation amplifier adopted advanced integrated circuit technology. When its input current, or the ratio of the two input currents, varied within a 100 dB range, total output error is always less than 0.37% of the satisfactory

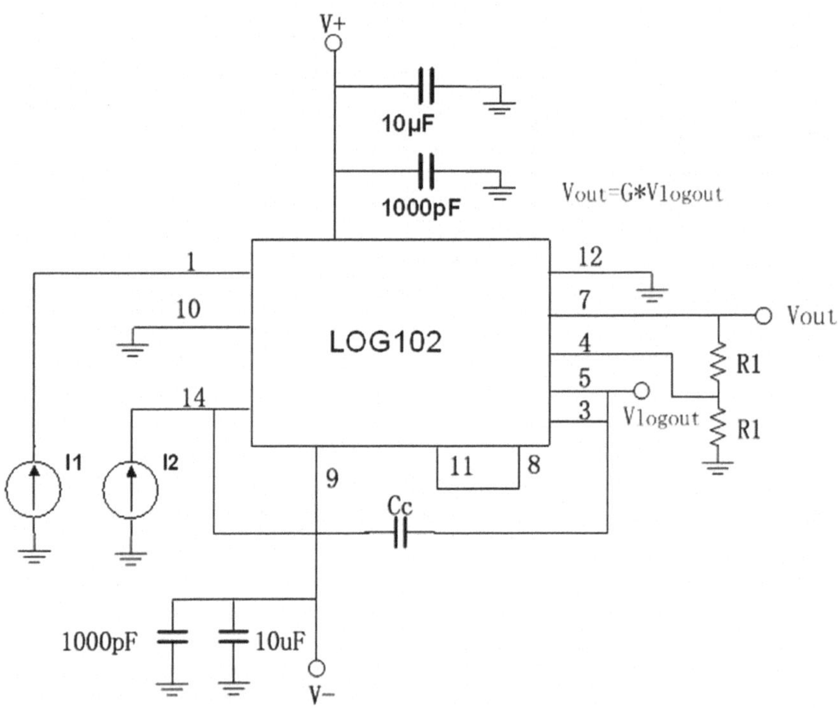

Fig. 10.25 Logarithm ratio calculation circuit

output voltage and the error from ideal logarithmic relationships will not be more than 0.1%. Since this chip integrates an amplifier, logarithm transistor, and low-drift membrane resistance, no external gain-setting resistors are necessary.

In operation, the azimuth information output will change between $-180°$and $+ 180°$, and resolutions are no less than 0.05°. So analog-to-digital conversion chips are required to be no less than 13 digits apiece. At the same time, analog-digital conversion chips themselves suffer from linear error and temperature drifting, which affects their accuracy. So analog-to-digital conversion chips are always one or two digits lower than their target values. Based on the analysis above, we adopted a USB7325, high-speed photoelectric isolation collection cards manufactured by ZTIC (Beijing Zhongtai Yanchuang Technology Co., Ltd) as data collection modules. SB7325, the high-speed photoelectric isolation interface modules, are composed of circuits enabled by multiple routes of simulated switches, high-accuracy amplified circuit, analog-to-digital conversion circuit, photoelectric isolation, and DC/DC circuit, FIFO buffering storage circuit, switch amount input and output circuits, timing/counter circuit and interface control logic circuit, power supply circuit, etc. Its functions mainly include: 16 digit transformation precision, $4\mu s$ transformation timing, embedded sampling-maintaining circuit, and a three-state serial data output.

Hardware installation of data collection modules is simple. When the modules are used, the only thing to do is insert the USB port of USB7325 into any one of computer USB ports. Its I/O signal, pulse input and output signals are all connected to an external signal origin and equipment through a J2 double-row needle plug on the modules. A/D transformed data will be read by USB general cable after FIFO storage buffering.

Angle calculation modules take PC machines as a processing platform, and conduct analysis and calculation of data sent from data collection modules based on navigation models: azimuth angles should be calculated for navigation program routines.

Based on the design above, we established an experimental platform for a polarization compass. The experimental platform adopts a sleeve structure to achieve its testing function. Sleeve structures can meet experimental system requirements on the one hand, and can ensure that an equal light radiation intensity is received by six channels at the same time, as shown in Fig. 10.26. There is a central axis at the centre of the sleeve. Inside the sleeve, it is divided into six uniform channels by separation boards. The angles between the polaroids in each channel are 60°. A polaroid with mutually perpendicular transmission axes is placed in the two channels opposite each other. On the plate under each polaroid, a photosensitive diode is placed. Electrical signal output from the photosensitive diode will be introduced after logarithmic calculation, which is equivalent to a direction analyser in this experimental system. We define the direction of one polaroid as the 0° polarization channel, and the other are 60° and 120° polarization channels, successively. At the same time, in the polarization direction of 0° polarization aligns with the main direction of the sleeve. Meanwhile, polarized light enters six light-sensitive amplifiers and produces six amplified voltage signal outputs. Meanwhile, the two light-sensitive signals in

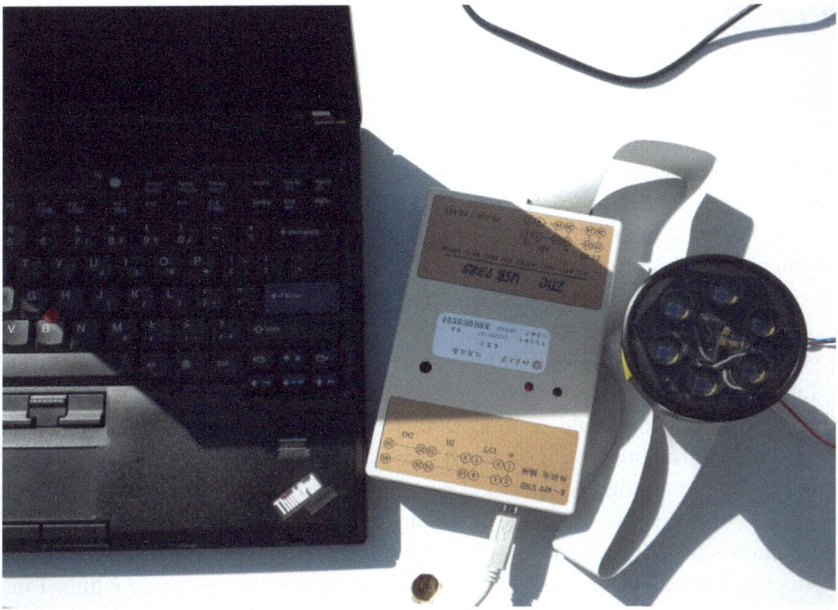

Fig. 10.26 Sleeve structure of the testing system

opposite directions should be output after logarithm calculation. Six light-sensitive signals and three logarithm signals are transformed into digital signals through 16 routes in the A/D collection modules. They should then be collected by USB and transferred to a computer. Based on the algorithm described in the theoretical analysis section, azimuth angle information can be obtained.

We have produced a working polarized navigation angle measurement device, an important part of bionic polarized navigation as established by a real-time hardware system. This helps to realise the basis for bionic polarized navigation at hardware level and laid a foundation for realising bionic polarized navigation in reality.

10.4 Bionic Polarized Navigation Models Based on Cloud Computing and Accuracy Measurement

Bionic polarized navigation may have results that are not accurate and cannot be repeated due to weather, timing issues, and other factors in practice. Bionic polarized navigation models based on cloud computing can realise a multi-object, multi-temporal, and multi-angle polarized navigation rationale that is stable and can be repeated. This section covered bionic polarized navigation and accuracy measurement under cloud computing theories.

10.4.1 How a Cloud Computing Support System for Multiple Navigation Units Was Proposed and Parallel Algorithm Analysis

To enable highly accurate automatic running of the bionic polarized navigation system on site, we adopted the research plan summarised as "Measure-Calculate-Search-Analyse-Describe-Improve" featuring multiple-object, long-term trials, and repeated operation, to analyse time zone and space zone error models of information origin and we then measured carrier error in a certain time zone and space zone. We carried out experiments relating to the information processing algorithm, which limits error and the effectiveness of this algorithm.

Taking polarized navigation with a precision of 10 m as an example, the information origin should contribute no more than 10% (less than 1 m) thereto in real use. Error in the information origin is mainly caused by changes of polarization patterns in the sky due to timing and spatial changes. The first issue is the total area where the change of polarization pattern of the sky can lead to less than 1 m error at the information source. The second issue is that time interval within which information origin error caused by changes in polarization pattern of the sky can be within 1 m.

Measurement error is the major error source here for two carriers. In certain areas and at certain intervals, information origin error can be viewed as a constant. So how does carrier measurement error change? How do we extract it? Firstly, 100 samples in the same direction should be placed in certain area (Area A). Within a certain interval T, data should be sampled multiple times and random error of single object should be eliminated through Kalman filtering. Then measured values containing system error can be obtained. The 100 samples were then placed in Area A. The measured values of 100 samples should be collected, but because the system errors of different samples are unrelated, Kalman filtering gives quasi-real values with measuring error eliminated therefrom. Then mathematical models of bionic polarized navigation should be simulated. By comparing simulated results and quasi-real values, system errors of all single objects can be obtained to form a foundation for fixing systemic error in future use.

This experiment is related to 100 individual object measuring systems, 10,000 sampling moments, 100 repeatable measurements and calculations, and recursive calculation using improved Kalman filtering. Extracting information source error and carrier measurement error all involve timing and space issues. Continuous changes in the spatial scope and sampling interval involves even more calculation, but each object needs Kalman filtering to eliminate random error. Multiple individual objects need statistical calculations at the same time. Those issues are intertwined and calculations are complex. These issues are utilised for experiments of multiple individual objects and multiple repeatable experiments in parallel or serial mode. On the other hand, there are various calculations, many structures, and many models, so task division, taking advantage of features of the calculation array, namely its ability to be grouped, can happen at the same time. This helps increases calculation efficiency and, calculating array structures through error analysis based on multi-object filtering has strong error tolerance capacity. Even calculation error, or missing information from some units, will not affect the accuracy. System restructuring can be done at any time, to ensure the timing effectiveness and validation of calculation. In such a system, involving a multi-unit, multi-process calculation array akin to a cloud computing structure, adopting cloud computing and parallel algorithms for joint resolution through varying networks of multiple navigation units is necessary.

10.4.2 Polarization Pattern Space and Time Resolution Model Based on Cloud Computing

To improve the precision of on-site bionic polarized navigation measurement which runs automatically based on cloud computing, we need to establish spatio-temporal error models. An example of group activities made up of 100 single objects, individual movements including 10 objects, and a 1 m system tolerant spatial error (namely a 10 m earth surface azimuth distance error) is discussed (Fig. 10.27). The time field error models for the information origin are shown in Fig. 10.28.

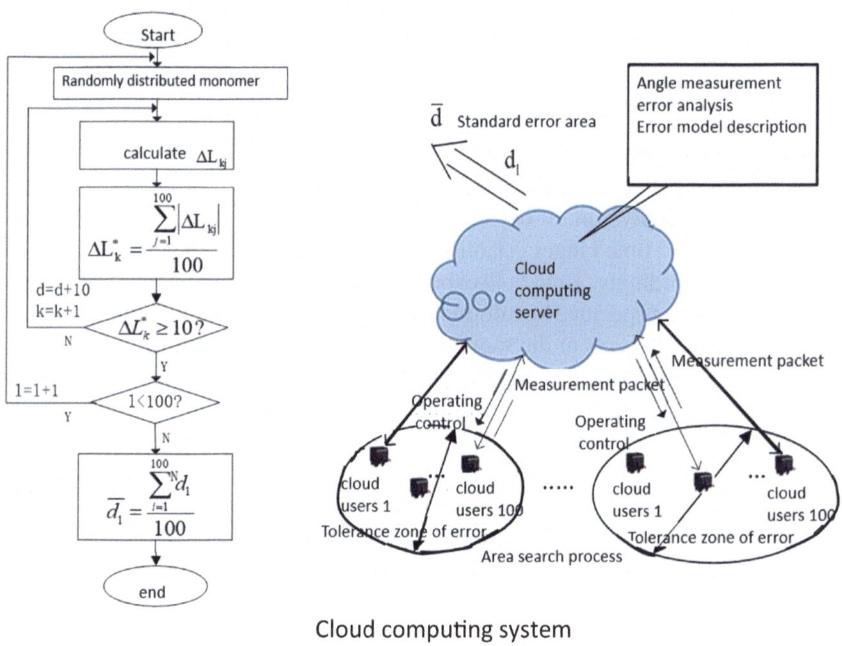

Fig. 10.27 Information origin space field error analysis experiment

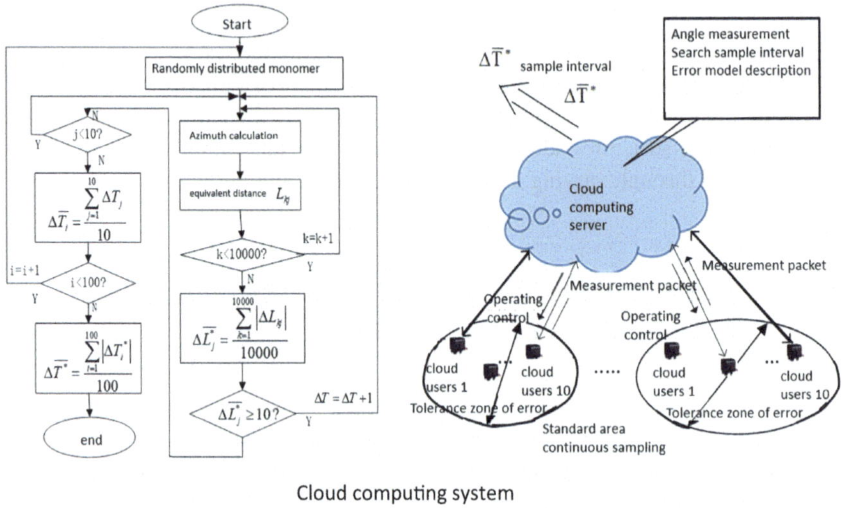

Fig. 10.28 Information origin time field error analysis experiment

10.4.3 Multi-object Measurement Error Analysis Within a Certain Space and Time Range Based on Cloud Computing

The measurement error of carriers is mainly caused by random error from noise and high-temperature effects, and the system error from mechanical structures, and other reasons. Here we analyse multi-object random time and space field errors and system errors in space and time ranges established in the experiments above, and establish error models accordingly. Time field experiments adopted sampling intervals and an area with a diameter specified by information origin error models to measure azimuth angles of multiple objects at the same time and to obtain statistical average values of equivalent length error. Then, through improved Kalman filtering and statistical calculation, analysis of systematic and random features of error was conducted to assist in final model building (Fig. 10.29). Space field experiments adopted sampling intervals and an area with a radius specified by information origin error models to measure azimuth angles of multiple objects through long periods of time. Then through improved Kalman filtering and statistical calculation, analysis of systemic and random errors was conducted to assist in final model building (Fig. 10.30).

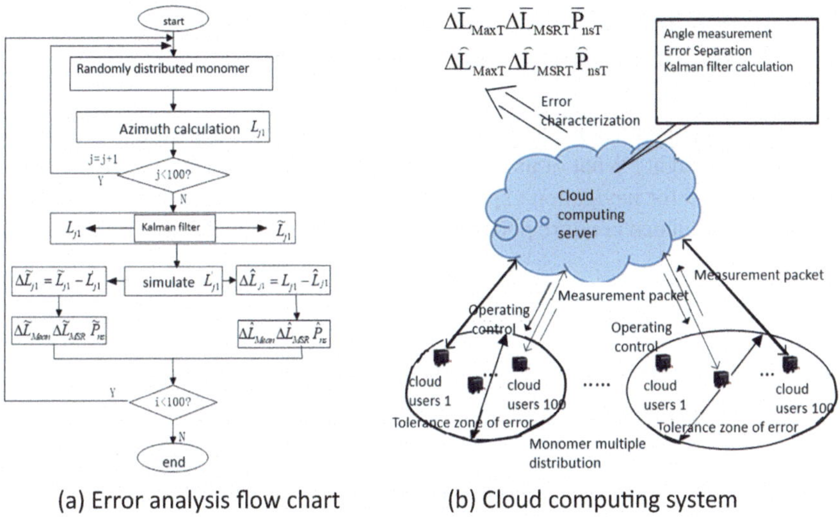

Fig. 10.29 Space field system measurement error analysis experiment

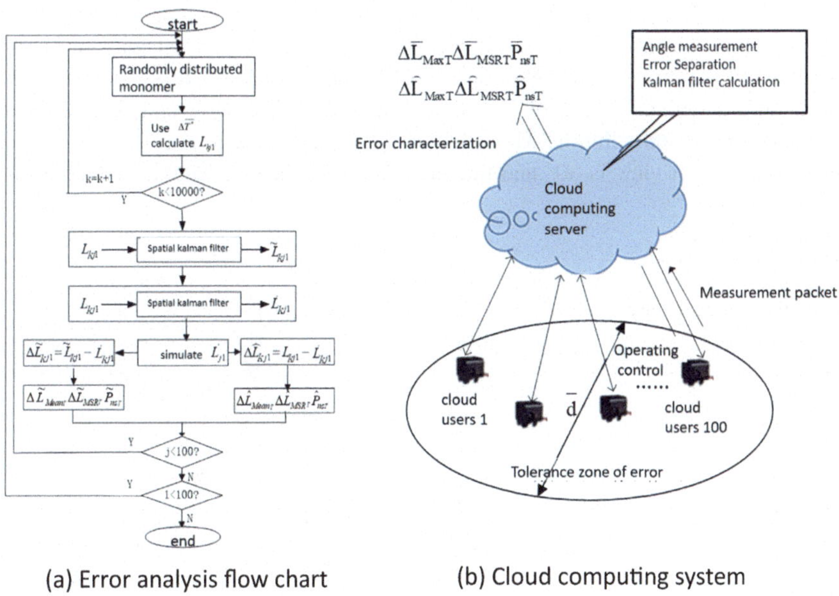

Fig. 10.30 Time field system measurement error analysis experiment

10.5 Experimental Validation of Bionic Polarized Navigation Models and Precision Analysis

Navigation is an important but how can humans realise bionic polarization navigation using polarized light? What of its navigation ability? Certain experimental methods must be adopted for navigation experiment validation and precision analysis. This section uses simulated creature polarized navigation experimental equipment to this end.

10.5.1 Bionic Polarized Navigation Angle Measurement Experiment

1. Single-channel polarization orthogonality unit testing

This experiment established optical and electrical models of polarization orthogonality units by simulating vision nerve cell of insects, and symmetrical distribution of sky polarization pattern was sensed through models to obtain a navigation reference direction. The model is shown in Fig. 10.31. The photoelectrical converter used a photosensitive diode with a blue optical filter. Three photosensitive diodes (with linear amplifier function) are placed on the metal box. The photosensitive diode is surrounded by metal columns whose inner wall is painted with a diffuse reflection layer. The field of view is 60° and the angle between the optical axis and zenith is

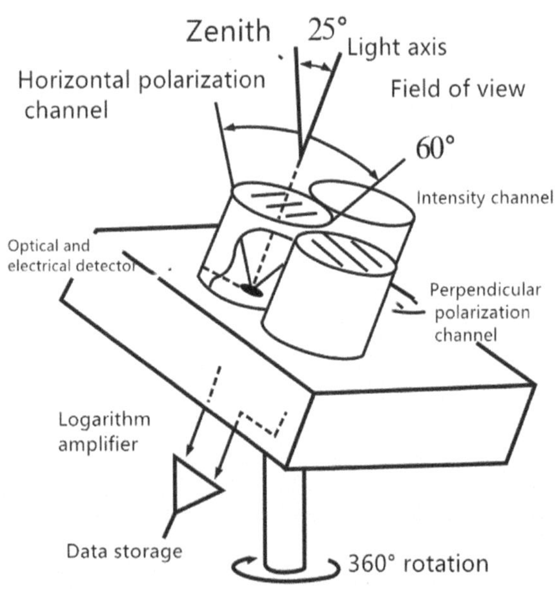

Fig. 10.31 Optical and electro models of polarization orthogonality units

25°. Two of the photosensitive diodes are equipped with a polaroid (horizontal polarization and perpendicular polarization) and are placed on top of the columns. The penetrations are mutually perpendicular. The other photosensitive diode serves as an intensity detector. The final output from the horizontal and perpendicular polarization channels after logarithmic amplification forms the curve of polarization orthogonality units and their response to polarized light.

We used a direct current motor to drive the rotating shaft and scanning rotation was 360°. Then the response curve of polarization orthogonality units compared to the sky polarization light distribution can be obtained. The curve responded most in sun meridian and anti-sun meridian directions. The output of the intensity detector can help to identify the sun meridian direction, which is the reference direction of navigation, as shown in Fig. 10.32.

In real measurement, a bracket is controlled by the direct current motor (ZGA20RU1100i) and rotates at a constant rate of 30°/s, driving the sleeve through 360° per scan in 12 s. Each rotation would start from a southerly direction. Figure 10.33 shows data collected at 10:18 a.m. on 1 April.

The measurement result shows that output response of a single channel is subject to circular changes and there are two maxima per cycle of output response, denoting the sun meridian, and anti-sun meridian, directions, respectively. This is in line with the theory, meaning that angle measurement models of bionic polarized navigation designed here are accurate and can generate responses at different intensities for light polarized in different directions and that a reference direction can be identified based on response intensity.

When a polariser is added, the waveform of three routes of signals generated by continuous rotation of the sleeve around the axis varies as shown in Figs. 10.34, 10.35 and 10.36.

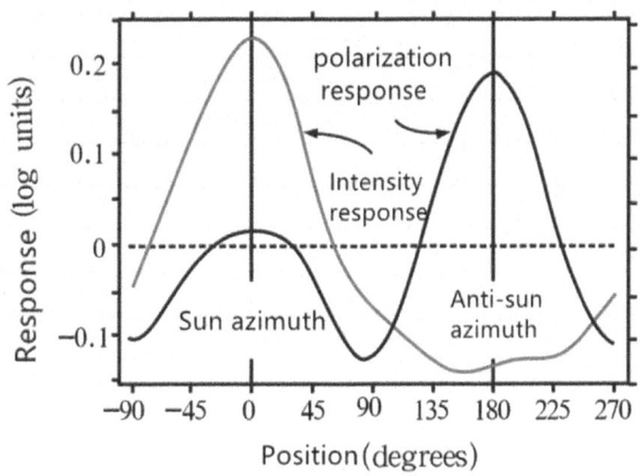

Fig. 10.32 Response curve of polarization orthogonality units

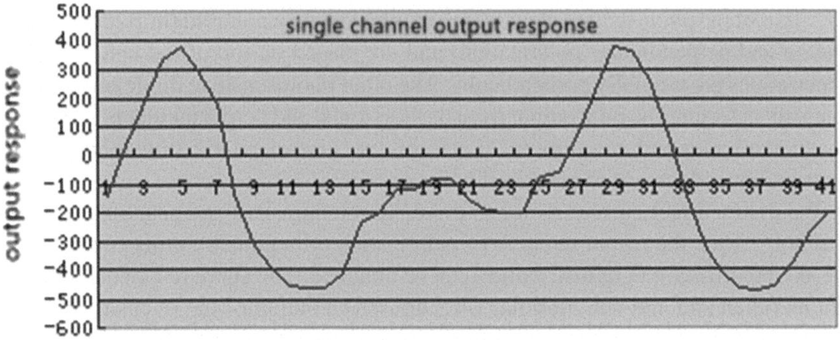

Fig. 10.33 Response curve of single-channel polarization orthogonality units

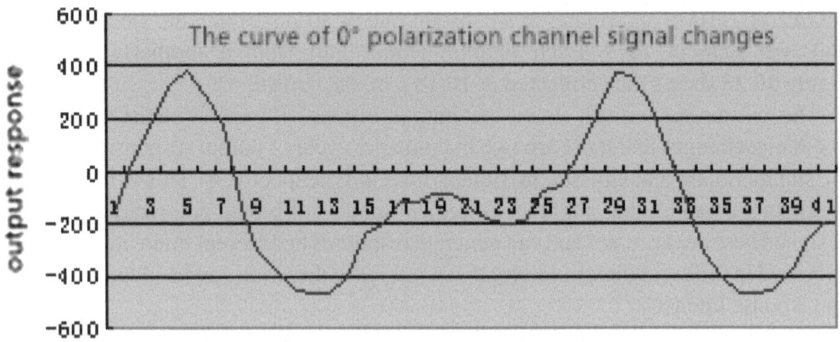

Fig. 10.34 Response curve of polarization orthogonality units at 0°

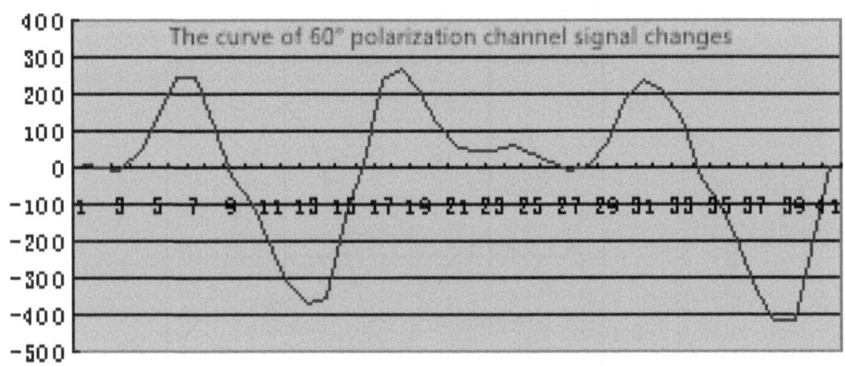

Fig. 10.35 Response curve of polarization orthogonality units at 60°

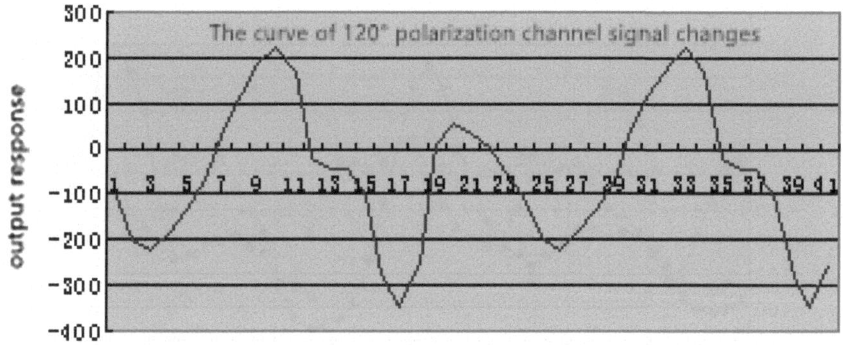

Fig. 10.36 Response curve of polarization orthogonality units at 120°

This figure shows that, when incident light is polarized, three routes of output signals generated when the angle measurement platform rotates around its axis generates a change in the trend of a certain sinusoidal response, but since conditions are limited and each time it can only rotate two rounds rather than undergoing continuous rotation, and since the sleeve rotates unevenly, the output response of different channels is not as smooth as ideally required. In general, the change in waveform output signals of the three channels match the response output from theoretical models: responses at different density levels for interference from different directions can thus be completed.

2. Three-channel polarization orthogonality unit testing

When a polariser is added, the incident light becomes linearly polarized. The waveform of three routes of signals generated by continuous rotation of the sleeve around its axis varies as shown in Figs. 10.37, 10.38 and 10.39. Figures 10.37, 10.38 and

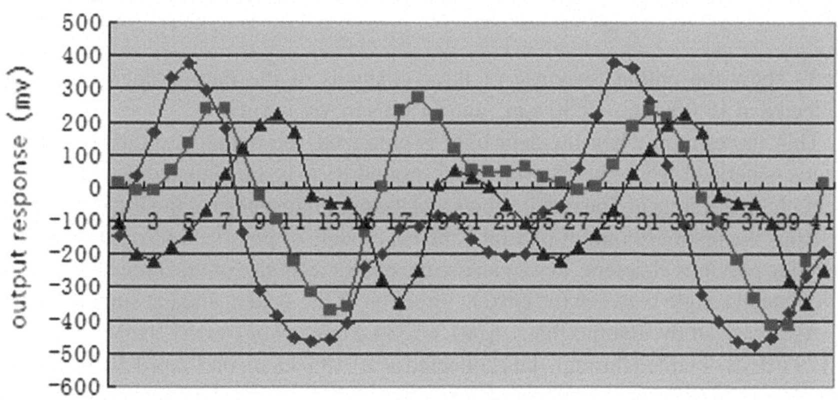

Fig. 10.37 Response curve of the three channels at the point of polarized light incidence

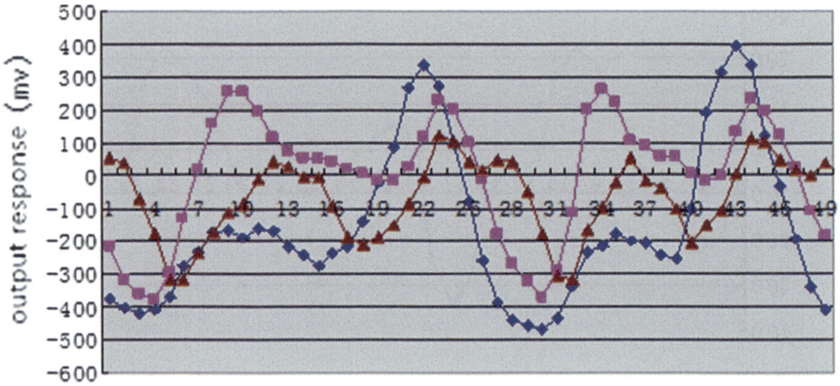

Fig. 10.38 Response curve of the three channels at the point of polarized light incidence

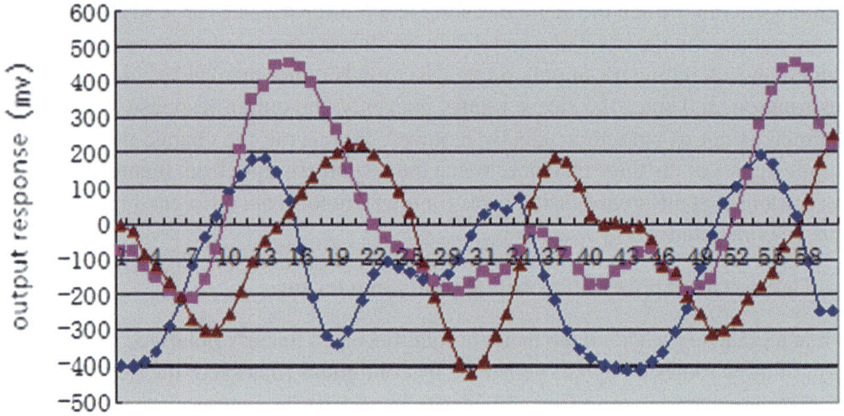

Fig. 10.39 Response curve of the three channels at the point of polarized light incidence

10.39 show the output response of three channels of the measurement platform collected at 10:00 a.m., 10:30 a.m., and 11:00 a.m. on 1 April.

This shows that, when incident light is polarized, the output responses of three routes generated when the sleeve rotates around its axis are sinusoidal with a constant phase difference. The result shows that the output response of the three channels matches the response output of simultaneous models of polarized navigation introduced in previous chapters, so we can use the output response of the three channels to calculate the angle between the current direction of the system and the sun meridian.

As a case study assume that: $p_1(\varphi) = 334.4\,\text{mV}$, $p_2(\varphi) = 51.9\,\text{mV}$, $p_3(\varphi) = -179\,\text{mV}$ as obtained through data collected at 10:18:04 a.m. on 1 April 2010. $\bar{p}_1(\varphi)$, $\bar{p}_2(\varphi)$, and $\bar{p}_3(\varphi)$ can be calculated. Then the angle between the current movement direction of the carrier and sun meridian can be calculated, thus $\varphi = 18.72°$.

The response of the polarization orthogonality unit models of three channels to polarized light basically conforms to the response rule such that an ant visual nerve-system is sensitive to the direction of sky polarization of light, thus validating the proposed angle measurement models. Of course, the precision of angle measurement models needs to be proved by considering other navigation methods, which warrants further research.

10.5.2 Error Analysis

Since this book is about exploratory studies to support a new theory, the current study remains in its initial experiment phase. The results from the experiment only initially reflected the basic trend whereby the angle measurement validation platform responds to sky polarization light. The more accurately to express sensitivity rules, certain questions must be resolved:

1. Models are crucial. Description of a polarization sensor as $s(\varphi) = KI[1 + d\cos(2\varphi - 2\varphi_{\max})]$ determines angle measurement precision in the whole system. The light received by two polarization sensors on a single channel differs. To reduce intensity differences, the only way is to do our best to reduce the volume of bionic polarized navigation sensors.
2. Mechanical structure of polaroid. In each polarization orthogonality unit, the polarization axes of the two polarization sensors are perpendicular to each other, to simulate insects who have perpendicular rhabdomes in their compound eyes. They are required to respond sinusoidally to polarized light in a certain direction and this is a rigid direction analysing device, so the mechanical design of any polaroid component therein should guarantee that the polarising directions of two opposed polaroids are mutually perpendicular, and that each polaroid is a relatively ideal linear polaroid. In addition, we also need to ensure that the angle between penetration directions of polaroids on neighbouring channels is 60°.
3. For the differences between light-sensitive components, each light-sensitive component should have as consistent a set of parameters as possible.
4. Since the sleeve rotates unevenly, we should avoid using switched power or a transformer to provide power, so as to avoid 50 Hz mains interference from alternating current.

References

Fent K, Wehner R (1985) Oceili: a Celestial compass in the desert ant Cataglyphis. Science 228(4696):192–194

Rossel S, Wehner R (1986) Polarization vision in bees. Nature 323(6084):128–131

Wehner R (2003) Desert ant navigation: how miniature brains solve complex tasks. J. Comp. Physiology A-neuroethology Sensory Neural Behavioral Physiology 189(8):579–588

Wehner R, Menzel R (1969) Homing in the ant Cataglyphis bicolor. Science 164(3876):192–194
Wehner R, Gallizzi K, Frei C, Vesely M (2002) Calibration processes in desert ant navigation: vector courses and systematic search. J. Comp. Physiology A-neuroethology Sensory Neural Behavioral Physiology 188(9):683–693
Wittlinger M, Wehner R, Wolf H (2006) The ant odometer: stepping on stilts and stumps. Science 312(5782):1965–1967

Chapter 11
New Areas in Polarization 2: Remote Sensing for Advanced Space Exploration and Global Change Research

The most typical application of polarization remote sensing is used on advanced space exploration and global change research although it is also an important breakthrough method used to solve significant international scientific problems. These include: objective conclusions drawn from astronomical polarization observation of stars and planets to complement and convert polarization observation methods in the fields of astronomy and remote sensing; the comparison of full sky polarization vector fields, the gravitational field, and geomagnetic field, to confirm their contribution to the observation of astronomical phenomena; to use polarization observation methods to prove the effectiveness of the moon as a radiance benchmark, and explain polarization's filtered saturation caused by bright light attenuation and the micro-fluctuation in radiation caused by weak light intensification; the international debate with regards to whether global vegetation biomass is positively or negatively related to C and N content and the singularity screening of polarization remote sensing to correct the inadequacies of polarization inversion theory; the theoretical outlook of the systemisation of polarization remote sensing physics to achieve concrete improvement in polarization remote sensing and increase acceptance of polarization remote sensing in practice.

11.1 Objective Conclusions Drawn from "Phase" Characteristics of Polarization V Component and Astronomic Polarization Observation

The development of polarization remote sensing techniques has allowed it to be applied objectively and widely in the observation of distal stars and planets, which shows how polarization observation methods in the fields of astronomy and remote sensing can be complemented and converted. Polarized light can be expressed using the Cartesian coordinate system (or the Cartesian coordinate system after rotation

L. Yan et al., *Polarization Remote Sensing Physics*, Springer Remote Sensing/Photogrammetry, https://doi.org/10.1007/978-981-15-2886-6_11

through a certain angle) and the Stokes vectors. In optical and radio electric experiments, it is most realistic to test for intensity. As the four Stokes vectors (I, Q, U, and V) are all intensity dimensions, so it is easy to realise experimentally and this has been used in polarization observation for distal stars.

11.1.1 Polarization Probe Theory of Stokes Vectors

Polarization characteristics in completely polarized images are often expressed as linear polarization, circular polarization or elliptical polarization. The polarization phenomenon produced by geophysical targets is generally linear polarization, and circular polarization is minimal. Stokes vectors are often used to express light's state of polarization in polarization remote sensing, in which

$$\begin{aligned} S &= [S_0, S_1, S_2, S_3]^T \\ S_0 &= \langle \tilde{E}_x^2(t) \rangle + \langle \tilde{E}_y^2(t) \rangle \\ S_1 &= \langle \tilde{E}_x^2(t) \rangle - \langle \tilde{E}_y^2(t) \rangle \\ S_2 &= \langle 2\tilde{E}_x(t)\tilde{E}_y(t) \cos\delta \rangle \\ S_3 &= \langle 2\tilde{E}_x(t)\tilde{E}_y(t) \sin\delta \rangle \end{aligned} \tag{11.1}$$

where $\tilde{E}_x$ and $\tilde{E}_y$ are the components of electric vectors in the x and y-directions in the selected coordinate system, δ is the phase difference between the two vibration components, and the $\langle\,\rangle$ sign denotes time-averaging. Using this four-vector model, one can show any polarized light's state, including its degree of polarization.

In polarization testing technology, using $S = [S_0, S_1, S_2, S_3]^T$ instead of $S = [I, Q, U, V]^T$, where I is the intensity of non-polarized light, Q and U, respectively, represent linearly polarized light on two directions, and V represents circularly polarized light. I represents the measure of radiance's total intensity (or the intensity of non-polarized light), Q is used to measure the linear degree of polarization in the horizontal direction, U is used to measure the linearly polarized component which makes a 45° angle with the horizon, and V represents circularly polarized light (Liu et al. 2008).

11.1.2 V-Component's Motion Object Probe Phase Characteristics

To conduct polarization probing of geophysical targets, one would need to obtain simultaneously the polarization images and their I, Q, U, and V-components. The Poincaré sphere (Fig. 11.1) considers the complete structure of the V-component.

Fig. 11.1 Poincaré sphere

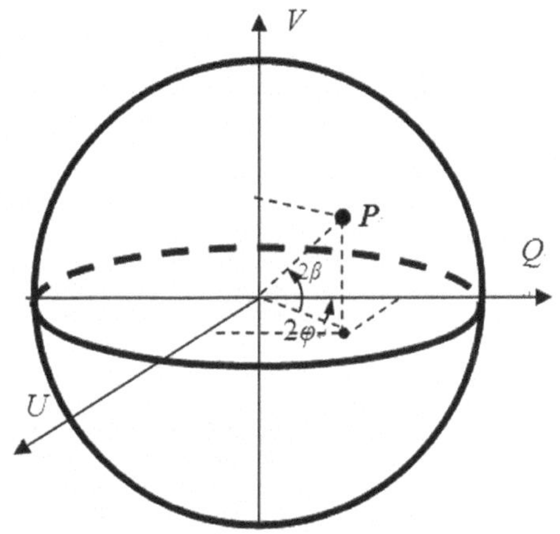

According to the distribution of P, the Stokes vector should be written as:

$$S = \begin{pmatrix} I \\ Q \\ U \\ V \end{pmatrix} = I \begin{bmatrix} 1 \\ P\cos 2\beta \cos 2\varphi \\ P\cos 2\beta \sin 2\varphi \\ P\sin 2\beta \end{bmatrix} \tag{11.2}$$

β is the ellipticity angle, φ is the inclination angle, and V is the value of the polarization component V.

It can be seen that the characterisation of the V-component is its distance P sin 2β which is either deviating upwards or downwards away from the equatorial plane, while it is also equivalent to the polarization phase value in any reflectance off the terrestrial feature of interest. To test the nature of polarization therefrom, we need to conduct comprehensive measurement of the degree of polarization P, ellipticity angle β, inclination angle φ, and natural light intensity.

From the perspective of four-component energy, the information contained by the V-component is sufficiently strong so there is a significant amount of entropy and the V-component image can highlight some details in the images that other components and their images are unable to highlight (Figs. 11.2 and 11.3), from this perspective, the V-component is useful.

To extract frame sequences from the video in Fig. 11.4, from frame 1 to 50, as the imaging order is Q, V, U, I, for every adjacent component, the motion target's position will be closer to the rear in the former component than latter. Comparing each adjacent frame, through four sets of adjacent components (Q, V), (V, U), (U, I), and (I, Q), each set with 50 frames of adjacent components, we get the following conclusion: the motion target in the latter component can overlap with the motion target in the former component every three frames. As such, the imaging time difference between adjacent components is three frames. Displaying this video in video-editing software that satisfies persistence of vision requirements, we see that the time difference

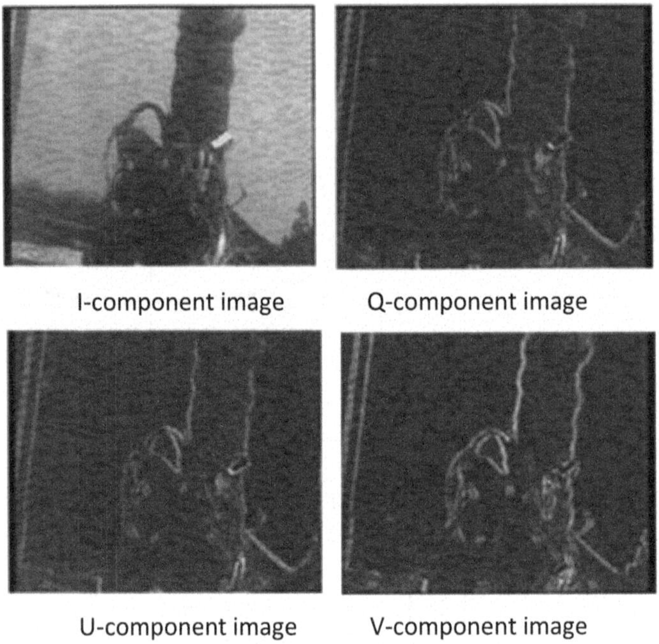

Fig. 11.2 Images of I, Q, U, and V-components

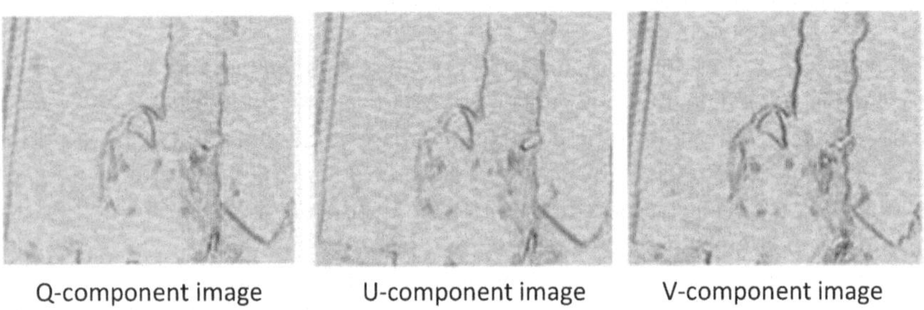

Fig. 11.3 Grey values of Fig. 11.2 after inversion

between frames is 1/24 s, which allows us to calculate the imaging time interval between two adjacent components as 3/24 s (or 0.125 s). The time taken for imaging to be completed for the four vectors is 3×0.125 s $= 0.375$ s, which allows us to achieve the time delay between different component images which is caused by the spectrophotometry in the device.

From the four components' images, we can see that the motion target not only experiences a change in position but also a change in behaviour. Conducting vector decomposition, we obtain results utilising all-component imaging software to analyse motion image characteristics and we also get the difference in action between the

Fig. 11.4 Motion target's I, Q, U, and V-component images

four components' images during the imaging process; in the image the position of the rider is different in each frame, which allows us to see that the rider's direction of motion is from right to left. According to polarization theory, when an incident beam experiences moderate reflection, as there are differences between different media and their surfaces, the polarization phase and polarization parameter of the emitted light will change; the V-component image show more visible effects as compared to the other components when detecting motion of a target; the V-component images of motion targets are clearer than that of stationary targets and can even eliminate the influence of bright light, so V-component images can effectively filter bright light (even reflected sunlight). A motion target is essentially a first derivative discretisation, and taking derivative of δ, the equation can be written as $S_3' = \langle 2\tilde{E}_x(t)\tilde{E}_y(t)\cos\delta \rangle$. Calculating the maximum value when $\delta = 0$ enables the V-component image to capture and reflect the characteristics of motion targets. The V-component is also good at portraying details with regards to velocity of variation and gradient. The phase image derived from image processing can reflect the outline of the original image after interpretation, contributing greatly to the extraction of motion target outlines.

11.1.3 Objective Conclusions Drawn from V-Component Probing of Stars and Planets

Domestically and internationally, polarization observation methods have been objectively applied to the probing of stars and planets. Qu Zhongquan's research team of Yunnan Astronomical Observatory, Chinese Academy of Sciences used their self-designed polarization light probe device to obtain the first polarization spectrum of sun glare in Gansu, Jinta on 1 August 2008. Their success was reported in the US magazine, Science, and the results were also published in the world-renowned scientific journal The Astrophysical Journal Letters (ApJ Letters). Nature magazine also reported on the European Space Agency's polarization probe device detecting "signs of life" from the Moon on March 2012 which is essentially data about the Earth that was filtered out from the polarization information of the Moon's reflected light. The extremely accurate results of the polarization detection showed gas absorption peaks clearly and can reflect the Earth's characteristic vegetation, which provides a new method of detecting signs of life from extragalactic planets. One of the top-ten most popular scientific research results as complied by Nature magazine in 2011 is the discovery made by Hungarian and Sweden scientists, through testing the nature of the sky's polarization light, that navigators from 750 C.E to 1050 C.E used polarized light to identify direction in the Atlantic Ocean, solving a mystery in scientific history.

The phase images that are obtained after image processing can reflect the original image's outline after interpretation but this is not the case for intensity images obtained after image processing. The objective conclusion that the V-component has phase characteristics can allow it to contribute greatly to the observation of distal stars and planets. Using the Sun as an example, the Sun is essentially a mass of plasma balls that burns and shines at high temperature and a high pressure. The Sun is a stable star which has solar activity that occurs in short periods of time. Solar activity is closely related to the magnetic field in the solar atmosphere, the vector magnetic field in the solar atmosphere is an important physical quantity which is critical to our understanding of all sorts of solar activity on the Sun's surface (Lites et al. 1988; Jeeffries et al. 1989). Applying the objective characteristics of phase data expressed by the V-component to the measurement of solar atmosphere's vector magnetic field will lead to objective conclusions drawn from the use of the polarized V-component to conduct probing of stars and planets.

11.2 Comparison of Global Attributes of Full Sky Polarization Vector Field, Gravitational Field, and Geomagnetic Field: Objective Conclusions Drawn Therefrom

It is widely known that there are two common fields that exist on our Earth: the gravitational and geomagnetic fields. Earth's gravitational field can reflect the distribution and change in Earth's internal quality and density and can also reflect the distribution, behaviour and change of the space between terrestrial matter. Earth's gravitational field is a type of physical field, which is distributed amongst the source field that causes it: the interior, surface, and surrounding space of the Earth. It is the gradient of gravitational potential energy and can be obtained through gravitational tests, astro-geodetic tests, and by observing disruptions in the orbit of artificial Earth satellites.

The geomagnetic field includes two components: the basic magnetic field and variable magnetic field. The basic magnetic field makes up the bulk of the geomagnetic field and its source lies within the interior of the Earth, it is relatively stable, and is a static magnetic field; the variable magnetic field includes all short-term changes of the geomagnetic field which mainly originates from the interior of the Earth and is relatively weaker; however, according to our research on polarization, there is in fact a little-known field that exists on Earth apart from the gravitational and geomagnetic fields: this is Earth's polarization field.

11.2.1 Earth's Gravitational Field

Earth's gravitational field exerts a force caused by an attraction from Earth and it is strictly speaking not simply caused by a singular force of Earth attracting objects. It is in fact created by the attraction from Earth to objects and the inertial centripetal force of its self-orbit. In it, attraction is the fundamental factor which decides the size of the gravitational force. On Earth, its size and direction is related to the location of the object. Earth's gravitational field is any existing gravitational effect or phenomenon in the space on and surrounding Earth that can reflect the distribution and change of Earth's interior quality and density. Earth's gravitational field is a type of physical field, which is distributed amongst the source field that causes it: the interior, surface, and surrounding space of the Earth.

In Earth's gravitational field, the size and direction of the gravitational force acting on any point are only related to that point's location. This is to say that the same object at different points on Earth will bear a different amount of gravitational force. Figure 11.5 shows the first image of Earth's gravitational field to be produced by the European exploration satellite Goce, which allows us to see that at different points on Earth, the gravitational potential energy differs.

At some points and places on Earth, such as when underwater, abnormalities often exist in Earth's gravity and we need to correct these abnormalities. As shown in Fig. 11.6, we conducted interpolation over a certain portion (60×60) of a map of gravitational anomalies. We extracted a 30×30 piece of map by extracting samples at intervals from the original image and then conducted interpolation using the fractal interpolation method and compared the interpolated image with the original image.

The research into the Earth's gravitational field can be used in geodesy to calculate the average ellipsoid shape of Earth and also to establish national geodetic, and levelling, networks; it can be used in spatial science to determine the orbit corrections of spacecraft caused by the Earth's gravitational field; it can be used in solid Earth physics to examine the internal structure and material distribution of the Earth.

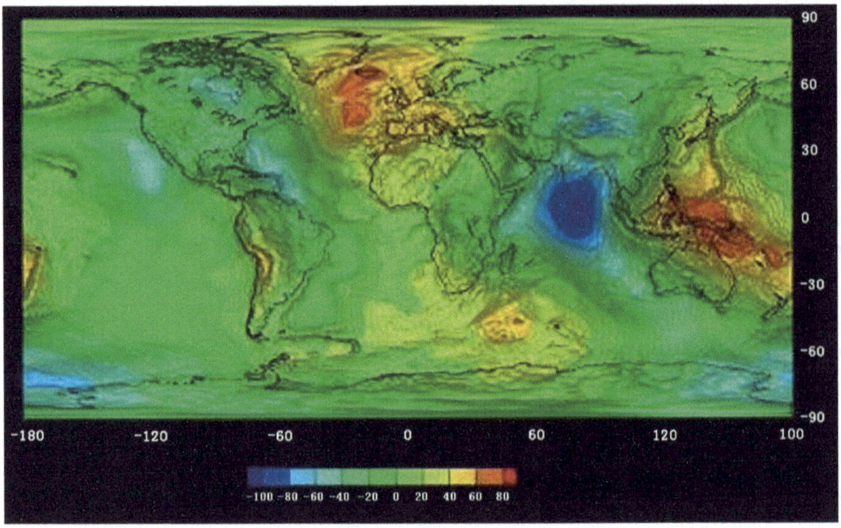

Fig. 11.5 The first image of Earth's gravitational field to be produced by Goce

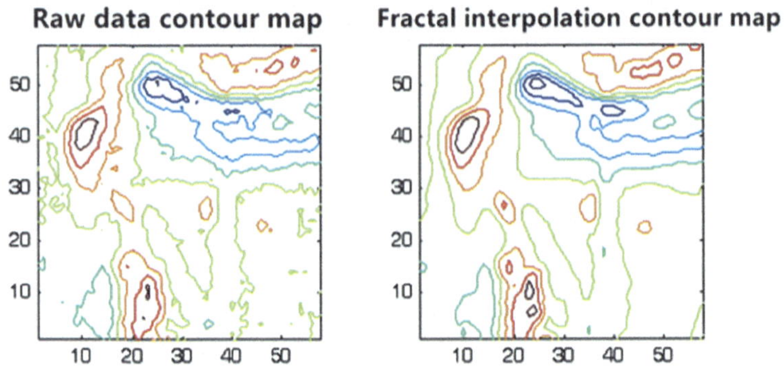

Fig. 11.6 Results of interpolation on samples of original data

11.2.2 Geomagnetic Field

The Earth can be regard as a magnetic dipole where one pole is located near the North Pole and another near the South Pole. An imaginary straight line that connects the two poles (the magnetic axis) and the Earth's axis of rotation form an angle of approximately 11.3° (Fig. 11.7).

At different portions on the Earth's surface, there is a different distribution of geomagnetism. Figures 11.8 and 11.9 show, respectively, the geomagnetic distribution over a part of an ocean and part of a flat land surface.

The geomagnetic field helps to provide directional information to many animals. For example, carrier pigeons are able to use the geomagnetic field to fly back from a

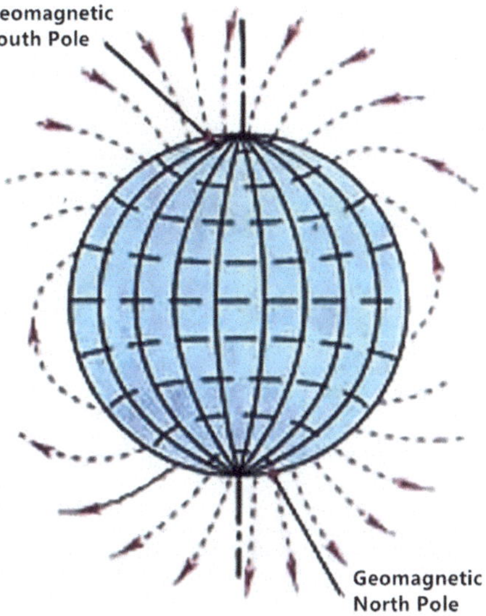

Fig. 11.7 Earth's geomagnetic field

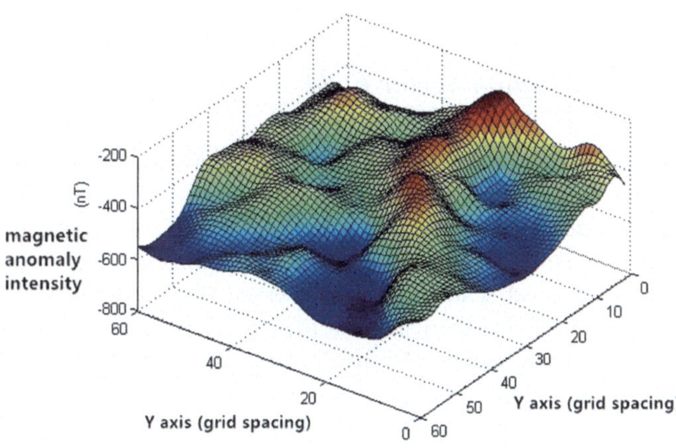

Fig. 11.8 Geomagnetic distribution over part of an ocean

distant location without getting lost, migrant birds use it to navigate their way around the world. In our earliest days, humans used the geomagnetic field to identify directions and now we are able to use abnormalities in the geomagnetic field to detect mine reserves, to search for iron, nickel, chromium, gold, oil, and other underground resources. We can even use the seismo-magnetic effect, a localised anomaly in the geomagnetic field, to predict earthquakes. When sunspots are very active, the geomagnetic field acts as a protective shield for the Earth and blocks energetic charged

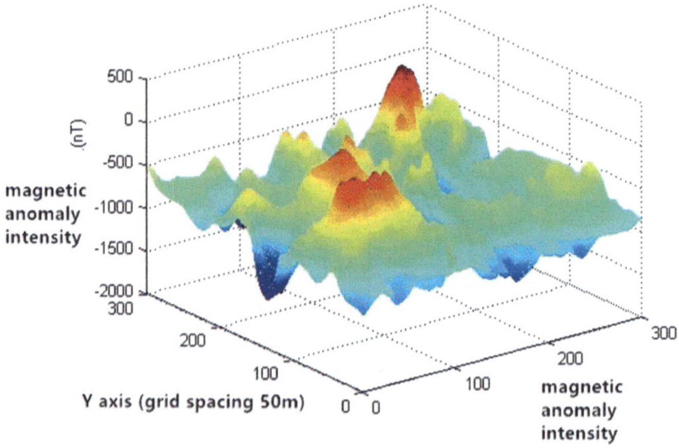

Fig. 11.9 Geomagnetic distribution over part of a flat land surface

particles from the Sun. The geomagnetic field is a natural umbrella of protection for the living organisms of Earth.

11.2.3 Sky Polarization Vector Field

In the process of atmospheric transmission, sunlight experiences scattering due to the air molecules, dust, and aerosol particles in the atmosphere. As such light in the sky has a polarization effect. The degree of polarization and its state are dependent on the size, shape, refractive index of the particles, and the polarization state of the incident beam and the angle at which scattered light is observed. If we only consider the single Rayleigh scattering that light experiences due to particles in the air, then at some time, in a certain location, there will be a relatively stable polarization pattern in the sky. This relatively stable polarization model can provide navigational information for some insects such as sand ants and crickets. In Fig. 11.10, a distribution model of sky

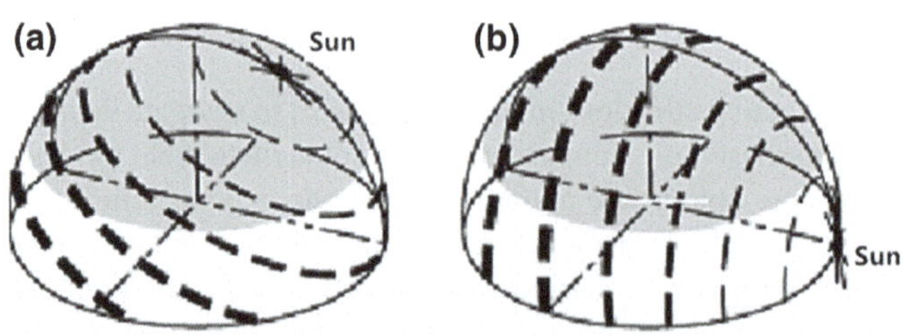

Fig. 11.10 Distribution of sky polarization light in ideal atmospheric conditions (the direction and width of the curve represent the direction and size of the degree of polarization)

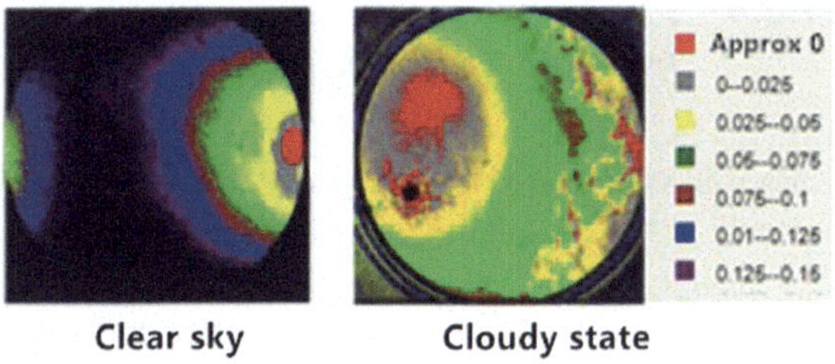

Fig. 11.11 Sky polarization model observed under clear (left) and cloudy (right) sky

polarization light in ideal atmospheric conditions is described and the atmospheric polarization model it simulates can roughly reflect the distribution of atmospheric polarization.

Understandably, there are numerous factors that influence the distribution of atmospheric polarization such as the number and size of atmospheric particles, the waveband at which weather conditions were being measured, solar altitude, and so on, all of which can influence the model image: however, in reality, there are differences between the actual polarization model of the sky and the idea model (as in Figs. 11.10 and 11.11 which are the sky polarization models observed under different weather conditions).

In Fig. 11.11, the weather conditions and solar altitude in the left, and right-hand images were different when the images were captured: the degree of polarization is higher in clear skies than in cloudy skies. We can also see red circular regions that are the neutral point zones of atmospheric polarization where the degree of polarization is close to zero. The sky's polarization model image is annularly distributed around the atmospheric polarization neutral point. As such, we can see that a quasi-stable polarization vector field exists in the sky and that it changes over time and differs under various conditions.

Similar to Earth's gravitational and geomagnetic field, the sky's polarization field's effect on research into major global issues should not be ignored. As there is indeed a stable polarization model image that exists in the sky, some organisms such as sand-ants use polarized light in the sky to navigate, accurately and speedily, their way to food sources and back to their habitat. As there is an atmospheric polarization neutral point that exists in the polarization model image, the degree of polarization in the sky is close to zero around this point, so if observation of the Earth's surface is undertaken from this point, then it is possible to eliminate the influence of polarization in light and improve the clarity and quality of images captured. This is a new breakthrough for those finding a location from which remote sensing satellites can observe the Earth, as observation can be done from the neutral point. As the sky's polarization light is mainly caused by aerosol scattering in the atmosphere, the sky's polarization field can even conduct inversion of the atmospheric aerosols. Currently, there are many researchers studying polarization field effects on human navigation and are hoping that one day we can utilise the polarization field for navigation of moving objects such as humans, cars, and ships.

11.2.4 The Comparisons and Conclusions Drawn from the Three Different Fields

Earth's gravitational and geomagnetic field as well as the sky's polarization field can not only provide a solid theoretical base as we study global problems and the way they change, but they are also very similar in many ways as listed in Table 11.1.

Similarities and differences of the three fields are summarised in Table 11.1. A centre point is present for each field: a geocentric point as the centre point in the gravitational field, an atmospheric polarization neutral point as the centre point in

Table 11.1 Areas of comparison for the three fields

Comparisons	Gravitational field	Geomagnetic field	Sky polarization field
Centre point	Earth's gravity	North and south magnetic poles	Atmospheric polarization neutral point
Stability	Very stable	Changes as N/S magnetic poles experience gradual but obvious change, can be influenced by sunspots	Takes the sun as its centre, changes regularly according to the solar altitude, can be influenced by weather conditions
Field strength difference	Relatively weak as the gradient of gravity weakens with gravity and is stronger or weaker according to the Earth's surface	Relatively strong, especially around the two poles, relatively weak around the Equator	Weak polarization at 0° incidence of the sun to Earth and 180° reflection of the Earth, stronger as it moves away from this to around ±90°
Sensitivity to surrounding environment	Insensitive	Influenced by man-made metallic objects	Easily affected by the atmosphere but the fact that it uses the Sun as its centre does not change
Observation profile	From up to down	From up to down	From down to up
Real world applicability	Establishing national levelling, and geodetic, networks, analysis of Earth's terrain, orbit correction of spacecraft, to research the internal structure and material distribution of the Earth	Provide directional cues for certain organisms such as carrier pigeons, to detect mineable reserves, to forewarn of earthquakes, and to protect Earth from energetically charged particles	To provide navigation for some organisms like sand ants, to observe and analyse astronomical catastrophes, conduct inversion of aerosol and atmospheric particles, to detect from neutral-point areas, and realise ground-atmosphere separation

the sky's polarization field, and north/south magnetic poles as centre points in the geomagnetic field. The stability of such centre points is also different: the geocentric pole is very stable; atmospheric neutral points vary with changes in solar altitude angle, both regularly and periodically, and are also affected by factors such as weather conditions; North and South Poles will change slowly with the variation in locus of the magnetic poles, and are also affected by factors such as sunspot activity. As for the intensity of the three fields, the gravitational field is relatively weak and stable, as the gravitational gradient weakens with decreasing gravitational acceleration and is stronger or weaker according to the Earth's surface elevation and geology; the Earth's magnetic field is relatively strong, especially around the two poles and is weaker around the Equator; in the sky polarization field there is weaker polarization at 0° incidence of the sun to Earth and 180° reflection of the Earth, which gets stronger as it moves away from this to around ±90°. The intensity of Earth's magnetic field and the polarization field are affected by different factors: Earth's magnetic field will undergo magnetic storms due to the impact of the flow of charged particles in space, and this produces a magnetic shock effect due to the influence of magnetised minerals; the polarization field will be affected by different weather conditions, solar elevation angle, and other factors to produce different degrees of polarization and polarization angles.

The three fields are closely related to the physical environment of the Earth, but their sensitivities are different: the gravitational field is generally insensitive to our surroundings; the geomagnetic field is easily affected by artificial metals resulting in local magnetic anomalies; the sky polarization field is easily affected by atmospheric effects, but its general trend of being centred around the sun is constant. While observing the three fields, the positions of observation are different, gravitational and geomagnetic fields are observed from up to down, while the sky polarization field is observed from down to up.

The gravitational field, geomagnetic field, and sky polarization field are able to provide theoretical bases for studies of global issues and their changes. Gravitational field data can be employed in geodesy to calculate the average ellipsoidal shape of the Earth for the establishment of a national network of geodetic and levelling stations: it can also be employed in space science to determine the corrective effects on orbits caused by the Earth's gravitational field acting on spacecraft. In solid earth physics, it can also be employed to study the internal structure of the Earth and its resource reserves.

The geomagnetic field can be used in sailing to determine directions; in resource exploration operations, it can be used to detect mines, looking for underground resources such as iron, nickel, chromium, gold, oil, *etc*. The shock magnetic effect of local geomagnetic anomalies can be used to predict earthquakes. Also, geomagnetic field data provides positional information for many creatures on Earth (for instance, carrier pigeons can be used to send letters using it). It also provides a protective shield when sunspot activity emits harmful energetic charged particles. Similarly, sky polarization field data is non-negligible in studying global issues. Some creatures, such as sand ants, navigate using the skylight polarization field to find food and their nests. As an atmospheric neutral point is present in the polarization pattern, the degree of polarization at this point is nearly zero, thus if this point is used to conduct observations on the Earth's surface, the polarization impacts of light can be erased and thus increase the clarity of images. This has provided a new idea for the positioning of remote sensing satellites during their observations of the Earth's surface, which is to observe at the atmospheric neutral point. As skylight polarization is mainly caused by the scattering of aerosols in the atmosphere, sky polarization field can be used to conduct inversion thereof. Sky polarization field data are also used to analyse and observe magnetic storms and other astronomical events.

From the analysis above, the sky polarization field, gravitational field, and geomagnetic field are all able to provide navigation field data for the globe, and their characteristics can be used to study ground-atmosphere separation, tomographic inversion, magnetic storms, and other astronomical events.

11.3 Density Calculation and Lunar Radiometric Calibration by Polarization

The Moon as the nearest celestial body from us, is valued in scientific circles. Polarization, as a new technical method, provides us new ideas for lunar exploration. This section will focus on the studies of moon density using polarization and the method of making reference to lunar radiometric calibration.

11.3.1 Using Polarization to Study Moon Density

Density is one of the important physical properties of the Moon; it changes with the surface and depth and affects the temperature, thermal conductivity and dielectric properties of the Moon. In 1976, astronauts on Apollo 11 and 12 brought back samples of moon rocks with densities of 3.2–3.4 g cm^{-3}, and it is generally seen as the surface density of the Moon. This is an anomaly as the surface density of the Moon (3.2–3.4 g cm^{-3}) shows little difference from its average density (3.33 g cm^{-3}). Normally, the surface density of a celestial body is much less than its average density, taking the Earth as an example, its surface density is 2.9 g cm^{-3} while with depth towards the core, the pressure increases and thus the density increases to as much as 12.0 g cm^{-3}, and the average density of the Earth is 5.5 g cm^{-3}.

The Moon's similar surface and average densities have induced guesses by some scientists that the Moon is hollow inside. Dr. Wilkins of the Royal Astronomical Society, in the book "Our Mysterious Spaceship" states that the moon is estimated to have a volume of about 14 million cubic miles of emptiness. Currently information on lunar surface density comes from samples brought back by Apollo 11 and 12, which are field probes with limited and fixed distributions of probe points thus the samples acquired are somewhat biased. Taking the Earth as an example, when the rocks have higher Fe_2O_3 contents, their density can reach 5.34 g cm^{-3} that far exceeds the average density of the Earth.

How to determine the surface density of a celestial body is a difficult question, as there is no better measurement method for it available at time of writing? The polarization characteristics of remote sensing spectroscopy provided a new method for measuring the density of the Moon's surface. Chinese scientists used polarization methods to simulate and verify the density of the Earth's surface and obtained a value of 2.824 g cm^{-3}, which is consistent with the well-recognised value of 2.9 g cm^{-3}. Through observing the spectral characteristics of polarization reflection of the minerals and rocks on Earth, the following four steps can be taken to measure and calculate surface density:

1. Measure and calculate the degree of polarization of the reflection spectrum of the rock surface, and then calculate the reflectivity;
2. Calculate the rock density from the reflectivity;
3. Calculate the average ratio of refraction value K of the overall surface rocks and minerals through the reflectivity and density of the rock;
4. Calculate the surface density using the average K value of the overall surface rocks and minerals.

From the methods mentioned above, accurate information on the surface density of the Moon can be deduced. In fact, whenever the measured celestial body does not emit light, the polarization characteristics of spectra can be used to deduce its surface density. For celestial bodies such as the Moon, the result is even better. Firstly, the Moon does not emit light by itself, and it fully reflects sunlight, which means that all the polarization spectral characteristics observed by the detector are the polarization characteristics of physical reflected light from the lunar surface. Secondly, there is no water on, or atmospheric surrounding to, the Moon. These two conditions ensure that all of the polarization reflection characteristics observed are generated by the surface rocks on the Moon, without the intervention of water and atmospheric effects.

As such, when the detector surrounds the moon, degree of polarization can be obtained by measuring the polarization characteristics of the Moon surface at different points, and thus estimate the reflectivity and density at different points among all lunar surface rocks to calculate the average K value thereof. Using this K value and the aforementioned methods, the surface density of the Moon can be calculated.

11.3.2 Using Polarization to Study the Radiometric Calibration with Reference to the Moon

Calibration is the prerequisite for quantitative remote sensing, and the accuracy directly affects the precision of data and image quality. The accuracy of radiometric calibration is always a difficulty: using polarization methods and making the Moon the reference for calibration provided a new idea with which to overcome this difficulty.

1. The advantages of using the Moon as a source of radiometric calibration

There are three methods of radiometric calibration: laboratory calibration, radiation field calibration, and air-borne/satellite-borne calibration. Accuracy and stability are the main goals of calibration devices in the laboratory design stage, but due to the problems of light wavelength shifts, different spectral channels, and the change in reflective properties that are present in the actual operating phase, the accuracy of calibration is affected. The accuracy of real-time radiation field calibration systems in orbit, as verified by surface water and other satellites, is largely affected by the atmosphere, the angular field of view and the degradation of mounted instruments in space cause errors. Air-borne/satellite-borne calibration refers to the source of radiation and the calibration of the optical system are both done in flight or with reference to a celestial body, thus using mounted radiation sources or nearby celestial bodies as the reference during calibration.

From the methods of radiometric calibration, it can be seen that the quality and the selection of calibration reference are important. Low-quality references will induce significant errors. The inaccuracy of the calibration of sensors in the laboratory design stage and in-orbit flight phase is a common problem in remote sensing, yet for many space probes, long-term continuous observations require the greater stability of radiometric calibration, such as the observation of global climate change in space, where the long-term stability of the solar spectrum has to be less than 1% within a decade. With the progress of science and technology, scientists have started to focus on the use of the Moon as a reference for the radiometric calibration of their sensors.

Lunar radiation observation provided a new method for the automatic radiometric calibration of the Earth observation satellites, its unprecedented stability of surface reflectivity makes it an excellent calibration source with high precision and accuracy. For the current reference of calibration, the manually attained stability can only reach 10^{-3}, while engineering attained stability should reach at least 10^{-6}. The brightness of the Moon is 1/4,000,000 of that of regular sunlight with radiation characteristics similar to that of its surface, and being the nearest natural celestial object, it has a photometric stability of 10^{-8} each year. The radiant flux of the moon is within the range of most imaging equipment and can be seen as being surrounded by the evenly low radiation target (cold space) in reflection and radiation bands, with soft-colour and monochrome. Thus the Moon is widely regarded as a potential ideal external calibration reference source (Kieffer and Robert 1996).

Comparing the Moon with the Sun, its phase and brightness are changing rapidly as shown in Fig. 11.12. The variation of the phases of the Moon will increase the difficulty of lunar calibration: however, the Moon is the only feasible calibration source with a certain brightness and a surface unaffected by an atmosphere, and its surface irradiance changes are small when in a certain phase, so no special hardware but the instrumental observations of the moon is needed for its radiometric calibration. The reasons stated above make the moon the ideal satellite sensor calibration source. For some sensor data that require high-precision calibration, lunar calibration is the only method able to reach the required standard in actual applications.

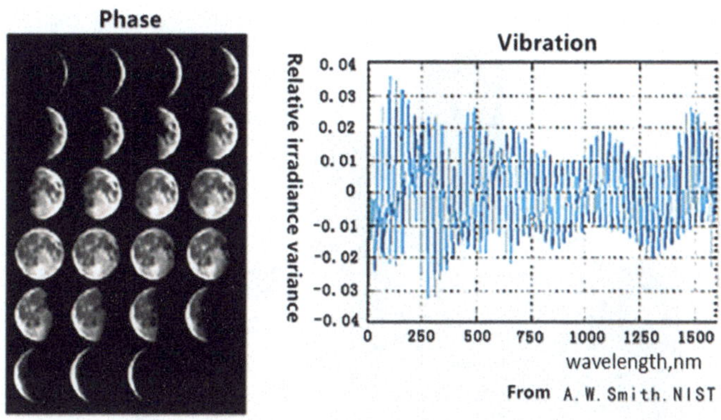

Fig. 11.12 Variation of the phases of the moon

2. The irreplaceability of polarization methods

Moon calibration experiments usually suffer from the saturation-precision dilemma: when a sensor receives lunar radiation energy, the problem of saturation will occur. To solve this problem, signal reduction has to be conducted which will hinder the collection of precise information about the lunar radiation changes. It is commonly believed that the radiation fluctuations of the Moon are of a magnitude of 10^{-8}, however after signal reduction, the fluctuations at this magnitude cannot be verified through observation, which lowers the precision of the reference. This dilemma is a long-term problem: to solve the problem, polarization provides a suitable method. First we filter the spectrum through a polarization paddle and use the property of the sensitivity of polarization to the change of signal-to-background ratio, or in other words the unique properties of "bright light attenuation" and "weak light intensification", that retains and highlights the differences of minor energy bands while supersaturated energy is weakened to acquire detailed information at high precision.

The current uncertainty of the energy radiated is around 7%, and polarization methods can improve the uncertainty to 2–3%. Taking the USA as an example, space probe organisations such as NASA and NOAA now recognise the importance and irreplaceability of such a method and invited a team of Chinese scientists to study radiometric calibration technologies using the Moon as a reference through polarization methods.

3. Establishment and operation of the relevant experimental conditions

The US Geological Survey (USGS) established a moon radiometric calibration programme and a Robotic Lunar Observatory (ROLO)observation station with reference to ground surface as shown in Figs. 11.13 and 11.14. ROLO has two telescopes that observe 23 visible and near-infrared bands (350–950 nm) and nine short-wave infrared (950–2350 nm) bands, respectively. The workstation (Fig. 11.14) has been continuously observing celestial bodies and carrying out extinction corrections for

Fig. 11.13 The ROLO station observation device

Fig. 11.14 The ROLO observation station

six years, and has allowed collection of 85,000 images of the moon and tens of thousands of celestial bodies to establish a large database. The ROLO workstation is able to observe phases of the moon, and build the polarization model of moon phases at different times and under various weather conditions. The observation workstation is high on a mountain with sparse atmospheric cover and suffers little perturbation from atmospheric aerosol and other atmospheric elements, thus precise observation can be conducted of the phases of the moon and its polarization effect can be more precisely described.

Polarization observation experiments on phases of the moon can be classified into ground and airborne-based trials. After using observation stations to observe the polarization of moon phases, ground-based simulation experiments must be conducted. The main idea is: place the sensor in areas of atmospheric neutral points (or area with a degree of polarization of near-zero) and non-polarization neutral points to conduct ground surface observations. At the atmospheric neutral point, the atmospheric polarization effect between the sensor and the ground surface is reduced to zero or a certain degree to achieve the effective removal of atmospheric polarization effects, while the purpose of the feature is to maximise the polarization information obtained. Using the atmospheric neutral point to observe, the atmospheric information between the Earth and the Moon can be blocked, thus enabling maximum acquisition of the original information contained in moonlight. Non-polarization neutral point observations are able to simulate the changes and attenuation of moonlight under atmospheric effects, while prolonged and continuous observations and data acquisition are able to collect much useful information for future experiments. Figure 11.15 shows the geometry of these aviation remote sensing polarization experiments, where the line connecting the ground surface and the neutral point has a degree of polarization of 0.

Traditional aerospace payloads mean it is unfeasible to conduct Earth observations and near-Earth simulations, thus problems such as being unable to fulfil normal working hours, and the high technical and economic cost, arise when an aerospace load is launched. A great amount of ground verification has to be carried out after launching the sensor. UAV platforms equipped with various types of aerospace loads can be employed to conduct simulation and verification of the actual ground observations to lower the operating costs and improve the stability of aerospace payloads.

As aviation remote sensors are on the platforms of satellites, their observations of the Moon are easier, however as there is no precise polarization model currently

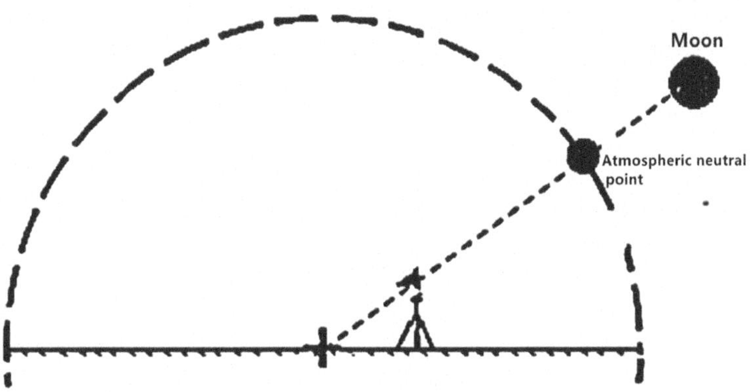

Fig. 11.15 Schematic diagram of ground-based observation experiments

available, precise calibration of the sensor cannot be undertaken. The precision of calibration is also unable to improve on that in the original model of the Moon. Airborne sensors can reach a certain height to simulate all kinds of conditions for which a sensor can conduct simulation experiments, and as a result, achieves the goal of verifying airborne sensor data when using the platform and experimental data obtained in this study.

11.3.3 Extension of the Density of the Lunar Surface and Polarized Estimation of the Radiance Reference

The density of rock is an important physical quantity that is meaningful to the measurement of celestial bodies. Density measurement of distal celestial bodies using current technology remains a pressing problem, while deducing the density of rock and minerals from the measurement of the degree of polarization of the rock surfaces is an unprecedented exploration and application advance in the field of polarization. This method has been trialled and verified on Earth, and will now be used on extra-terrestrial celestial bodies in the future, and in particular, on the Moon. If the precision of calculation is high enough, polarization methods will provide a remarkable means with which to study the surface density of any non-luminous celestial body, and greatly promote the application and development of polarization remote sensing. Remote sensing has four resolutions: temporal, spatial, spectral, and radiometric. Radiometric resolution is a measurement of radiation energy and photometric accuracy. The temporal resolution can reach 10^{-15} to 10^{-16} s with high precision, while spatial and spectral resolutions are unable to satisfy application requirements. Spatial resolution is related to the energy conversion capability of each cell while spectral resolution is related to the energy band within a narrow spectrum. The accuracy of the two is directly related to the radiometric resolution. Therefore, using the Moon as the radiance reference through polarization means can improve the radiometric resolution, and is critical to the improvement of temporal and spectral resolution. The improvement of radiometric resolution is the core reference for the breakthrough of radiation, spatial, and spectral accuracy, so using polarization methods provided a new idea for the improvement of remote sensing resolutions.

The measurement of Moon surface density and the calculation of a Moon radiance reference using polarization form a key future research direction. Accurate determination of the lunar surface density is an important part of lunar research for the better understanding of the material composition and physical properties of the moon and the promotion of progress in deep-space exploration. The establishment of natural objects, especially the Moon as absolute radiometric calibration of radiation reference is key to high-accuracy radiometric calibration. In the lunar radiation reference, the use of polarization provides a good solution for problems such as the oversaturation of energy and the preservation of accuracy for detail. Polarization observation with the Moon as radiance reference remains at the global research frontier, while aviation and space agencies in different countries are experimenting and have obtained certain progress, it is anticipated that knowledge of areas such as remote sensing, electronics, and devices will be applied to the exploration of more advanced space probes using polarization methods, so as to provide new ideas and solutions for radiometric calibration.

11.4 The Unique Means of Polarization Screening of the Remote Sensing Reversion of the Vegetation Biochemical Content and Global Climate Change Theory

The remote sensing of radiology and photometry has been carried out for years, and the emergence of polarization remote sensing is a new direction for the development of space remote sensing. Polarization remote sensing, as compared to conventional remote sensing, has its specialties, being able to solve some problems that common photometric sensing is unable to solve. According to the Fresnel principle, specular reflection causes polarization effects. At the visible and near-infrared band of remote sensing, sensors mainly receive surface reflectivity. As the ground surface is not ideal and Lambertian, when radiation interacts with the ground surface, the reflected radiation contains partially polarized light. The properties of radiant transfer of energy allow polarization's special role in the inversion of ground surface properties.

11.4.1 Remote Sensing Inversion of Vegetation Biochemical Composition

The special application of polarization remote sensing is demonstrated in the screening of international remote sensing controversy. Knjazihhin published three articles in 2013 in the Proceedings of the National Academy of Sciences of the United States of America (PNAS), stating that, other than the optical properties of the blade and the incident/observation direction, the structure of vegetation canopy is another hyperspectral aspect of vegetation Knyazikhin et al. (2013a, b, c). Before that, the article published by Ollinger et al. (2008) illustrating the linear correlation between the bidirectional reflectance factor (BRF) of vegetation and nitrogen content may be partially applicable because the study on the distribution of broadleaf and coniferous of the samples satisfies a basic linear relationship. If such structural conditions are not satisfied, the linear relationship will no longer exist. In Ollinger's article, a large amount of data proves a positive correlation may exist between the visible and near-infrared BFR and the nitrogen content of vegetation as shown in Fig. 11.16. If this relationship is correct, then it can be said that polarization data can be used to estimate and calculate global, or regional, nitrogen contents of vegetation, and

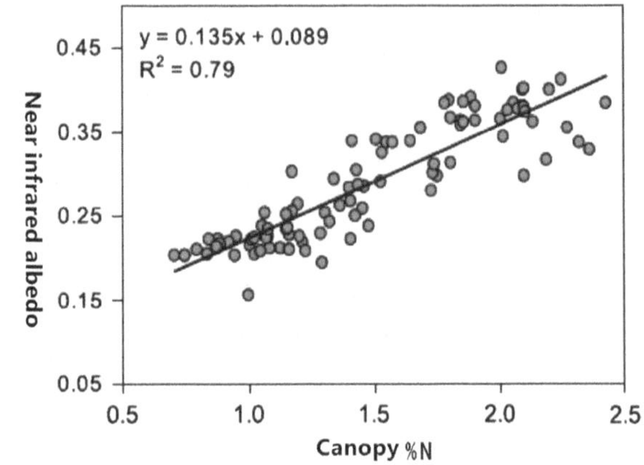

Fig. 11.16 Positive correlation between near-infrared BRF and the nitrogen content of the vegetation canopy

may even describe the current reflectivity information of remote sensing from the nitrogen content.

There are, however, flaws in Ollinger's conclusion: as in the infrared channel, the increase in nitrogen content will lead to the increase in absorbed radiation, while the increase in absorbed content will lead to plants absorbing more, and reflecting less, radiation which will eventually lower the BRF. Also, there is a saturation effect between the BRF and the structure of vegetation (Fig. 11.18), in other words, assuming that the leaf area index LAI is used to represent the structural information of the plant, with the increase in LAI, the BRF detected by sensors will increase or decrease but when LAI reaches a certain level, BRF remains unchanged and thus lead to the lack of any linear correlation in Fig. 11.18. The BRF acquired is actually the result of internal scattering when light passes through the surface. To understand the chemical content of leaf's interior, it is necessary to remove the effect of surface specular scattering using polarization methods.

11.4.2 *The Special Nature of Polarization Remote Sensing in Assessing the Controversy*

Professor Knjazihhin expounded on this problem in detail, and based on the existing achievements used to solve this problem, used leaf scale to understand the polarization characteristics of the leaf. Then, he increased the scaling to foliage-scale and canopy-scale to construct respective vegetation polarization models corresponding to different scalings (Fig. 11.17).

Initial experiments to examine the polarization models of vegetation, including how to acquire the polarization reflectivity of vegetation and how to establish its relationship with the constant spectral index were undertaken. This process forms the basis for a vegetation polarization model, and laid the theoretical foundation for

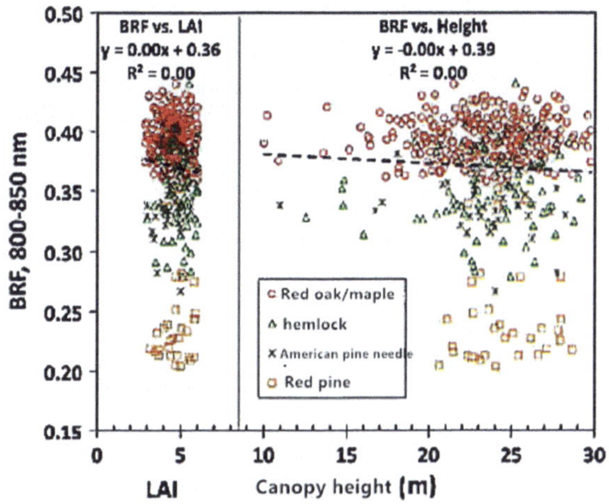

Fig. 11.17 Relationship between the BRF of vegetation and the three-dimensional structure of the canopy

subsequent experiments. Based on the above theoretical research, a series of polarization remote sensing experiments of vegetation are designed and implemented. It is found that the polarization reflectance of vegetation may be unrelated to the viewing angle as the polarization reflectance is approximately constant in all directions (Fig. 11.18). In addition, the study of polarization remote sensing of vegetation can provide a basis for effectively expanding the latest constant spectral index composition of the spectrum. After considering the polarization properties of vegetation, the stability of the constant spectral index has been improved:

$$\frac{BRF_i(\alpha, \Omega, \lambda)}{\omega_0(\lambda)} = p_1 BRF_i(\alpha, \Omega, \lambda) + R \tag{11.3}$$

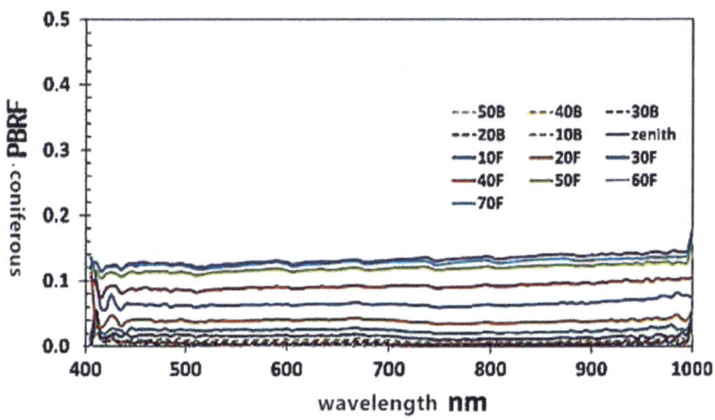

Fig. 11.18 Variation in polarization reflectivity of vegetation with measured wavelength

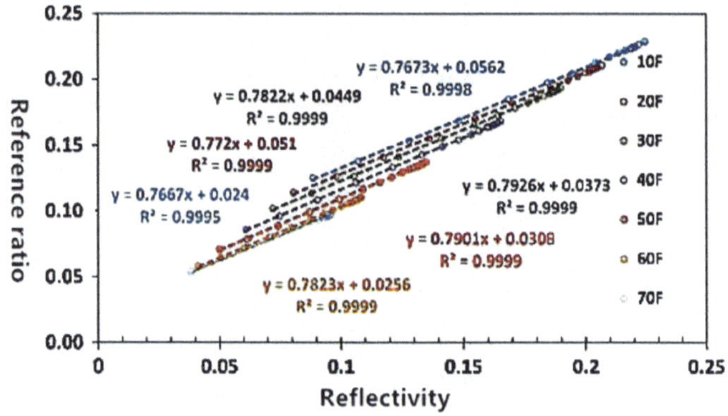

Fig. 11.19 A constant spectral index satisfies a linear correlation

where, $\omega_0(\lambda)$ represents the single scattering albedo, p_1 is the probability of a second collision, $BRF_i(\alpha, \Omega, \lambda)$ is the interior scattering of the leaf that is non-polarized, while R represents the escape factor.

From Fig. 11.19, target polarization information measured by the sensor is only related to the target's surface optical properties (such as complex refractive index and roughness) and the relative observation and incidence orientation. Polarized reflectance should be held constant for targets within the sensor range of 400–1000 nm. In Fig. 11.19, the horizontal axis represents the wavelength, the vertical axis represents the polarized reflectivity, the dotted line represents back scattering (in terms of "B") while full lines represent forward scattering (in terms of "F"), and the zenith represents vertical detection.

From Fig. 11.19, the linear correlation obtained from Eq. 11.3 proves the spectral invariance of the target information probed.

Until now, international research on this issue is still in its preliminary stage, since surface polarization information itself is relatively weak with a relatively large uncertainty; however, based on the polarization information used to help identify or analyse such an international dispute, such errors may be important. In fact, the nature of such polarization effects is determined by its radiation transmission capacity, on the basis of removing the impact of surface specular reflection on the inversion biomass, and it is able to describe the interaction between radiation and ground surface more accurately to impact on the studies of global radiation budget and global climate change.

11.5 Fundamental Status of Polarization Remote Sensing Parameters and the Prospects for the Theory of Polarization Remote Sensing System

The four high-resolution remote sensing types include radiation, spatial, spectral, and temporal resolution, wherein radiometric resolution depends on the intensity of received energy. The adequate amount of energy ensures sensitivity of detection of ground objects and also directly affects the hyperspectral intensity and the sensitivity of high-spatial-resolution pixels. Currently, conventional remote sensing is only able to obtain high quality images in one third of all cases with moderate light, while others are too bright (e.g., during solar magnetic storms, stellar objects, or water flares) or too dark (e.g., major natural disasters or distal extreme low light planetary explorations) and data are undetectable, which created a major obstacle to the advancement of space probes and surface remote sensing of major geological disasters. Polarization observation was based on the strong contrast ratio of different physical and chemical characteristics of observation targets, or the physical mechanism of bright light attenuation and weak light intensification to provide a possibility of solving these problems.

Through examples in Chaps. 2–6, the four major targets and five major characteristics of polarization remote sensing are analysed. The four major targets are four typical ground objects: rock, water, soil, and vegetation. The five characteristics are: multi-angle physical properties, the multi-spectral chemical characteristics, the structural characteristics of roughness and density, the information related to background high contrast ratio filter characteristics and the radiant transfer of energy characteristics of polarization by reflection from ground objects. In other words, polarization is multi-angular, multi-spectral, dependent on roughness and density, with an information background contrast ratio and radiative energy transfer corresponding with the physical, chemical, structural, filtering, and energy characteristics of the propagating medium and target, respectively.

In essence, the five polarization characteristics have a certain degree of correspondence with Einstein's mass-energy equation. The equation $E = mc^2$ reflects the relationship between mass and energy of matter. From a dimensional point of view, the spatial unit being m, the spectrum being $1/m$, time being t, and with radiation generalised as g, they cannot fully represent E. Maybe that is the reason why four resolutions of remote sensing are unable to cover, or probe, the required remote sensing features. Polarization included the related dimensionless form of the energy equation without limiting the strength of energy, making it the fifth dimensional component after geometry, radiation, and spectral and temporal resolutions, which can represent E. From the definition of the electromagnetic wave itself, it is mainly described by an amplitude, frequency, and phase, thus would polarization become the fourth component of electromagnetic wave description? Subsequent polarization studies may be able to answer this question.

The future theoretical development of polarization concentrates on the following areas: stellar observations (such as our solar system) where information is too strong while for distant planets information is too weak, restricting distal target observation. Polarization uses the characteristics of bright light attenuation and weak light intensification to achieve effective distal planet observations. Polarization can be used to prove the Moon is a stable radiance calibration reference, while also demonstrating that radiometric calibration is a pressing scientific problem to be solved. In addition, polarization can also be used to study the estimation and calculation of lunar rock properties, and to assess the effect of particulate matter and $PM_{2.5}$ air pollution probes. Detailed information includes:

1. The polarized information extraction technology of extremely distal or extreme light signals in advanced space probes where,
   (a) Study of polarization V-component features can achieve sensitivity towards extremely weak signals;
   (b) Study of bright light attenuation can achieve the removal of solar or planet flare effects and the acquisition of effective information; and
   (c) Study polarization radiation energy transfer to obtain polarization biological information of the Earth from the Moon.
2. Effective observations and polarization verifications of Moon radiance reference as well as the polarization calibration technology of high-resolution satellite-based sensors:
   (a) Observation of lunar radiance polarization to obtain Moon brightness variations;
   (b) Study strong polarization information using background ratio filter relationships to prove that the uncertainty of moon fluctuation is of a magnitude of 10^{-8}; and
   (c) Study polarization calibration methods to achieve Moon radiance calibration of satellite-based sensors.
3. Polarization probe and use of detection conversion technology among the densities of water-ice-rock and soil
   (a) Study the relationship between surface roughness of rocks and their multi-angle polarization spectral data to provide parametric mechanisms underpinning the aforementioned relationships;
   (b) Study the mechanism of bright light attenuation in water and ice;
   (c) Explore the relationship between rock composition, density, and degree of polarization to estimate the density of the Moon and its geological composition.
4. The delineation of atmospheric attenuation based on the sky polarization vector field and the search for technologies able to measure atmospheric pollution:
   (a) Observe the full sky polarization pattern and the region of the polarization neutral point to explore methods able to reduce the atmospheric window attenuation of sun-synchronous polarization observations;

(b) Atmospheric perspective chromatography of the full-sky polarization field to explore different variations caused by air pollutants and the root cause of $PM_{2.5}$ and soot formation;
(c) Dual characterisation of the magnetic-gravity-sky polarization fields to determine the Earth's third global vector field and significant application breakthroughs therewith.

The four observation directions of polarization mentioned above are going to be widely applied in areas such as space probes, autonomous navigation, and air pollution monitoring and will have major repercussions; however currently the international progress on polarization studies is relatively slow due to the inability of scientists to prove its objectivity, wide domain of application, irreplaceability, uniqueness, and repeatability. This chapter illustrates the astronomical polarization observation of stars and planets, compares the full-sky polarization vector field, gravitational field, and geomagnetic field, and delineates the employment of polarization observation methods to prove the Moon as a calibration reference, and describes inversion of vegetation biochemical content and the polarization remote sensing screening approach. As a reflective study of the most advanced systematic research results nationally and internationally, this study introduces the trends and theoretical references for applications in related fields.

References

Jeeffries J, Lites BW, Skumanich AP (1989) Transfer of line radiation in a magnetic field. Astrophys J 343:920–935
Kieffer HH, Robert LW (1996) Establishing the moon as a spectral radiance standard. J Atmoshp Ocean Technol 13:360–375
Knyazikhin Y, Lewis P, Disney MI (2013a) Reply to Ollinger et al. Remote sensing of leaf nitrogen and emergent ecosystem properties. In: Proceedings of the national academy of sciences of the United States of America, 110, E2438–E2438
Knyazikhin Y, Lewis P, Disney MI (2013b) Reply to Townsend et al. Decoupling contributions from canopy structure and leaf optics is critical for remote sensing leaf biochemistry. In: Proceedings of the national academy of sciences of the United States of America, 110, E1075–E1075
Knyazikhin Y, Schull MA, Stenberg P (2013c) Hyperspectral remote sensing of foliar nitrogen content. In: Proceedings of the national academy of sciences of the United States of America, 110, E185−E192
Lites BW, Skumanich AP, Rees DE et al (1988) Stokes profile analysis and vector magnetic fields. IV. Synthesis and inversion of the chromospheric Mg I b line. Astrophys J 330:493–512
Liu BL, Shi JM, Zhao DP et al (2008) Mechanism of infrared polarization detection. Infr Laser Eng 37(5):777–781
Ollinger SV, Richardson AD, Martin ME et al (2008) Canopy nitrogen, carbon assimilation, and albedo in temperate and boreal forests: functional relations and potential climate feedbacks. Proc Natl Acad Sci USA 105:19336–19341

The manufacturer's authorised representative in the EU is Springer
Nature Customer Service Centre GmbH, Europaplatz 3, 69115 Heidelberg,
Germany. If you have any concerns regarding our products, please
contact ProductSafety@springernature.com

Printed and bound by CPI Group (UK) Ltd, Croydon, CR0 4YY
30/09/2026
02225507-0015